ANTENNA ENGINEERING

McGRAW-HILL ELECTRONIC SCIENCES SERIES

EDITORIAL BOARD

**Ronald Bracewell, Colin Cherry, Willis W. Harman
Edward W. Herold, John G. Linvill, Simon Ramo, John G. Truxal**

ABRAMSON · Information theory and coding
BATTIN · Astronautical guidance
BLACHMAN · Noise and its effect on communication
BREMER · Superconductive devices
BROXMEYER · Inertial navigation systems
GELB AND VANDER VELDE · Multiple-input describing functions and
                                  nonlinear system design
GILL · Introduction to the theory of finite-state machines
HANCOCK AND WINTZ · Signal detection theory
HUELSMAN · Circuits, matrices, and linear vector spaces
KELSO · Radio ray propagation in the ionisphere
MERRIAM · Optimization theory and the design of feedback control systems
MILSUM · Biological control systems analysis
NEWCOMB · Linear multiport synthesis
PAPOULIS · The fourier integral and its applications
STEINBERG AND LEQUEUX (TRANSLATOR R. N. BRACEWELL) · Radio astronomy
WEEKS · Antenna engineering

ANTENNA ENGINEERING

W. L. WEEKS

Professor of Electrical Engineering
Purdue University

McGRAW-HILL BOOK COMPANY

New York, St. Louis, San Francisco
Toronto, London, Sydney

ANTENNA ENGINEERING

Copyright © 1968 by McGraw-Hill, Inc. All Rights Reserved.
Printed in the United States of America. No part of this
publication may be reproduced, stored in a retrieval system,
or transmitted, in any form or by any means, electronic,
mechanical, photocopying, recording, or otherwise, without
the prior written permission of the publisher.

Library of Congress Catalog Card Number 68-13106

ISBN 07-068970-9

7890KPKP79876

TO MY FAMILY

PREFACE

The purpose of this book is to provide a broad introduction to antenna engineering; it is intended primarily for electrical engineering graduates or seniors who have completed courses in electromagnetic fields and transmission lines. The manuscript is an enlargement of a set of notes prepared for a course taught by the author at Purdue University.

It was decided at the outset that the book should be relatively short, one that could be covered in a single semester of study. It was necessary, therefore, to choose carefully from the myriad of topics that might be covered, and to keep the discussions brief. Considerable breadth of coverage was deemed necessary, however, to provide a good introduction to the subject. It should be kept in mind (and this book reflects the fact) that an antenna is primarily a necessary part of some specific system, and the types of applications vary widely. The job of the antenna engineer is to select and design the antenna type which best fits the particular application.

The performance of antennas is governed by Maxwell's equations for the electromagnetic field; where feasible this theory is presented in some detail. However, the book is not and is not intended to be a book on the mathematical theory of antennas. Physical arguments or factual discussions, rather than mathematical theories, are given when such presentations are economic and/or illuminating. Every effort has been made to provide concrete and realistic examples.

The introductory chapter includes a brief history of the development of antennas and reviews fundamental theory, such as the determination of the fields of a point dipole, superposition, and reciprocity. With this background, the discussion turns naturally to the subject of small antennas, a subject which has been and which will continue to be of great practical interest. The third chapter consists mainly of the theory of linear arrays; one section is devoted to modern signal-processing antennas. A chapter on conventional wire antennas includes more details than the typical text contains on balance and feeding problems, mainly because these topics are not well covered in earlier literature. Chapter 5 introduces some of the mathematical theory of antennas; however, the book is arranged so that if the reader finds the mathematics beyond his level of accomplishment at a given time, this material may be skipped without serious loss of continuity in the subsequent chapters. (Moreover, in most cases sufficient mathematical detail is presented that the reader can, with sufficient effort, raise his level of mathematical accomplishment.) Standard material is presented on aperture antennas, while more recent material is included on frequency-independent antennas and receiving antennas. The book concludes with a chapter on antenna measurements, the material being largely based on the IEEE Standards publication on the subject.

In the early course notes, the reciprocity theorem and reaction concept were used a great deal in formulating and solving all types of antenna problems because these methods are not widely described in the literature and they are usually interesting and efficient. Most of the students enrolled in the course were relatively inexperienced in electromagnetic theory. Although they were able to learn these methods easily and well, the author, in watching their performance in oral exams and elsewhere, became concerned that they did not appear to be well schooled in the methods used by the "rest of the world" in solving electromagnetics problems. Consequently, though vestiges of the early versions remain, many of the discussions were rewritten and formulated in terms of the conventional magnetic vector potential or other potentials.

The author is indebted to Ross Bell of the Collins Radio Company and Olin Ely of the Marshall Space Flight Center at Huntsville for making available some of the photographs which appear in the book. Thanks are also due to Professor W. H. Hayt of Purdue University, who read most of the manuscript and made helpful suggestions for its improvement.

W. L. Weeks

CONTENTS

Preface *vii*

Chapter 1 Introduction 1

1.1 Brief History of Antenna Development *1*
1.2 Applications for Antennas *4*
1.3 Maxwell's Equations and the Boundary Conditions *5*
1.4 Measurement of Dynamic Field Quantities *7*
1.5 Solution of Maxwell's Equations *8*
1.6 Solution for the Vector Potential Function *12*
1.7 Field of a Point Current or Current Element-radiation Pattern *16*
1.8 Superposition *18*
1.9 The Role of the Currents in Antenna Theory *21*
1.10 Auxiliary Theorems and Principles *22*

Chapter 2 Small Antennas 27

2.1 The Capacitor-plate Antenna *27*
2.2 Radiation Efficiency *31*
2.3 Small Antennas with Transmission-line Loading *31*
2.4 Other Types of Transmission-line Loading *37*
2.5 Images *39*
2.6 Small Antennas above Ground *41*
2.7 Small Loop Antennas *56*

Chapter 3 Arrays 62

3.1 Basic Notions—The Array Factor *62*
3.2 Polynomial Representations for Linear Arrays *84*

ix

- 3.3 Fourier Representations and Antenna Pattern Synthesis *92*
- 3.4 Chebyshev Arrays *101*
- 3.5 Summation Procedures for Array Factors *115*
- 3.6 Tolerance Problems and the Effects of Unequal Element-to-element Spacing *117*
- 3.7 Signal-processing Antennas *122*

Chapter 4 Wire Antennas and Related Forms 136

- 4.1 Radiation Patterns of Wire Antennas *137*
- 4.2 Bent- and Curved-wire Structures *140*
- 4.3 The Biconical Transmission Line *145*
- 4.4 Self-impedance of Wirelike Structures *151*
- 4.5 Balance Problems in Transmission Lines and Antennas *167*
- 4.6 Folded Dipoles *180*
- 4.7 Mutual Effects *184*

Chapter 5 Introduction to the Mathematical Theory of Antennas 199

- 5.1 Theory of the Biconical Antenna *199*
- 5.2 Idealized Cylindrical Structures *214*
- 5.3 Cylindrical Antenna Theory *222*

Chapter 6 Aperture Antennas 228

- 6.1 Slot Antennas *230*
- 6.2 Slotted Cylinder Antennas *235*
- 6.3 Horn Antennas *240*
- 6.4 Other Waveguide Radiators *249*
- 6.5 Circular Aperture Antennas *254*

Chapter 7 Antennas for Multiple Frequencies 263

- 7.1 Spot-band Antennas *263*
- 7.2 Frequency Independent Antennas *267*

Chapter 8 Receiving Antennas 292

- 8.1 Power Delivered by a Receiving Antenna *292*
- 8.2 Effective Area of a Receiving Antenna *297*
- 8.3 Power Transfer between Elliptically Polarized Antennas *299*
- 8.4 Noise Considerations *301*
- 8.5 External Noise *303*

8.6 Internal Noise *307*
8.7 Receiver Noise *313*
8.8 Diversity Reception *317*

Chapter 9 Antenna Measurements 320

9.1 Scale Models *322*
9.2 Measurement of Radiation-field Amplitude *323*
9.3 Measurement of Other Pattern Characteristics *332*
9.4 Measurement of Power Gain and Directivity *341*
9.5 Input Impedance Measurements *349*
9.6 Power-handling Measurements *350*
9.7 Noise Temperature Measurements *352*

Bibliography *354*
Glossary of Symbols *355*
Index *361*

1. INTRODUCTION

It is the purpose of this book to provide a relatively brief, up-to-date introduction to antenna engineering. To place the study in proper perspective, this introductory chapter begins with a short review of the highlights in the history of antenna technology. At the present time, the theoretical basis for antenna engineering is, and should be, found in the Maxwell field equations. Consequently, a review of these equations and of a method to solve them in order to find the field of an elemental current is presented next. Some of the auxiliary theorems and principles that are frequently employed in antenna work are also reviewed.

1.1 BRIEF HISTORY OF ANTENNA DEVELOPMENT

The first antennas were employed by Hertz in 1887 in his classic demonstrations of the electromagnetic waves that had been predicted earlier by Maxwell. The form of one of his antennas is indicated in Fig. 1.1a. It consisted of two flat plates, 40 cm square, each attached to a rod 30 cm in length. The rods were terminated at their other ends by small balls and lined up with a spark gap between them about 7 mm in length. The spark gap was excited by an induction coil. Today, this antenna would be called a capacitively loaded dipole, although in today's technology the loading would be accomplished by a different physical arrangement of the plates. Hertz's receiving antenna was a circular loop of wire broken by a microscopic gap. The radius of the loop was 35 cm, a dimension that had been found by experiment to put the loop into resonance with the transmitter. Hertz later placed his rod antenna in the focal plane of a cylindrical mirror.

About 8 years after the early work of Hertz, a Professor Popoff of Kronstadt was engaged in a study of atmospheric electricity, and in con-

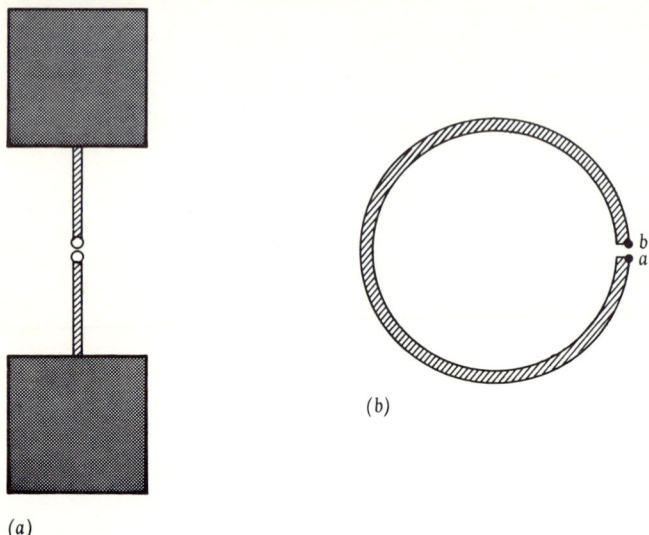

Fig. 1.1 (a) Hertz's transmitting antenna; (b) Hertz's receiving antenna (1887).

nection with this study, he placed a receiving antenna, consisting of a metallic rod, above his housetop. The receiver ("coherer") was connected between this rod and the earth.

In 1897, Marconi described a complete system for wireless telegraphy. In this system, one terminal of the spark transmitter was connected to an elevated wire and the other terminal was connected to the earth. Apparently, Marconi was the first to realize the importance of elevating the transmitting antenna.

The transmitting antenna for the first wireless transatlantic communication (from Poldhu in Cornwall to Newfoundland, in 1901) was a vertical fanlike structure consisting of 50 vertical copper wires supported by a horizontal wire, as indicated in Fig. 1.2. The horizontal wire was stretched between two masts about 150 ft high and 200 ft apart. The receiving antenna in Newfoundland was supported by kites. By 1907, commercial telegraph services had been established, and the advantage of top-loaded antennas was widely recognized. The frequency of operation of these early communications systems was usually in the range from 50 to 100 kHz, and consequently the antennas were small compared to the wavelength.

Unlike the transmitters of today, the antennas that were employed in the early systems usually strongly influenced the operating frequency. It is not surprising therefore that the point of view of the early theorists was somewhat different from that of the present. One of the earliest papers on the subject was that of Abraham,[1] who, in extending some earlier work on spheres by J. J. Thomson, studied the natural oscillations of a conducting prolate spheroid. He determined the natural frequencies and calculated the fields of a half-wave dipole.

Hertz himself had studied the fields of point dipoles. His work was carried on further by Sommerfeld and others, and by 1914, Hertz potentials and vector potentials had been employed extensively in calculations of the radiation patterns of known currents. Further, Poynting's theorem had been employed to calculate the total power radiated from antennas together with their radiation resistances.

Interest in resonant length antennas (half-wavelength dipoles or quarter-wavelength monopoles above ground) began to grow about 1920, after the discovery that the De Forest triode tube could be made to produce continuous wave oscillations at the higher frequencies (hundreds or even thousands of kilohertz). At these higher frequencies, it became practical to construct resonant length antennas or even arrays of these. By about 1930, the theory and practice of simple linear arrays had been developed and applied to broadcast transmitters for interference control.

As antennas that were of the order of a wavelength came into use, the need for a better understanding of the interaction of the antennas with transmission lines and transmitters grew more pressing. Outstanding

[1] M. Abraham, *Ann. Physik*, **66:**435 (1898).

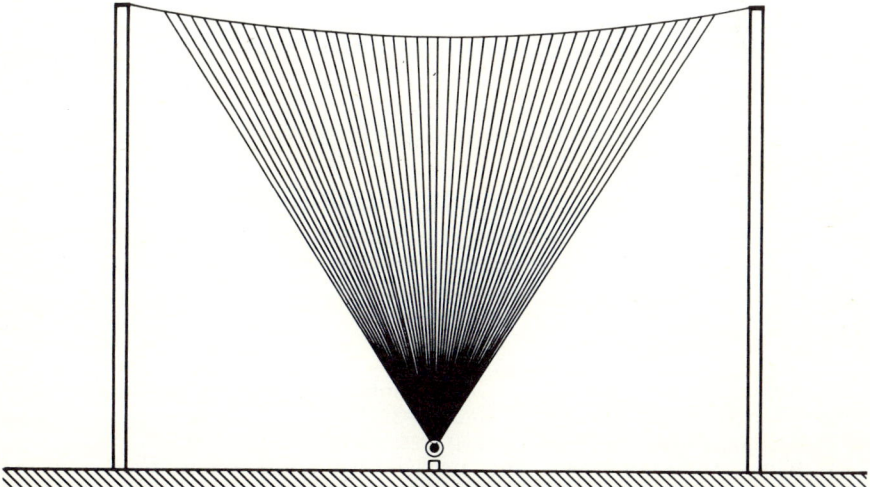

Fig. 1.2 Marconi's first transatlantic transmitting antenna (1901).

contributions on this subject were made by J. R. Carson,[1] who presented a generalization of the reciprocity theorem, and by P. S. Carter,[2] who published antenna terminal-impedance definitions and calculations. Later in the thirties the treatment of the antenna as an electromagnetic boundary-value problem was revived. King[3] and Hallen[4] formulated the linear-antenna problem as an integral equation. Stratton and Chu[5] employed a prolate spheroidal model for the linear antenna and deduced some of its properties by means of spheroidal wave functions. Perhaps the most illuminating model was introduced by Schelkunoff.[6] According to his initial model, the straight-wire antenna was regarded as a limiting case of a biconical horn antenna. With such a model, the solutions may be expressed in spherical coordinates.

In the mid-thirties, a new branch of antenna technology began to develop. The developments went hand in hand with the development of generators in the microwave frequency range and the use of metallic pipes as waveguides. These waveguides were flared out into horns, a rather natural step by analogy with the corresponding acoustic problem. Later, radiating slots were introduced into the walls of the waveguides. Still later, the World War II requirements for special high-gain antennas led to the development of large parabolic reflectors and lenses (the general subject of microwave optics). Schelkunoff[7] also published a beautiful generalization of linear array theory.

This brief account of the history of antenna development will be stopped arbitrarily at this point. The more recent history of antenna development will be covered incidentally as the detailed subject matter is considered.

1.2 APPLICATIONS FOR ANTENNAS

Antennas help to convey electromagnetic energy from one place to another. A sensible question to ask is: "Under what circumstances should the energy be transferred by antennas, rather than by wires, transmission lines, or waveguides?" An obvious answer is: "In those situations in which the running of wires or waveguides is impossible, impractical, or uneconomic." Examples of such situations are (1) aircraft or spacecraft communications, (2) ship communications, (3) other

---

[1] J. R. Carson, *Bell System Tech. J.*, **3**:393 (July, 1924); **9**:325 (1930).
[2] P. S. Carter, *Proc. IRE*, **20**:1004 (June, 1932).
[3] L. V. King, *Trans. Roy. Soc. (London)*, **236**:381 (November, 1937).
[4] E. Hallen, *Nova Acta Upsala*, **11**:1 (1938).
[5] J. Stratton, and L. J. Chu, *J. Appl. Phys.*, **12**:230 (March, 1941).
[6] S. A. Schelkunoff, *Proc. IRE*, **29**:493 (September, 1941).
[7] S. A. Schelkunoff, *Bell System Tech. J.*, **22**:80 (1943).

mobile-vehicle communications, (4) broadcasting, in which the object is to send energy in literally every direction.

A less obvious application is the transmission of high-frequency energy over long distances. The reason for this application can be seen by examining the details concerning attenuation. The basic transmission loss for free-space transmission is the 6-db decrease that accumulates each time the distance from the source is doubled (the inverse-square decrease of power with distance). Some of the modes of propagation in the actual earth environment have losses that are only slightly greater than this. For such modes, when the distance between source and receiver is great enough, the average number of decibels lost per mile is quite small. On the other hand, the number of decibels lost per mile in transmission lines or waveguides is the same for each mile. For example, in coaxial cables typical losses are about 0.05 db/100 ft at 1 MHz, 0.3 db/100 ft at 100 MHz, and 1 db/100 ft at 1,000 MHz, at best. Attenuation in standard waveguide is typically about 0.2 db/100 ft at 1,000 MHz, 1.5 db/100 ft at 5,000 MHz, and 5 db/100 ft at 10,000 MHz. The choice between antennas and transmission lines is made on the basis of which has the least total loss over the desired distance, with due attention given to the economic factors. Small distances and low frequencies tend to favor transmission lines and waveguides. Large distances and high frequencies tend to favor antennas.

Another question that is frequently in the minds of budding (or even flowering) antenna engineers has to do with the choice of an antenna type for a given application. The detailed answer to this question is not usually simple. The specific requirements must be compared with the properties of the different types of antennas. Mastery of the material in the remainder of this book will help an engineer to find an intelligent answer to the question and also to design the required type.

1.3 MAXWELL'S EQUATIONS AND THE BOUNDARY CONDITIONS

The characteristics and performance of antennas are primarily governed by the pair of equations known as Maxwell's equations

$$\nabla \times \mathbf{H}^t = \frac{\partial \mathbf{D}^t}{\partial t} + \mathcal{J}^t \qquad \nabla \times \mathbf{E}^t = -\frac{\partial \mathbf{B}^t}{\partial t}$$

where the superscript t simply indicates that the symbols are functions of time. However, much of ordinary antenna theory and practice applies to situations in which the sources and therefore the fields vary sinusoidally in time at essentially a single frequency. (When transient solutions are required, the best analytical technique is to carry out a Fourier or Laplace transformation on the Maxwell equations, in which case the single-frequency steady-state solutions are of great importance and utility.) Thus most antenna work is based on the Maxwell equations written in complex vector form with the time eliminated:

$$\nabla \times \mathbf{H} = j\omega\epsilon\mathbf{E} + \mathcal{J} \tag{1}$$

$$\nabla \times \mathbf{E} = -j\omega\mu\mathbf{H} \tag{2}$$

This form of the equations, in which the quantities **E**, **H**, and $\mathcal{J}$ are complex vectors independent of the time, may be obtained either by (1) a Fourier transform process or by (2) assuming the sources vary as $e^{j\omega t}$. With such single-frequency sources, the fields will likewise vary as $e^{j\omega t}$ and this factor, containing the time variation, may be canceled from each term. After Eqs. (1) and (2) are solved for **E** and **H**, the explicit time functions may be found, if required, by (1) taking the inverse transforms or (2) taking the real part of a product, as follows:

$$\mathbf{E}^t = \mathrm{Re}(\mathbf{E}e^{j\omega t}) \qquad \mathbf{H}^t = \mathrm{Re}(\mathbf{H}e^{j\omega t})$$

In the equations, the term $\mathcal{J}$ will ordinarily represent a known electric current density and will be considered to be the source of the electromagnetic field. As we shall use these equations, ϵ is the complex permittivity, $\epsilon = \epsilon' - j\dfrac{\sigma}{\omega}$, which includes the conductivity σ of the medium, if any. The permeability μ may also be complex. The term $j\omega\epsilon\mathbf{E}$ therefore includes any conduction currents that may exist in the region, $j\omega\epsilon\mathbf{E} = j\omega\epsilon'\mathbf{E} + \sigma\mathbf{E}$.

In certain types of antenna structures, the electric current distributions become quite complicated to handle. Sometimes, the work can be simplified by introducing a fictitious magnetic current source distribution $\mathfrak{M}$ that produces the same fields as the complicated electric current distribution. An example of this is the replacement of a small current loop by a magnetic dipole, but this is by no means the only example, nor is it the most interesting. As far as the mathematical theory is concerned, the Maxwell equations can be solved as well with magnetic current sources as with electric current sources. It is therefore economic to rewrite Eq. (2) so as to include the possibility of magnetic current sources from the outset

$$\nabla \times \mathbf{E} = -j\omega\mu\mathbf{H} - \mathfrak{M} \tag{2a}$$

Simply remember that $\mathfrak{M}$ is zero unless otherwise stated, for convenience.

The theoretical descriptions of antenna performance must be extracted from Eqs. (1) and (2) by methods to be described throughout this book. In these processes, the conditions on the field components at material boundaries will often be required. It will be assumed that the reader is generally familiar with these boundary conditions. (The details may be found in all the books in the bibliography.) The most important of the boundary conditions are reproduced here for reference:

$$\mathbf{n} \times (\mathbf{H}_2 - \mathbf{H}_1) = \mathbf{K} \qquad \mathbf{n} \times (\mathbf{E}_2 - \mathbf{E}_1) = -\mathfrak{K} \tag{3}$$

In words, the first of these says that the tangential component of the magnetic field **H** is continuous at a boundary unless it has a discontinuity equal to the magnitude of the density of an electric surface current **K**. The second says that the tangential component of the electric field **E** is continuous across a boundary, except for a specified discontinuity equal to the negative of the magnetic surface current density $\mathfrak{K}$ at the boundary. Furthermore, it must be recalled that the electromagnetic fields inside a perfect conductor are zero; thus tangential **E** and normal **H** are zero at the surface of a perfect conductor of electricity.

1.4 MEASUREMENT OF DYNAMIC FIELD QUANTITIES

To understand fully the electrodynamic field quantities **E** and **H**, it is necessary to understand how they can be measured. Their measurement at frequencies below optical frequencies involves some type of antenna, and this provides the connection to the subject of our main interest. Field probes are small receiving antennas.

The magnetic field quantity **H** may be measured with the aid of a small perfectly conducting loop, not unlike the one first used by Hertz except much smaller, into which a point, infinite-impedance voltmeter has been inserted. The connection between the voltmeter reading and the H field may be obtained by integrating Eq. (2) over the area of the loop and by applying Stokes' theorem

$$\oint \mathbf{E} \cdot \mathbf{dl} = -j\omega\mu H_n \, \Delta S$$

where ΔS is the area of the small loop and the line integral is around the perimeter of the loop. Since the wire is perfectly conducting, **E** is nonzero only at the position of the voltmeter. Thus the line integral reduces to the voltage at the voltmeter: $\oint \mathbf{E} \cdot \mathbf{dl} = \int_a^b \mathbf{E} \cdot \mathbf{dl} = V_{ab}$ where V_{ab} is the voltmeter reading. The equation relating **H** to the voltmeter reading is thus

$$H_n = -\frac{V_{ab}}{j\omega\mu \, \Delta S} \tag{4}$$

where H_n is that component of the magnetic field which is perpendicular to the plane of the loop. Of course, to obtain anything like a point measurement of **H**, the loop must be quite small. This means that amplification will usually be required. Nevertheless, the requirement for amplification does not invalidate the fundamental connection between the magnetic field and the voltmeter reading given by Eq. (4).

Frequently, from a measurement of **H** the electric field **E** may be deduced. For example, if the field is a plane traveling wavefield, $\mathbf{E} = (\mu/\epsilon)^{\frac{1}{2}}\mathbf{H} \times \hat{\mathbf{a}}$ where $\hat{\mathbf{a}}$ is a unit vector in the direction of propagation. Moreover, in an arbitrary field, if all components of **H** are measured at all points in a region, then **E** can be deduced from Maxwell's equations. Sometimes, however, it is more convenient to measure the electric field directly. For this purpose a straight-wire probe is helpful. For example, the probe might resemble Hertz's original transmitting antenna (Fig. 1.1), although it would usually be much smaller. Or, it might consist of the wires alone, without the end plates. The electric field to be measured induces a voltage in the probe, and this appears across the terminals. A measurement of this voltage may be made by means of an ideal voltmeter connected to the terminals. This voltage is proportional to the component of the electric field that is parallel to the probe. The proportionality constant depends on the length of the probe in wavelengths and the extent of the end-plate loading. With a very short probe having no loading, the equation relating E to the terminal voltage is $E_p = 2V_{ab}/h$ where h is the overall length of the probe.

Problem 1.1 In practice, the input impedance of the test loop is often more complicated than just a series inductance with a little resistance. A shunt capacitance is often introduced by the lead wires or by the loop itself. (The latter effect is more pronounced if the loop has many turns.) In such a case, a little more analysis is required to relate the magnetic field strength to the voltmeter reading. (1) Employ a circuit-type analysis and obtain a formula which shows how a shunt capacitance affects the voltage at the accessible terminals. (2) Some voltage measuring devices have relatively low input impedances (50 ohms is common). Derive a formula for the voltage indicated by such a meter when the loop is placed in a given field.

1.5 SOLUTION OF MAXWELL'S EQUATIONS

In the previous section it was pointed out that measurements of **H** and/or **E** may be made with the aid of loops and probes and sensitive rf meters. It is suggested that the reader imagine himself moving around making such measurements, as we deduce the field configurations associated with the various radiating systems. Such imaginative experiments serve to eliminate some of the abstractness of the theory.

Our task is to find solutions to Eqs. (1) and (2). That is, the object is to find the fields **E** and **H** that are generated by a prescribed source function $\mathcal{J}$ in a prescribed environment. The reader should understand and remember that the Maxwell equations, (1) and (2), are partial differential equations that have infinitely many solutions, but that our task is only to find the particular solutions that result from the prescribed sources and boundary conditions.

Mathematical solutions to Maxwell's equations are commonly obtained with the help of potential functions. The particular potential functions we shall employ may be called vector potentials or Hertz potentials, according to personal preference. In this book we shall usually refer to them as vector potentials. In order to introduce the potentials and deduce the equations they must satisfy with a minimum of special mathematics, we shall present here a slightly unconventional derivation of conventional results. The two unknowns in Eqs. (1) and (2) are **E** and **H**. To solve, one or the other must be eliminated. Here, we choose to eliminate **H**. But before eliminating anything, let us note that from the first equation, if we apply the divergence operator to both sides, since the divergence of any curl is zero, we obtain

$$0 = j\omega\epsilon \, \nabla \cdot \mathbf{E} + \nabla \cdot \mathbf{\mathcal{J}}$$

but by the equation of continuity of charge

$$\nabla \cdot \mathbf{\mathcal{J}} = -j\omega\rho$$

Therefore

$$\nabla \cdot \mathbf{E} = \frac{\rho}{\epsilon}$$

a result that is familiar in electrostatics. Now using the second Maxwell equation to eliminate **H** from the first, we have, assuming that μ is a constant,

$$\nabla \times \nabla \times \mathbf{E} = -j\omega\mu(j\omega\epsilon\mathbf{E} + \mathbf{\mathcal{J}})$$

or with the vector identity for the curl of a curl,

$$\nabla(\nabla \cdot \mathbf{E}) - \nabla^2\mathbf{E} = \omega^2\mu\epsilon\mathbf{E} - j\omega\mu\mathbf{\mathcal{J}}$$

or

$$-\nabla^2\mathbf{E} = \omega^2\mu\epsilon\mathbf{E} - j\omega\mu\mathbf{\mathcal{J}} - \nabla\left(\frac{\rho}{\epsilon}\right)$$

or

$$\nabla^2\mathbf{E} + \omega^2\mu\epsilon\mathbf{E} = j\omega\mu\mathbf{\mathcal{J}} - \nabla\left(\frac{\nabla \cdot \mathbf{\mathcal{J}}}{j\omega\epsilon}\right)$$

As in the original equations, the source of the field in these differential equations is the current density $\mathbf{\mathcal{J}}$. However, the equation for E appears to have two sources, one proportional to $\mathbf{\mathcal{J}}$ and the other related to the gradient of the divergence of $\mathcal{J}$, which is to say that it is proportional to the gradient of the charge density associated with the current density $\mathcal{J}$. This suggests that it might be simpler to split **E** into two parts, $\mathbf{E} = \mathbf{E}_1 + \mathbf{E}_2$, associating the $\mathbf{E}_1$ part of the field with the partial source $j\omega\mu\mathcal{J}$, and the remainder $\mathbf{E}_2$ with the partial source $\nabla(\rho/\epsilon)$. That is, instead of trying to solve the equation for **E** as it stands, let us solve the two simpler equations

$$\nabla^2 \mathbf{E}_1 + \omega^2 \mu\epsilon \mathbf{E}_1 = j\omega\mu\mathcal{J}$$

$$\nabla^2 \mathbf{E}_2 + \omega^2 \mu\epsilon \mathbf{E}_2 = \nabla\left(\frac{\rho}{\epsilon}\right)$$

This is possible because of the linearity of the equations; adding the latter two equations gives the original equation, and adding the solutions to the equations gives a solution to the original equations, since $\mathbf{E} = \mathbf{E}_1 + \mathbf{E}_2$.

To cast the analysis into the form usually found in the literature, we define the partial field $\mathbf{E}_1$ in terms of a new variable **A**, and the partial field $\mathbf{E}_2$ in terms of a new variable V, as follows:

$$\mathbf{E}_1 = -j\omega\mu\mathbf{A} \qquad \mathbf{E}_2 = -\nabla V$$

Thus

$$\mathbf{E} = \mathbf{E}_1 + \mathbf{E}_2 = -j\omega\mu\mathbf{A} - \nabla V$$

By so doing, we split the total electric field **E** into a part $\mathbf{E}_2$ that is conservative, electrostatic-like, and that vanishes for divergenceless source distributions, plus a part $\mathbf{E}_1$ that is negligible when the frequency is close to zero. Note also that $\mathbf{E}_1$ is a nonconservative field in general; the latter type of field is required by the Maxwell equation (2). In terms of the new variables **A** and V, the pair of equations to be solved are

$$\nabla^2 \mathbf{A} + \omega^2 \mu\epsilon \mathbf{A} = -\mathcal{J} \tag{5}$$

$$\nabla^2(\nabla V) + \omega^2 \mu\epsilon(\nabla V) = -\nabla\left(\frac{\rho}{\epsilon}\right) \tag{5a}$$

Note that the potential V does not enter into the calculation for the magnetic field **H**. That this is so can be seen from the second Maxwell equation and the nature of the split in the E field

$$\mathbf{H} = -\frac{1}{j\omega\mu}\nabla \times \mathbf{E} = -\frac{1}{j\omega\mu}\nabla \times (-j\omega\mu\mathbf{A} - \nabla V)$$

or

$$\mathbf{H} = \nabla \times \mathbf{A} \tag{6}$$

since the curl of any gradient is zero. This means that to find **H**, we need only solve Eq. (5) for **A** and then use Eq. (6). Historically, the quantity **A** is termed a magnetic vector potential. Commonly, the magnetic vector potential is defined such that **B** = ∇ × **A**, but in most antenna work the magnetic flux density **B** need not enter the theory explicitly. Thus the Eq. (6) will suffice: our **A** and the **A** frequently found in the literature differ only by a factor of the permeability μ (usually μ_0 in practice).

In the case of the single-frequency harmonic fields that we are studying, since ρ and $\mathcal{J}$ are not independent, the partial fields $\mathbf{E}_1$ and $\mathbf{E}_2$ are not independent, and the potential quantities **A** and V are not independent. Let us find the connection between **A** and V. Recall that the divergence of E is equal to ρ/ϵ; thus

$$\nabla \cdot \mathbf{E} = \frac{\rho}{\epsilon} = -j\omega\mu \nabla \cdot \mathbf{A} - \nabla^2 V$$

or

$$\nabla \cdot \mathbf{A} = -\frac{1}{j\omega\mu}\left(\nabla^2 V + \frac{\rho}{\epsilon}\right)$$

Therefore

$$\nabla(\nabla \cdot \mathbf{A}) = -\frac{1}{j\omega\mu} \nabla\left(\nabla^2 V + \frac{\rho}{\epsilon}\right)$$

or by Eq. (5a)

$$\nabla(\nabla \cdot \mathbf{A}) = -j\omega\epsilon \nabla V$$

This is the connection we were seeking. The value of this relationship is much greater than it might seem at first glance. The value stems from the fact that, since $\nabla V = -\nabla(\nabla \cdot \mathbf{A})/j\omega\epsilon$, we can find the total electric field **E** from the function **A** alone:

$$\mathbf{E} = -j\omega\mu\mathbf{A} + \frac{\nabla(\nabla \cdot \mathbf{A})}{j\omega\epsilon} \tag{7}$$

This means that, as was the case for the H field, it is not necessary to solve Eq. (5a) to find E.[1] This is not really so surprising, since if **H** = ∇ × **A**, then from Maxwell's equation (1) we have

$$\mathbf{E} = \frac{1}{j\omega\epsilon}(\nabla \times \mathbf{H} - \mathcal{J}) = \frac{1}{j\omega\epsilon}(\nabla \times \nabla \times \mathbf{A} - \mathcal{J}) \tag{7a}$$

which gives **E** in terms of only **A** and the known source density.

To sum up, to solve Maxwell's equations for the field generated by an electric current density $\mathcal{J}$, we solve Eq. (5) for the vector potential **A**, then use Eq. (6) to find **H** and Eq. (7) or (7a) to find **E**.

---

[1] It is interesting to note that Eq. (5a) shows the modification which Poisson's equation from electrostatics must undergo to cover single-frequency fields: $\nabla^2 V + \omega^2\mu\epsilon V = -\rho/\epsilon$. The extra term $\omega^2\mu\epsilon V$ tends to zero as the frequency tends to zero.

Problem 1.2 As an exercise and to obtain the results for future reference, the reader should consider the solution of Eqs. (1) and (2a) with the term $\mathcal{J} = 0$ but with the magnetic current term $\mathfrak{M} \neq 0$. He should show that in this case the electric field **E** may be represented by a vector potential function **F**, that is, $\mathbf{E} = \nabla \times \mathbf{F}$. The reader should work through an analysis similar to that presented in the text and should show that the potential function **F** satisfies the equation

$$\nabla^2 \mathbf{F} = \gamma^2 \mathbf{F} + \mathfrak{M}$$

(where $\gamma^2 = -\omega^2 \mu \epsilon$) and that given **F** the magnetic field **H** may be found from

$$\mathbf{H} = j\omega\epsilon \mathbf{F} - \frac{1}{j\omega\mu} \nabla(\nabla \cdot \mathbf{F})$$

1.6 SOLUTION FOR THE VECTOR POTENTIAL FUNCTION

We must now consider the solution of Eq. (5) for the vector potential. First of all, note that the vector equation may be written in terms of its rectangular components and that the resulting three equations all have the same form. For example, introducing $\gamma^2 = -\omega^2\mu\epsilon$, the equation for the z component is

$$\nabla^2 A_z = \gamma^2 A_z - \mathcal{J}_z \tag{8}$$

Although this equation is in terms of the z components of vectors, it is an equation relating scalars (the magnitude of a vector is a scalar). To emphasize this point and to help prevent the formation of a mental block in the path to the solution, let us change variables as follows: $A_z = \psi$, $\mathcal{J}_z = p$; thus

$$\nabla^2 \psi = \gamma^2 \psi - p \tag{8a}$$

In this equation, p is the known forcing function and we wish to find ψ. Both are functions of position. The general procedure for solving an equation of this type is first to find the solution for a particular simple p—usually a point source—and from this to construct the solutions for more complicated source distributions by superposition. That is, the field produced by the whole distribution of sources is just the sum of the fields produced by the incremental parts of the distribution.

In the sense that it is a fundamental building block, the field produced by a *point* source is the most fundamental. We shall now find the field produced by a point source. In particular, let the point source be located at the origin, and let us represent it by impulse functions*

---
* The use of the impulse function $\delta(t)$ in circuit analysis is familiar to all electrical engineers. Here we employ the same idea to represent a source density. Thus,

$p = \delta(x)\,\delta(y)\,\delta(z)\,p_0$. In this case Eq. (8a) becomes

$$\nabla^2\psi - \gamma^2\psi = 0$$

at all points except the origin. Moreover, since p is a point source at the origin, the solution ψ should have spherical symmetry and be independent of the spherical coordinate variables θ and φ. This suggests that we should solve (8a) in spherical coordinates. The laplacian operator expressed in spherical coordinates gives

$$\nabla^2\psi = \frac{1}{r^2}\frac{\partial}{\partial r}\left(r^2\frac{\partial\psi}{\partial r}\right) + \frac{1}{r^2\sin\theta}\frac{\partial}{\partial\theta}\left(\sin\theta\frac{\partial\psi}{\partial\theta}\right) + \frac{1}{r^2\sin^2\theta}\frac{\partial^2\psi}{\partial\varphi^2}$$

Because of the spherical symmetry, the equation for ψ simplifies, except at the origin, to the form

$$\frac{1}{r^2}\frac{\partial}{\partial r}\left(r^2\frac{\partial\psi}{\partial r}\right) = \gamma^2\psi \tag{9}$$

We shall now attempt to find a solution to this source-free (homogeneous) equation that is consistent with the source condition at the origin. After a little preliminary pencil scratching, it becomes evident that a change of variable $\psi = G/r$ will change the form of (9) into a very familiar differential equation, as follows:

$$\frac{\partial\psi}{\partial r} = \frac{\partial}{\partial r}\left(\frac{G}{r}\right) = \frac{1}{r}\frac{\partial G}{\partial r} - \frac{1}{r^2}G$$

$$r^2\frac{\partial\psi}{\partial r} = r\frac{\partial G}{\partial r} - G$$

$$\frac{\partial}{\partial r}\left(r^2\frac{\partial\psi}{\partial r}\right) = r\frac{\partial^2 G}{\partial r^2} + \frac{\partial G}{\partial r} - \frac{\partial G}{\partial r}$$

$$\frac{1}{r^2}\frac{\partial}{\partial r}\left(r^2\frac{\partial\psi}{\partial r}\right) = \frac{1}{r}\frac{\partial^2 G}{\partial r^2}$$

or

$$\frac{\partial^2 G}{\partial r^2} = \gamma^2 G$$

This latter equation is recognized to be the equation of any simple oscil-

---

the quantity $\delta(x - x_0)$ represents a source that is infinitely dense on the x_0 plane. This is usually called a surface density and the strength of the source is given by $\int_{-\infty}^{\infty}\delta(x - x_0)\,dx = 1$. Similarly, the product $\delta(x - x_0)\,\delta(y - y_0)$ represents a line density along the line of intersection of the planes $x = x_0$ and $y = y_0$. Finally, the product $\delta(x - x_0)\,\delta(y - y_0)\,\delta(z - z_0)$ represents an infinite density at a point. The source strength is $\iiint_{-\infty}^{\infty}\delta(x - x_0)\,\delta(y - y_0)\,\delta(z - z_0)\,dx\,dy\,dz = 1$.

latory system and has the solution
$$G = C_1 e^{-\gamma r} + C_2 e^{\gamma r}$$
where C_1 and C_2 are arbitrary constants. It follows from the change of variables introduced above that the function ψ is then
$$\psi = \frac{G}{r} = \frac{C_1}{r} e^{-\gamma r} + \frac{C_2}{r} e^{\gamma r}$$
where $\gamma = j\omega \sqrt{\mu \epsilon}$. Note that if the medium has even the slightest loss, then the propagation constant, $\gamma = j\omega \sqrt{\mu \left(\epsilon - j\frac{\sigma}{\omega} \right)}$, has a positive real part. This means that the second term in the solution would become infinite at large enough distances from the source. Thus we conclude that the constant C_2 must vanish. Next, we must make certain that the solution $\psi = C_1 e^{-\gamma r}/r$ satisfies the equation
$$\nabla^2 \psi = \gamma^2 \psi - \delta(x)\, \delta(y)\, \delta(z)\, p_0 \tag{10}$$
at the origin. The solution certainly becomes infinite there, but it must become infinite in the proper way. Notice that the integral of the source density $\delta(x)\, \delta(y)\, \delta(z)$ over a negligibly small sphere about the origin is unity: $\iiint \delta(x)\, \delta(y)\, \delta(z)\, dv = 1$. Thus if we integrate (10) over a small sphere at the origin, we obtain
$$\iiint \nabla^2 \psi\, dv - \gamma^2 \iiint \psi\, dv = -p_0 \tag{11}$$
Now we proceed step by step to evaluate the integrals over a small sphere at the origin. First consider the integral of ψ:
$$\iiint \psi\, dv = C_1 \iiint \frac{e^{-\gamma r}}{r} r^2 \sin \theta\, d\theta\, d\varphi\, dr$$
In the limit as r tends to zero, this integral likewise tends to zero. Next consider the integral of $\nabla^2 \psi$. First recall the identity $\nabla^2 \psi = \nabla \cdot \nabla \psi$, and then apply Gauss' theorem from vector analysis to change the volume integral into a surface integral (letting $\hat{r}$ be a unit vector in the direction of the spherical coordinate r)
$$\iiint \nabla \cdot \nabla \psi\, dv = \iint \nabla \psi \cdot d\mathbf{S} = \iint \nabla \psi \cdot \hat{r} r^2 \sin \theta\, d\theta\, d\varphi$$
$$= \iint \nabla \left(C_1 \frac{e^{-\gamma r}}{r} \right) \cdot \hat{r} r^2 \sin \theta\, d\theta\, d\varphi$$
$$= C_1 \int_0^{2\pi} \int_0^{\pi} \left(\frac{-\gamma e^{-\gamma r}}{r} - \frac{e^{-\gamma r}}{r^2} \right) r^2 \sin \theta\, d\theta\, d\varphi$$
and in the limit as $r \to 0$
$$\iiint \nabla \cdot \nabla \psi\, dv = -2\pi C_1 \int_0^{\pi} \sin \theta\, d\theta$$
$$= -4\pi C_1$$

Thus to satisfy the source condition given by (11), we must have $-4\pi C_1 = -p_0$ or $C_1 = p_0/4\pi$. Finally, then, the solution to (8a) that is satisfied everywhere is

$$\psi = p_0 \frac{e^{-\gamma r}}{4\pi r}$$

Since (8a) differs from (8) only in symbols ($A_z = \psi, p = \mathcal{J}$) then it follows that the potential

$$A_z = p_0 \frac{e^{-\gamma r}}{4\pi r} \tag{12}$$

satisfies Eq. (8), provided the source density is a point current $\mathcal{J}_z = \delta(x)\,\delta(y)\,\delta(z)\,p_0$. The function A_z given in Eq. (12), produced by a unit point current, is known as a Green's function. The fields of more complicated source distributions may be found by superposition.

Without doubt we have in (12) a mathematical solution for the potential due to a point current. However, physically, the idea of a point current may be repugnant at first since it is somewhat difficult to visualize charge flowing at a single point but nowhere else. We can construct a physically more palatable current distribution by superposition. As a first step, to illustrate the idea, let us construct the solution for a pair of z-directed point currents on the z axis p_1 and p_2, one on either side of the origin. The fields of the two simply add:

$$A_z = \frac{p_1 e^{-\gamma R_1}}{4\pi R_1} + \frac{p_2 e^{-\gamma R_2}}{4\pi R_2}$$

where R_1 and R_2 are the distances from the individual point currents to the observation point (see Fig. 1.3). Now let us impose the restriction

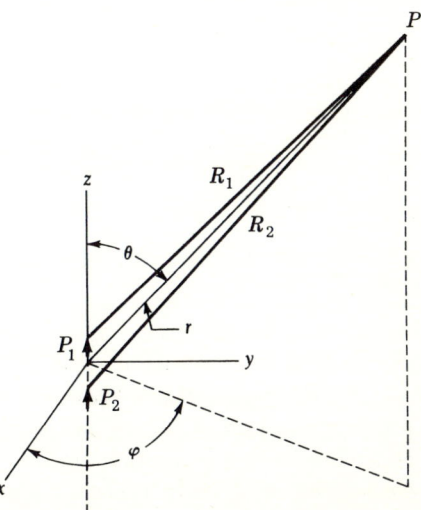

Fig. 1.3 Geometry for point dipoles close to the origin.

that the point currents are to lie so close to the origin that the approximation $R_1 \doteq r \doteq R_2$ is sufficiently good for all observation points of interest. The vector potential for such a pair of point currents is

$$A_z = \frac{p_1 e^{-\gamma r}}{4\pi r} + \frac{p_2 e^{-\gamma r}}{4\pi r}$$

Similarly, if there are N point currents, all close to the origin, the potential function is

$$A_z = \sum_{n=1}^{N} \frac{p_n e^{-\gamma r}}{4\pi r}$$

If there are so many of these point currents that they make up a continuous function $p(z) = I(z)\, dz$ along the z axis, then the potential is given by an integral. In particular, if $I(z)$ extends along the z axis from $-\Delta z/2$ to $+\Delta z/2$, the potential is

$$A_z = \int_{-\Delta z/2}^{\Delta z/2} \frac{I(z')}{4\pi R} e^{-\gamma R}\, dz'$$

Here we will assume that the overall length Δz of the elemental current is small enough so that $R \doteq r$ for all observation points of interest. Then since the quantity r is independent of the source points z', we have for the potential function of a *uniform current* I_0, of length Δz,

$$A_z = \frac{e^{-\gamma r}}{4\pi r} \int_{-\Delta z/2}^{\Delta z/2} I(z')\, dz' = \frac{I_0 \Delta z\, e^{-\gamma r}}{4\pi r} \tag{13}$$

To sum up, in Eq. (13) we have the vector potential function for an incremental or elemental current. Note that this potential is the potential of a unit point current times the product $I_0\, \Delta z$. This latter product is known as the *current moment*.

1.7 FIELD OF A POINT CURRENT OR CURRENT ELEMENT–RADIATION PATTERN

Now that we have the z component of the vector potential of a z-directed point or elemental current, we have in fact a solution to Eq. (5) for the vector potential $\mathbf{A}$. Since the x and z components of $\mathbf{A}$ satisfy similar equations, then if there are neither x- nor y-directed currents, the components A_x and A_y have no sources and are consequently zero. The complete vector potential for this case is $\mathbf{A} = A_z \hat{\mathbf{z}}$. The fields $\mathbf{E}$ and $\mathbf{H}$ associated with this vector potential may be found by the methods described in Sec. 1.5. Thus, from Eq. (6), the field **H** is given as follows:

$$\mathbf{H} = \nabla \times A_z \hat{\mathbf{z}} = \nabla A_z \times \hat{\mathbf{z}}$$

$$= \frac{I \, \Delta z}{4\pi} \nabla \left(\frac{e^{-\gamma r}}{r} \right) \times \hat{\mathbf{z}}$$

$$= \frac{I \, \Delta z}{4\pi} \left(\frac{-\gamma e^{-\gamma r}}{r} - \frac{1}{r^2} e^{-\gamma r} \right) \hat{\mathbf{r}} \times \hat{\mathbf{z}}$$

where I is the current level in the uniform current element. We can make use of the relationships between unit vectors,

$$\hat{\mathbf{r}} \times \hat{\mathbf{z}} = \hat{\mathbf{r}} \times (\hat{\mathbf{r}} \cos\theta - \hat{\boldsymbol{\theta}} \sin\theta) = -\hat{\boldsymbol{\phi}} \sin\theta$$

to find the following expression for the magnetic field of a point current or elementary dipole

$$\mathbf{H} = \hat{\boldsymbol{\phi}} \frac{I \, \Delta z}{4\pi} \left(\frac{\gamma}{r} + \frac{1}{r^2} \right) e^{-\gamma r} \sin\theta \tag{14}$$

The associated electric field can be found from Maxwell's equation $\mathbf{E} = \frac{1}{j\omega\epsilon} \nabla \times \mathbf{H}$, or from Eq. (7), $\mathbf{E} = -j\omega\mu A_z \hat{\mathbf{z}} + \frac{1}{j\omega\epsilon} \nabla \left(\frac{\partial A_z}{\partial z} \right)$:

$$E_\theta = \frac{I \, \Delta z}{4\pi} e^{-\gamma r} \left(\frac{j\omega\mu}{r} + \sqrt{\frac{\mu}{\epsilon}} \frac{1}{r^2} + \frac{1}{j\omega\epsilon r^3} \right) \sin\theta \tag{15}$$

$$E_r = \frac{I \, \Delta z}{2\pi} e^{-\gamma r} \left(\sqrt{\frac{\mu}{\epsilon}} \frac{1}{r^2} + \frac{1}{j\omega\epsilon r^3} \right) \cos\theta \tag{16}$$

Equations (14) to (16) give the complete electromagnetic field generated by an elemental current at the origin. These equations, along with Eq. (13) for the potential function, are invaluable in the solution of more complicated problems. In the application of these equations, the terms that vary as $1/r$ are commonly called the distant fields or radiation fields, and the terms in the higher inverse powers of r are called the near-field terms. Sometimes, the terms in $1/r^2$ are called the induction field terms, and the terms in $1/r^3$ are called the quasi-static field terms.

In antenna work one of the quantities of primary interest is called the *radiation pattern*. To understand precisely what is meant by this, the reader should imagine himself performing an experiment as suggested at the beginning of Sec. 1.5. In particular, he might imagine himself at a distance r_0 from the antenna making a measurement of E in the distant or radiation field. Thus he sets up an electric field measuring probe, with voltmeter, oriented in the θ direction. Then, keeping the distance fixed, he moves about, noting the different field strength or voltmeter readings as a function of angle. (It might be easier to accomplish the same thing by staying in one place and rotating the transmitting

18 ANTENNA ENGINEERING

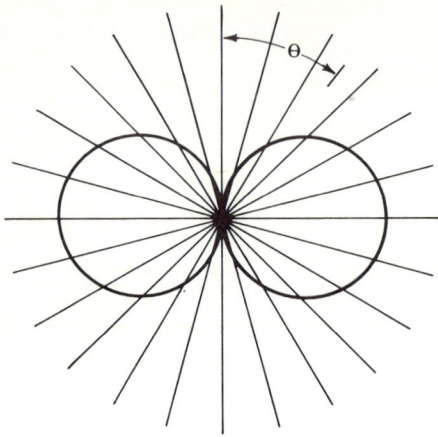

Fig. 1.4 Radiation pattern of elemental current: $E_\theta(\theta)$ in a polar plot.

antenna.) To obtain the radiation pattern, he plots the relative field strength or voltage readings as a function of angle. Of course, if he were to measure H and make a similar plot, it would look the same. For example, suppose the transmitter is an elemental current with fields given by Eqs. (14) to (16). Then, in any plane through the z axis, the electric field E_θ varies sinusoidally with θ. In the xy plane, there is no variation of the field with angle (φ). These are the *principal plane* radiation patterns for the elemental current—the principal planes being the plane containing E and that containing H. Radiation patterns are frequently plotted on polar-coordinate paper for ease in visualization. Figure 1.4 is such a display of the radiation pattern for the elemental current. Sometimes, the square of E or H is plotted as a function of angle; if this is done, it is referred to as a *power* pattern.

1.8 SUPERPOSITION

One idea that is central to antenna theory is that of superposition. The field produced by a collection of sources is the sum of the fields of the individual sources. If the sources all have the same current direction, the total potential can be found by a scalar summation of the individual potentials. Let R be the variable representing the distance between the observation point and an arbitrary point on a source distribution (Fig. 1.5). R is then a function of the point of observation (r,θ,φ) and the source point (r',θ',φ'). Let us symbolize the dependence as follows: $R(r,r')$. The potential function generated by a source distribution $\mathcal{J}_z(r')$ is then given by the integral over the source points

$$A_z = \frac{1}{4\pi} \iiint \frac{e^{-\gamma R(r,r')}}{R(r,r')} \mathcal{J}_z(r') \, dv' \qquad (17)$$

A simple but interesting special case is that of a current confined to a line, $\mathcal{J}_z(r') = I(z')\,\delta(x)\,\delta(y)$:

$$A_z(r,\theta,\varphi) = \frac{1}{4\pi} \int \frac{e^{-\gamma R(r,z')}}{R(r,z')} I(z')\,dz' \tag{18}$$

If there are x and y components to the source currents, then other components of the vector potential A_x and A_y may be calculated in similar fashion.

The superposition integrals for the potential components as given above are restricted to currents in free space, but they hold in the near field as well as the distant field. The field vectors **E** and **H** also super-

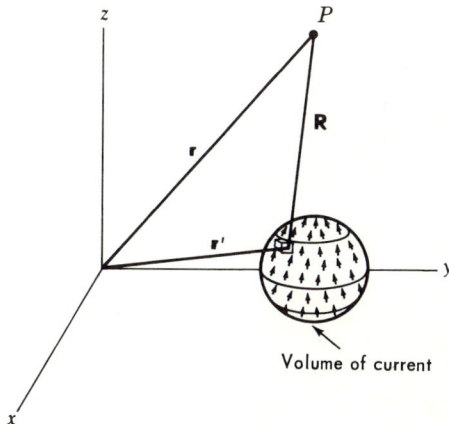

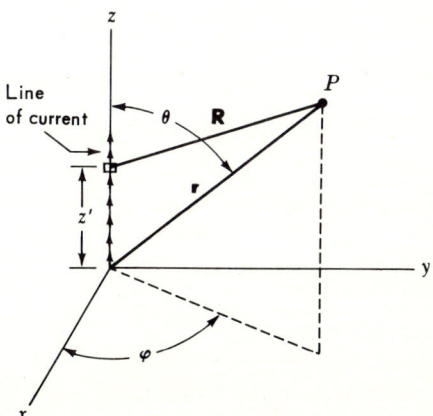

Fig. 1.5 Geometry for superposition integrals.

pose; in practice, however, the process of superposing the components of the vectors is complicated by differences in current directions and positions with respect to a common origin. Nevertheless, at distant-field points, if currents have the same direction, a convenient approximation may be employed to find the spherical components of the field vectors. A typical approximation is

$$E_\theta \doteq j\omega\mu A_z \sin\theta \tag{19}$$

$$H_\varphi \doteq \sqrt{\frac{\epsilon}{\mu}} E_\theta \tag{20}$$

The justification for this approximation is as follows: Making use of Eq. (7) for **E**, we find the θ component

$$E_\theta = \mathbf{E} \cdot \hat{\theta} = -j\omega\mu A_z \hat{\mathbf{z}} \cdot \hat{\theta} + \frac{1}{j\omega\epsilon} \hat{\theta} \cdot \nabla\left(\frac{\partial A_z}{\partial z}\right)$$

Note that $\hat{\mathbf{z}} \cdot \hat{\theta} = -\sin\theta$; also since $\hat{\theta} \cdot \nabla(\partial A_z/\partial z) = (1/r)(\partial/\partial\theta)(\partial A_z/\partial z)$, the second term in the expression for E_θ has a higher inverse power of r than does the first term. Thus, in the distant-field region, the second term may be dropped in comparison to the first; and in the distant-field region, the local fields become essentially plane waves. The ratio of E to H is then the intrinsic impedance of the medium; therefore for example H can be found from E by a relationship like Eq. (20).

Equations (17) to (20) provide the method for solving for the distant fields produced by any z-directed current distribution.

As a simple example, consider a source distribution consisting of a pair of z-directed point currents on the z axis. This distribution was considered in Sec. 1.6, but there the restriction was imposed that the currents were very close to the origin. Now let us remove that restriction. The current distribution may be represented by the equation $\mathcal{J}_z = L_1 I_1 \delta(x)\delta(y)\delta(z-z_0) + L_2 I_2 \delta(x)\delta(y)\delta(z+z_0)$, from which it follows that the potential integral of Eq. (17) reduces to

$$A_z(r,\theta,\varphi) = \frac{L_1 I_1 e^{-\gamma R_1}}{4\pi R_1} + \frac{L_2 I_2 e^{-\gamma R_2}}{4\pi R_2}$$

The distance R_1 is given by $R_1 = (r^2 + z_0^2 - 2rz_0\cos\theta)^{1/2}$. In the distant field $R_1 \doteq r - z_0\cos\theta$, $R_2 \doteq r + z_0\cos\theta$. In air or free space, the propagation constant γ is $\gamma = j\beta = j\omega(\mu\epsilon)^{1/2}$; therefore in the distant field the potential is

$$A_z = \frac{e^{-j\beta r}}{4\pi r}(I_1 L_1 e^{j\beta z_0 \cos\theta} + I_2 L_2 e^{-j\beta z_0 \cos\theta})$$

and the E field is

$$E_\theta(r,\theta,\varphi) = j\omega\mu \frac{e^{-j\beta r}}{4\pi r} \sin\theta (I_1 L_1 e^{j\beta z_0 \cos\theta} + I_2 L_2 e^{-j\beta z_0 \cos\theta})$$

If the two current moments are equal, $L_1 I_1 = I_2 L_2$, then since

$\exp(j\beta z_0 \cos\theta) + \exp(-j\beta z_0 \cos\theta) = 2\cos(\beta z_0 \cos\theta)$

the distant fields are

$$E_\theta(r,\theta,\varphi) = I_1 L_1 j\omega\mu \frac{e^{-j\beta r}}{2\pi r} \sin\theta \cos(\beta z_0 \cos\theta)$$

$$H_\varphi = \left(\frac{\epsilon}{\mu}\right)^{1/2} E_\theta$$

For spacings $2z_0$ of many wavelengths, there are many directions in which the distant field is zero or null.

Problem 1.3 Calculate the electric field at points that are at a great distance from a uniform line of current. The current has length $2h$ and lies along the z axis.

1.9 THE ROLE OF THE CURRENTS IN ANTENNA THEORY

Let us pause briefly to relate the hypothetical source currents that we have been considering to the currents on an actual antenna structure. The typical antenna is a metal and dielectric structure which at some point connects to a pair of circuit terminals (Fig. 1.6). Commonly, a generator of the circuit type is connected to the terminals. In theory, this generator may be represented by a constant-current source or a constant-voltage source in shunt or series with the generator impedance. This generator is the true source of the field in the sense that when it is switched off, everything stops. Such a source is often referred to as the *primary* source and is presumed to be known. If the source is of the constant-current type, it may be regarded as a source of the type we have been considering. The electromagnetic field which is established follows from (1) Maxwell's equations with this primary source, together with (2) the boundary conditions on the antenna, and (3) the boundary conditions at infinity (or on any other closed surface about the antenna).

The natural question arises: "Is there any more or less direct way we

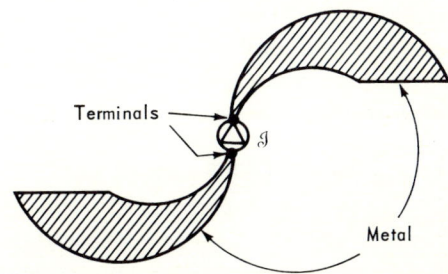

Fig. 1.6 Antenna structure.

can apply the theory developed so far to the calculation of the fields of actual antennas?" To see the answer, consider the currents that flow on the actual antenna structure. These currents contribute to the field—in fact, their contributions are usually more important to the distant fields than is that of the primary source. (That is, the same primary sources give vastly different fields with different types of antenna structures connected to their terminals.) Nevertheless, these antenna currents are resulting currents, sometimes called induced currents, and not source currents. Mathematically, this means that they are not a part of the known currents $\mathcal{J}$ which appear in the fundamental equations. Rather, they are a part of the term $j\omega\epsilon\mathbf{E}$. To find the antenna currents exactly, there is no alternative to solving the boundary-value problem. Nevertheless, sometimes these currents can be found approximately; if this is possible, then for a given antenna the induced current distribution can be split off from the $j\omega\epsilon\mathbf{E}$ term and treated as if it were a source $\mathcal{J}_i$. The effect of this is to remove the antenna structure from consideration so that the antenna currents behave as if they were sources acting in free space. Another way of saying the same thing is as follows: Suppose we know the current distribution on the antenna; if we were to arrange a distribution of primary sources $\mathcal{J}$ in free space in such a way that they had the same spatial distribution as the currents on the actual antenna structure, the external field would be the same as that produced by the actual antenna. That is, we can find the fields produced by an actual antenna by calculating the fields of a fixed, hypothetical distribution of currents acting in free space.

Keeping the latter point of view in mind will help keep our thinking straight. When we apply Eq. (13) or (17) to calculate fields, we can say that we are calculating the fields of a distribution of currents in free space that has the same distribution (and therefore the same external fields) as the currents on the actual antenna.

1.10 AUXILIARY THEOREMS AND PRINCIPLES

Although the fields generated by unidirectional electrical currents are readily found by means of a vector potential, different types of radiators are more readily handled by other techniques. We shall collect and review the main additional theorems and principles here, for use as the need for them arises.

1.10.1 DUAL SOLUTIONS

Given the fields that satisfy the Maxwell equations (1) and (2) with source currents $\mathcal{J}_1$, and with certain boundary conditions, the solutions to the equations with a similar magnetic current distribution (that is,

$\mathfrak{M}_2 = \mathfrak{J}_1$) and dual boundary conditions (that is, $\mu_2 = \epsilon_1$, $\mu_1 = \epsilon_2$) may be obtained from the first solutions by the transformation of variables $\mathbf{E}_2 = -\mathbf{H}_1$, $\mathbf{H}_2 = \mathbf{E}_1$, $\epsilon_2 = \mu_1$, $\mu_2 = \epsilon_1$. In free space the transformations

$$\mathbf{E}_2 = -\left(\frac{\mu_0}{\epsilon_0}\right)^{1/2} \mathbf{H}_1 \qquad \mathbf{H}_2 = \left(\frac{\epsilon_0}{\mu_0}\right)^{1/2} \mathbf{E}_1 \qquad \mathfrak{M}_2 = \left(\frac{\mu_0}{\epsilon_0}\right)^{1/2} \mathfrak{J}_1 \qquad (21)$$

are sufficient to change the solutions that hold with a source distribution $\mathfrak{J}_1$ into those that hold with a source distribution $\mathfrak{M}_2$.

For example, in Sec. 1.7, Eqs. (14) to (16), the free-space fields of an electric current element (electric dipole) are given. Using the duality notion, we may write down by inspection the fields of a magnetic dipole (magnetic current element) having a magnetic moment $M\,\Delta z$ in free space. From Eqs. (14) and (21), after replacing source amplitude $I\,\Delta z$ by $(\epsilon_0/\mu_0)^{1/2}$ times the magnetic moment $M\,\Delta z$, we obtain the result

$$\mathbf{E} = -\left(\frac{\mu_0}{\epsilon_0}\right)^{1/2} \hat{\boldsymbol{\phi}} \left(\frac{\epsilon_0}{\mu_0}\right)^{1/2} M\,\Delta z \left(\frac{\gamma}{r} + \frac{1}{r^2}\right) e^{-\gamma r} \sin\theta$$

and from Eqs. (15) and (21)

$$H_\theta = \frac{M\,\Delta z}{4\pi} e^{-\gamma r} \left[\frac{j\omega\epsilon_0}{r} + \left(\frac{\epsilon_0}{\mu_0}\right)^{1/2} \frac{1}{r^2} + \frac{1}{j\omega\mu_0 r^3}\right] \sin\theta$$

1.10.2 THE RECIPROCITY THEOREM

Consider a set of sources $\mathfrak{J}_a$, $\mathfrak{M}_a$ (labeled for identification purposes only) that generate a field $\mathbf{E}_a$, $\mathbf{H}_a$. These sources act together with a second set of sources $\mathfrak{J}_b$, $\mathfrak{M}_b$, operating at the same frequency, that generate the fields $\mathbf{E}_b$, $\mathbf{H}_b$. These sources all act in an environment for which the permittivity and permeability are scalars. The generalized reciprocity theorem is the following integral equation among the field quantities:

$$\int_{\text{vol}} (\mathbf{E}_b \cdot \mathfrak{J}_a - \mathbf{H}_b \cdot \mathfrak{M}_a)\,dv = \int_{\text{vol}} (\mathbf{E}_a \cdot \mathfrak{J}_b - \mathbf{H}_a \cdot \mathfrak{M}_b)\,dv \qquad (22)$$

This theorem can be derived directly from the field equations,[1,2] and other forms of it may be found in the references cited. The reciprocity theorem becomes a powerful problem-solving tool if one of the sources is taken to be a point source. For example, suppose the source b is a hypothetical one, defined as follows:

$$\mathfrak{M}_b = 0 \qquad \mathfrak{J}_b = \hat{\mathbf{p}}\,\delta(x - x_p)\,\delta(y - y_p)\,\delta(z - z_p)$$

where $\delta(x - x_p)$ is the impulse, or Dirac delta, function, and $\hat{\mathbf{p}}$ is a unit vector. With such a current distribution, the integral on the right-hand

---

[1] W. L. Weeks, "Electromagnetic Theory for Engineering Applications," pp. 318ff., John Wiley & Sons, Inc., New York, 1964.

[2] R. F. Harrington, "Time-harmonic Electromagnetic Fields," pp. 118–120, McGraw-Hill Book Company, New York, 1961.

side of Eq. (22) reduces to the component of the electric field $\mathbf{E}_a$ in the p direction (produced by $\mathcal{J}_a$, $\mathfrak{M}_a$) at the observation point (x_p, y_p, z_p):

$$\int_{\text{vol}} (\mathbf{E}_b \cdot \mathcal{J}_a - \mathbf{H}_b \cdot \mathfrak{M}_a) \, dv = \mathbf{E}_a(x_p, y_p, z_p) \cdot \hat{\mathbf{p}} \tag{23}$$

This equation is then an explicit formula for the computation of the fields produced by the known currents $\mathcal{J}_a$, $\mathfrak{M}_a$. The quantities $\mathbf{E}_b$, $\mathbf{H}_b$ are the fields produced by the point source $\mathcal{J}_b$. These fields are presumed to be known; for sources in free space, they are derived in detail in Sec. 1.7.

Problem 1.4 Repeat Prob. 1.3, this time doing the calculation by means of the reciprocity theorem.

Problem 1.5 Use the reciprocity theorem to show that the distant field of any finite current distribution in free space can have no radial component.

As in ordinary circuit theory, the reciprocity theorem says, among other things, that the positions of source and meter may be interchanged in a system with no change in the meter reading. To see this, suppose $\mathcal{J}_a$ is a unit constant-current source connected to the terminals of an antenna-circuit structure a. This source produces a voltage at a pair of terminals $t_{b,1}$, $t_{b,2}$ located in antenna-circuit structure b. Call this voltage $V_a(b)$

$$V_a(b) = \int_{t_{b,1}}^{t_{b,2}} \mathbf{E}_a \cdot \mathbf{dl}_b$$

Now suppose a constant-current source $\mathcal{J}_b$ is connected across the terminals of structure b (the same terminals $t_{b,1}$, $t_{b,2}$). This in no way disturbs the fields produced by source a, because of the linearity of the system. $\mathcal{J}_b$ extends along the incremental line $\mathbf{dl}_b$ between $t_{b,1}$ and $t_{b,2}$. Its integral over the cross-sectional area $\mathbf{dA}_b$, perpendicular to $\mathbf{dl}_b$, gives the current

$$I_b = \iint \mathcal{J}_b \cdot \mathbf{dA}_b$$

The reciprocity integral, Eq. (22), applied between sources a and b gives

$$\begin{aligned} \int \mathcal{J}_a \cdot \mathbf{E}_b \, dv &= \int \mathcal{J}_b \cdot \mathbf{E}_a \, dv \\ &= I_b \int_{t_{b,1}}^{t_{b,2}} \mathbf{E}_a \cdot \mathbf{dl}_b \\ &= I_b V_a(b) \end{aligned} \tag{24}$$

(If I_b is taken to be unity amperes, we have in Eq. (24) a formula for $V_a(b)$, the voltage at b produced by the source a.) But $\mathcal{J}_a$ is also connected to a set of terminals $t_{a,1}$ and $t_{a,2}$ in structure a. When I_b is operating, it produces a voltage $V_b(a)$ at the terminals of structure a: Note,

$$V_b(a) = \int_{t_{a,1}}^{t_{a,2}} \mathbf{E}_b \cdot \mathbf{dl}_a \qquad I_a = \iint \mathbf{J}_a \cdot \mathbf{dA}_a$$

Therefore

$$\int \mathbf{J}_a \cdot \mathbf{E}_b \, dv = I_a V_b(a)$$

Thus we have the result

$$I_a V_b(a) = I_b V_a(b) \tag{25}$$

Let us examine this latter equation in relation to the reciprocity statement concerning the interchange of source and meter. Recall that the quantity I_a represents the current source that is connected to the terminals of structure a and that it produces at the terminals of b the voltage $V_a(b)$; similarly, I_b represents the current source at the input of b that produces the voltage $V_b(a)$ at the terminals of a. If we are talking about an interchange, then $I_a = I_b$ so that

$$V_b(a) = V_a(b)$$

This equation tells us that the voltage at the terminals of structure a produced by a current source at the terminals of structure b is equal to the voltage at the terminals of b produced by the current source at the terminals of a.

In antenna work, this is a very important and interesting result. Note that there is no restriction on the type of antenna structure, orientation, environment, or any other physical restriction (except the linearity and isotropicity required for the derivation of the reciprocity theorem). This means, for example, that *the transmitting radiation pattern and the receiving pattern of an antenna are the same*. To see this, let the structure b consist of a pair of short wires forming a small dipole; let this dipole be located in the distant field of another arbitrary antenna structure a. In this case, as the structure b is moved around to different points in the distant field of a, the quantity $V_a(b)$ describes the transmitting pattern of the antenna structure a. But, even as the dipole b is moved around, the result $V_a(b) = V_b(a)$ is satisfied, point for point. Now $V_b(a)$ is the signal received by the structure a from the small dipole at a great distance; that is, for the different angular bearings of the dipole, $V_b(a)$ is the receiving pattern of antenna a.

We note also from Eq. (25) the result

$$\frac{V_b(a)}{I_b} = \frac{V_a(b)}{I_a} \tag{26}$$

which says that the transfer or mutual impedances of the structures are the same.

1.10.3 POYNTING VECTOR AND POWER TRANSFER

Electromagnetic fields may transfer power. When the time variation of the fields is sinusoidal at a single frequency, the power that leaves the sources or crosses the surface bounding a region of space may be found by integrating the complex Poynting vector $\mathbf{E} \times \mathbf{H}^*$ over the closed surface in question. In particular, the time-average power flow through the surface is

$$P_{av} = \tfrac{1}{2} Re \oint \mathbf{E} \times \mathbf{H}^* \cdot \mathbf{dS} \qquad (27)$$

where $\mathbf{dS}$ is the vector normal to the surface, the asterisk means complex conjugate, and $|\mathbf{E}|$ and $|\mathbf{H}|$ are the peak values of the fields.[1,2]

As an exercise, the reader should compute the power radiated from a current element (Sec. 1.7) and check his results by comparison to the material in Sec. 2.1.1.

1.10.4 HUYGENS' PRINCIPLE

Huygens' principle for electromagnetic waves, sometimes known as the equivalence theorem, is vital to the subject of antenna aperture theory and aperture-type antennas. It will be developed and described in detail in Chap. 6.

[1] *Ibid.*, p. 21.
[2] Weeks, op. cit., pp. 315–318.

2. SMALL ANTENNAS

The subject of small antennas provides a good starting point for the study of antennas. The word "small" herein means small relative to the wavelength of operation. The currents on small structures are pretty much determined by circuit considerations or, at most, transmission-line considerations. Thus, the current distributions on small antennas are often easily determined or approximated, and this in turn means that the fields may be calculated by the simpler methods outlined in Chap. 1. However, although the mathematical representations are relatively simple, the subject is by no means trivial or lacking in practical interest. Antennas for very low frequencies are almost inevitably small for structural reasons. Many vehicular antennas must be small because of space and mobility restrictions. Reasons of security or aesthetics also often dictate the use of small antennas.

In this chapter, the essentials of small antennas are covered. The material will be presented in a way that should also help the reader to obtain an intuitive feeling for the behavior of larger antennas.

2.1 THE CAPACITOR–PLATE ANTENNA

The point current source, discussed in Secs. 1.6 and 1.7, is of course a nonphysical entity. In fact, even a *short* linear element having a uniform current along its length is nonphysical because by the charge conservation law, $\nabla \cdot \mathcal{J} = -j\omega\rho$, an infinite charge density would be required at the end points of the current. Mathematically, this is so because the divergence of the pulselike function $\nabla \cdot f(z)\hat{\mathbf{z}}$ where

$$f(z) = \begin{cases} 1 & z_0 < z < z_1 \\ 0 & \text{elsewhere} \end{cases}$$

is infinite at the ends of the pulse. Thus, in order to have a physical structure with the characteristics of the point or elemental current model, it is necessary to incorporate some feature, such as a large plate, that provides a place for the charge to accumulate.

To better appreciate this basic problem, let us labor the point a little and examine in some detail the physical processes in an ordinary linear dipole antenna. This basic antenna consists of a pair of wires or rods with a pair of input terminals. The excitation at the input terminals is provided by a generator of the electronic-circuit type. This excitation may be visualized as a voltage, perhaps delivered by a transmission line. With the stimulus provided by the generator, currents flow in the antenna wires. But since the currents must be zero at the free ends of the wires, charge must build up along the wire. To have a *uniform* current, the current must fall abruptly to zero at the free ends of the wire. But if the current density were to fall rapidly to zero, this would imply a very large charge density in that region. Such a large charge density in a small conducting region is an unnatural situation, since the charges mutually repel one another. Consequently, on an actual wire, the current falls to zero at a leisurely space rate. The charge distributes itself along the wire, providing termination points for the electric field lines that extend from the upper half of the structure to the lower. To say it in another way, the current flows into the capacitance between the upper wire and the lower, and the current along the wires is nonuniform.

One structure that provides a place for the charge to accumulate, and consequently permits an essentially uniform current on the wires, is variously known as the capacitor-plate antenna or the top-hat-loaded dipole (Fig. 2.1). Such a structure can be very nearly the practical realization of the hypothetical point dipole or elemental current. The current will be essentially uniform if the length of the linear wires Δz is small in comparison with the radius of the plates as well as to the wavelength. The fields produced by the current on the z-directed wires are given by Eqs. (14) to (16), Chap. 3. The currents in the loading plates flow in the radial direction. Hence, the distant fields that would be produced by the currents in the plates may be effectively canceled by oppositely directed currents.

One item of interest in nearly all antennas is the input impedance.

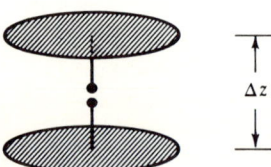

Fig. 2.1 Capacitor-plate antenna or top-hat-loaded dipole.

The input impedance is defined as in circuit analysis, with reference to the input terminals of the antenna, $Z_{in} = V_{in}/I_{in}$. In the case of the capacitor-plate antenna, it is clear by inspection that the input impedance will be primarily capacitive reactance, the capacitance being primarily the capacitance between the plates. There will be some ohmic resistance in the wires and plates, but this can be quite small, so for the moment let us ignore it. Even so, the real part of the input resistance will not be zero. The reason is that there is some power that is *radiated* (power that leaves the generator never to return), as we shall now demonstrate. The fact that power leaves the generator implies that there must be a real part to the input impedance, even though the structure has no ohmic losses.

2.1.1 RADIATED POWER AND RADIATION RESISTANCE

We shall suppose that the antenna operates in free space. In order to calculate the power flow from the antenna, we make use of the distant-field terms in the field expressions of a uniform current, and Poynting's theorem. According to this theorem as it applies to single-frequency fields, the average power flow out of a closed region of space may be found by integrating the normal component of the Poynting vector, **E** × **H***, over the surface of the region. Thus, in terms of peak values, the average power radiated by the uniform linear current in a capacitor-plate antenna is given by the following expression:

$$P_{rad,av} = \tfrac{1}{2} \operatorname{Re} \iint_{\text{large sphere}} (\mathbf{E} \times \mathbf{H}^*) \cdot d\mathbf{S}$$

$$= \frac{1}{2}\left(\frac{I\,\Delta z}{4\pi}\right)^2 \operatorname{Re} \iint \left(\frac{e^{-j\beta r}}{r} j\omega\mu \sin\theta \hat{\boldsymbol{\theta}} \times \frac{(-j\beta)e^{+j\beta r}}{r} \sin\theta \hat{\boldsymbol{\varphi}}\right)$$
$$\cdot \hat{\mathbf{r}} r^2 \sin\theta\, d\theta\, d\varphi$$

$$= \frac{1}{2}\left(\frac{I\,\Delta z}{4\pi}\right)^2 \beta^2 \left(\frac{\mu}{\epsilon}\right)^{1/2} 2\pi \int_0^\pi \sin^3\theta\, d\theta$$

$$= \left(\frac{I\,\Delta z}{4\pi}\right)^2 \beta^2 \left(\frac{\mu}{\epsilon}\right)^{1/2} \pi \cdot \frac{4}{3}$$

or

$$P_{rad,av} = I^2 \left(\frac{\Delta z}{\lambda}\right)^2 40\pi^2 \tag{1}$$

since $\beta = 2\pi/\lambda$ and $(\mu/\epsilon)^{1/2} = 120\pi$.

Equation (1) gives the average power radiated. This much power must be supplied by the generator. The generator can do so only if the input impedance has a real part. This required part of the input impedance is called the *radiation resistance* R_r. The input power is given by the familiar equation $P_{in} = \frac{1}{2}|I|^2 R_r$ (still working with peak values), in the absence of other losses. Equating the input power to the radiated power as given by Eq. (1) gives a value for the radiation resistance

$$R_r = 80\pi^2 \left(\frac{\Delta z}{\lambda}\right)^2 \doteq 800 \left(\frac{\Delta z}{\lambda}\right)^2 \qquad (2)$$

This will be the input resistance of a short antenna with a uniform current, in the absence of ohmic loss.

The power radiated and radiation resistance of other linear antennas may be found similarly.

2.1.2 DIRECTIVE GAIN

Another quantity of interest in the description of antenna performance is called the *directive gain* G_d. This quantity is defined as the ratio of the power density produced by an antenna, at a certain range in a certain direction, to the average radiated power density at that range. The average radiated power density is the total radiated power divided by the total area into which the power is radiated, that is, $P_{rad}/4\pi r^2$. Thus, the definition implies the following equation

$$G_d = \frac{\frac{1}{2} \text{Re}\,(E_\theta H_\varphi^* - E_\varphi H_\theta^*)}{P_{rad}/4\pi r^2} = \frac{\frac{1}{2}(\epsilon_0/\mu_0)^{\frac{1}{2}}(|E_\theta|^2 + |E_\varphi|^2)4\pi r^2}{P_{rad}} \qquad (3)$$

For the specific case of the capacitor-plate antenna or other uniform linear current, the directive gain is

$$G_d = \frac{1}{2}\frac{(I\,\Delta z/4\pi r)^2 \sin^2\theta \beta^2(\mu/\epsilon)^{\frac{1}{2}}}{(I\,\Delta z/\lambda)^2\,40\pi^2/4\pi r^2} = \frac{12}{8}\sin^2\theta \qquad (4)$$

The directive gain in the direction in which it is a maximum is thus $G_d(\pi/2) = 1.5$. This means that there is 1.5 times as much power density in the direction $\theta = \pi/2$ from this small antenna as there would be if the power were radiated uniformly in all directions. For this reason the gain quantity is sometimes defined in terms of a hypothetical isotropic (or omnidirectional) radiator. However, the latter does not exist for coherent electromagnetic fields; so it seems better to exclude the notion of an isotropic radiator from the definitions.

The value of the directive-gain function in the direction in which it is a maximum is called the *directivity*. It can be shown that nearly all small antennas have a directivity of 1.5. This is so because they all have essentially the same radiation pattern (a sinusoidal function of angle). This is not to say that all small antennas are equally efficient, however.

Exercise As a simple example of the directive-gain concept consider the following: Suppose that it is known that at a point in the direction of maximum field strength, a short lossless antenna produces a field of 10 mv/m per kw of input power. To provide a greater field strength at the point, without increasing the input power, a second antenna having a directive gain G_d is contemplated to replace the first. Find the field (at the corresponding position in space) that would be produced by the second antenna, per kilowatt of input power. Recall $|\mathbf{E} \times \mathbf{H}^*| = |E|^2/120\pi$.

2.2 RADIATION EFFICIENCY

In physical structures, besides the radiation resistance, the ordinary ohmic, dielectric, and/or magnetic losses contribute to the input resistance. One figure of merit for antennas is known as the *radiation efficiency*, η, defined as the ratio of the radiated power to the input power. For small antennas at least, the definition implies that the radiation efficiency is determined by the following formula:

$$\eta = \frac{|I|^2 R_r}{|I|^2 R_{\text{total}}} = \frac{R_r}{R_{\text{total}}} \tag{5}$$

The quantity R_{total} is the total input resistance. It is the sum of the radiation resistance plus all the resistance arising from the dissipative losses. The dissipative losses are calculated in standard fashion, with due regard for skin effect and other high-frequency phenomena.

2.3 SMALL ANTENNAS WITH TRANSMISSION-LINE LOADING

As was mentioned in Chap. 1, one of the problems of antenna theory is the determination of the current distribution on the antenna. In many cases, a good approximation for the current distribution may be obtained from a transmission-line model of the structure. Moreover, a transmission-line load on a short wire is another way to bring about a nearly uniform current distribution on the wire.

As a simple illustration, consider a parallel-wire line of length L, spacing Δz, excited by a voltage generator at one end and terminated at the other end by an impedance Z_L (Fig. 2.2). The current and voltage at

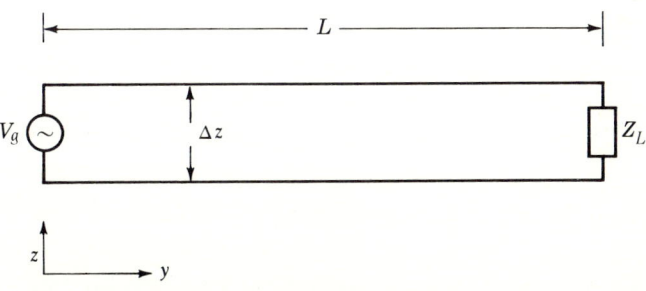

Fig. 2.2 Transmission-line-loaded antenna.

every point on the line may be obtained with transmission-line theory. For example, to find the current at the input, the first step is to find the transmission-line input impedance, since $I_{in} = V_g/Z_{in}$. The equation that shows the transformation of load impedance by a length of transmission line is well known:

$$Z_{in} = Z_0 \frac{Z_L \cosh \gamma L + Z_0 \sinh \gamma L}{Z_0 \cosh \gamma L + Z_L \sinh \gamma L}$$

or alternatively,

$$Z_{in} = Z_0 \frac{1 + \rho e^{-2\gamma L}}{1 - \rho e^{-2\gamma L}}$$

where

$$\rho = \frac{Z_L - Z_0}{Z_L + Z_0}$$
$$\gamma = (\mathcal{Y}\mathcal{Z})^{1/2}$$
$$Z_0 = (\mathcal{Z}/\mathcal{Y})^{1/2}$$
$$\mathcal{Y} = \mathcal{G} + j\omega \mathcal{C}$$
$$\mathcal{Z} = \mathcal{R} + j\omega \mathcal{L}$$

The symbols $\mathcal{Z}$, $\mathcal{R}$, $\mathcal{L}$, $\mathcal{Y}$, $\mathcal{G}$, and $\mathcal{C}$ represent the impedance, resistance, inductance, admittance, conductance, and capacitance, all per unit length of line.

The currents in the two sides of the transmission line, being oppositely directed, tend to cancel each other's effect in the distant field. Also, if the length of line L is only a small fraction of a wavelength, the current in the load and that in the input wire (generator) tend to be oppositely directed, so their distant fields also tend to cancel. The radiation per unit input current can therefore be enhanced by taking Z_L to be an open circuit, since in this case there is a vertical current only at the input. The resulting structure is a common type of transmission-line-loaded antenna.

Let us consider some of the electrical properties of the antenna consisting of a short wire loaded by an open-circuited transmission line. The input impedance, to a first approximation, is that of the open-circuited transmission line: $Z_{in} = Z_0 \coth \gamma L$. It follows that the input current is

$$I_{in} = \frac{V_g}{Z_0 \coth \gamma L} = V_g \frac{\tanh \gamma L}{Z_0} \tag{6}$$

The current distribution along the line is readily found from known forms of transmission-line solutions together with boundary values of the current at the open-circuited end and at the input. Some of the mathematical details are as follows:

$$I(y) = I_1 e^{\gamma y} + I_2 e^{-\gamma y}$$

$$I(L) = 0 = I_1 e^{\gamma L} + I_2 e^{-\gamma L} \qquad I_1 = -I_2 e^{-2\gamma L}$$

$$I(0) = V_g \frac{\tanh \gamma L}{Z_0} = I_1 + I_2 = I_2(1 - e^{-2\gamma L})$$

Thus

$$I_2 = V_g \frac{\tanh \gamma L}{Z_0(1 - e^{-2\gamma L})} \qquad I_1 = -I_2 e^{-2\gamma L}$$

$$I(y) = V_g \frac{\tanh \gamma L(-e^{\gamma(y-2L)} + e^{-\gamma y})}{Z_0(1 - e^{-2\gamma L})}$$

$$= -V_g \frac{\tanh \gamma L}{Z_0} \frac{\sinh \gamma(y - L)}{\sinh \gamma L}$$

or

$$I(y) = -V_g \frac{\sinh \gamma(y - L)}{Z_0 \cosh \gamma L} \tag{7}$$

If the line losses are small, a good approximation to the current distribution may be obtained by making the approximation $\gamma = \alpha + j\beta \doteq j\beta$, so that

$$I(y) = -jV_g \frac{\sin \beta(y - L)}{Z_0 \cos \beta L} \qquad \beta L \neq \frac{\pi}{2} \tag{8}$$

As long as the line separation Δz is small compared with the length L and the wavelength, the current in the vertical section (input wire) is essentially uniform along its length and equal to $I(0)$. At worst, the variation of current in the input wire is a form like $\sin \beta(L + \frac{1}{2}\Delta z - |z|)$. That is, the current distribution in the vertical wire is simply the extension of the transmission-line current distribution.

Next, consider the fields produced by this structure. The fields generated by the vertical section are given by Eqs. (13) to (16), Chap. 1, as long as the current distribution in this wire is essentially uniform. The fields generated by the transmission-line currents in the horizontal wires can be found from a vector potential having a y component only. The details of the field calculation are left as a problem for the reader (Prob. 2.1).

Exercise Investigate the effect in the distant field of a slightly nonuniform current in the vertical section of an open-circuited transmission-line-loaded antenna.

King and Harrison[1] have published an analysis of transmission-line-type antennas for missile applications. They give results for the radiation resistance as shown alongside the models in Fig. 2.3. Sketches of practical forms of such antennas are also shown in Fig. 2.3 (half-structure over ground—see Sec. 2.5). Similar types with appropriate structural changes are also useful in fixed installations on earth, particularly in the HF and lower bands of frequencies.

Problem 2.1 A parallel-wire transmission line has a spacing D, a length $L \leq \lambda/4$, and a propagation constant such that $\alpha/\beta = 0.01$. The terminal end is open-circuited. Compare the radiation field from the vertical wire (here assumed along the z direction), in the direction $\theta = \pi/2$ (that is, in the direction of its maximum), with the radiation field of the transmission-line pair in the direction $\theta = 0$ (that is, in the direction of their maximum). Also, consider the overall radiation pattern.

Problem 2.2 A transmission line of length L is shorted at one end and excited at the other end. The feed wire and the shorting wire are each of length D (line separation) and $D \ll L$. The characteristic impedance is 325 ohms and the propagation constant is such that $\alpha/\beta = 0.01$. Consider three cases: (1) $L = \lambda/4$, (2) $L = \lambda/2$, (3) $L \ll \lambda/4$. In each case find the currents everywhere, the radiation pattern due to the vertical currents, and a good approximation for the input impedance. In each case estimate the radiation efficiency. Finally, in the case that $L = \lambda/2$, consider the result of bending the transmission line around so as to bring the shorting wire into close proximity to the feeding wire.

Problem 2.3 A relatively short transmission line is excited at one end. The line is to be loaded at the other end with a reactance such that the currents in

[1] R. W. P. King and C. Harrison, *IRE Trans.*, **AP-8**:88 (January, 1960).

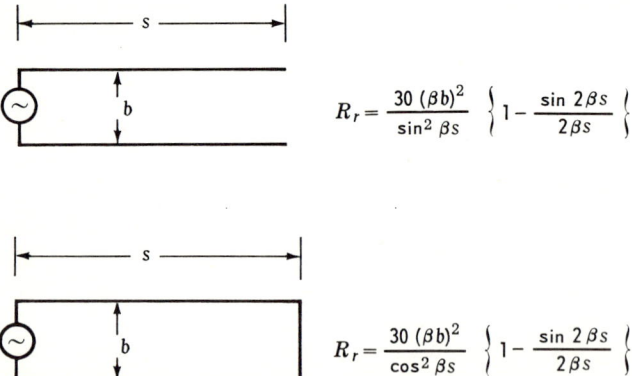

Fig. 2.3a Models for transmission-line-loaded antennas.

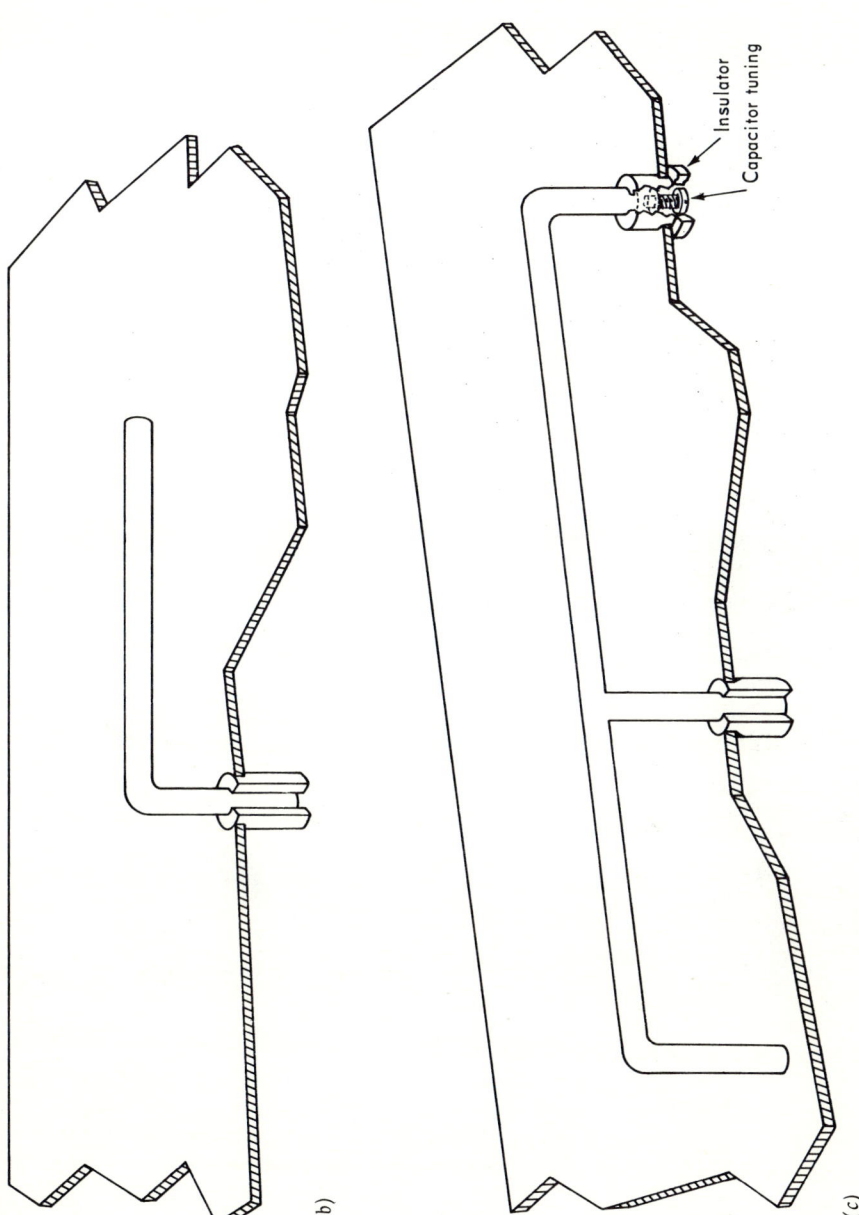

Fig. 2.3b, c Practical forms of transmission-line-loaded antennas for vehicles, aircraft, or other mobile applications.

35

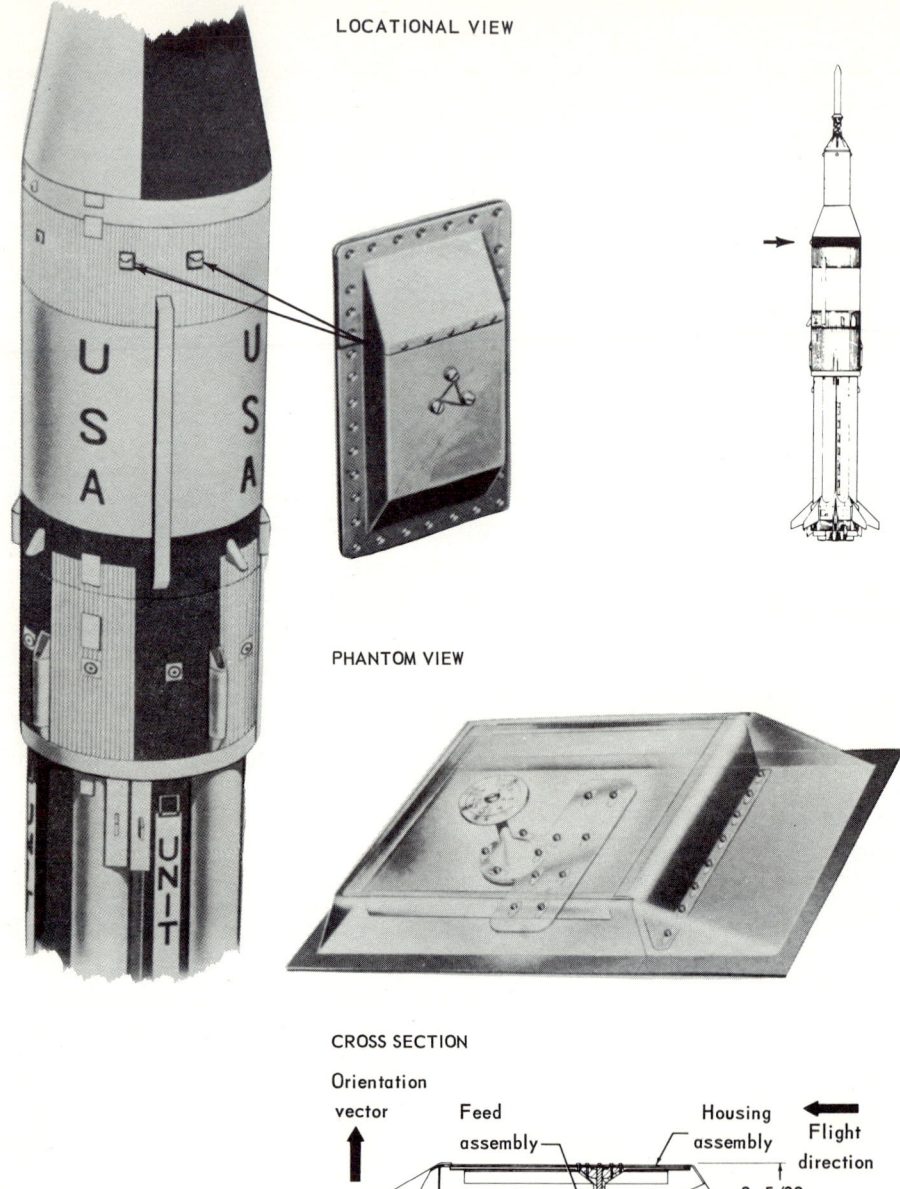

Fig. 2.3d Telemetry antenna system, 225–260 MHz.

this terminating wire and in the feeding wire have the same phase. Find the required reactance. This process has been called "multiple tuning." Consider the addition of another short length of line so as to obtain three in-phase vertical radiators.

2.4 OTHER TYPES OF TRANSMISSION–LINE LOADING

Having seen that a transmission line may be employed to load a vertical radiator, a fairly obvious next step is to load the vertical wire with more than one transmission line. The transmission lines may extend in more than one direction from the input. If, for example, lines are extended the same distance to the right and to the left (as in Fig. 2.4), then the input feeds two lines that are essentially in parallel. The method of finding the currents and radiation is basically the same as that illustrated above for an antenna with a single transmission-line load. However, the input reactance is only half that of the one-line structure.

Two additional lines, at right angles to the first two, may be added with little change, except for a further decrease in the input reactance. (Of course, the radiation from the horizontal wires is even less important than before.) Still more lines may be added to permit further division of the input current and thereby a still smaller input reactance (see Fig. 2.5). However, when there are a very large number of these radial wires, the structure (and the solutions) begins to resemble a radial transmission line or a large capacitor-plate antenna, according to point of view.

Problem 2.4 If the line is extended in two directions, it is not imperative that the extensions be the same length, nor that they have the same terminating impedance. With these factors at the disposal of the designer, it is possible to control the value of the input impedance at a specified frequency. This is often desirable for impedance-matching purposes. For example, suppose the line is extended to the left a distance d and to the right a distance t such that the overall length, $d + t$, is a quarter wavelength. The left line is shorted, and the right line is open-circuited. Let the characteristic impedance of both lines be 600 ohms and $\alpha/\beta = 0.02$. Find the distance d that results in an input impedance of 300 ohms, with zero reactance. Using the Smith chart to obtain values, plot a graph of the input admittance (or impedance) versus d. For a given d, plot also the input admittance (or impedance) over a small frequency range about the resonant frequency. This type antenna is shown in Fig. 2.3c, with capacitance added at the open end.

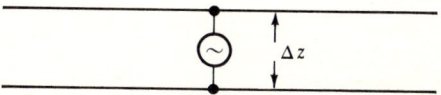

Fig. 2.4 Double-loaded transmission-line antenna.

38 ANTENNA ENGINEERING

A little more thought devoted to the behavior of transmission-line-loaded antennas can give insight into the impedance characteristics and current distributions of other antenna types. As a starting point in an evolution, consider the antenna as having open-ended transmission lines that extend in several directions. Then suppose that the lines diverge somewhat (as in Fig. 2.6), becoming nonuniform lines in the process. The change in the physical structure is not a major one, and therefore we expect the solutions to be similar. The solutions are more difficult to find, but certainly we expect on physical grounds that the main features of the input impedance and current distribution will be preserved. However, we note that the currents in the slanting transmission lines have components in the same direction as the current in the vertical feeding wire; we therefore expect that the radiation per unit input current should be increased. In the limit that the number of loading lines is very large, the structure tends toward a biconical antenna. We shall carry out a detailed analysis on the biconical antenna in Chaps. 4 and 5 and shall confirm the similarities noted. Having evolved the transmission-line-loaded structure into a biconical antenna, we can next pass to a further limit in which the angle of the biconical structure becomes very small.

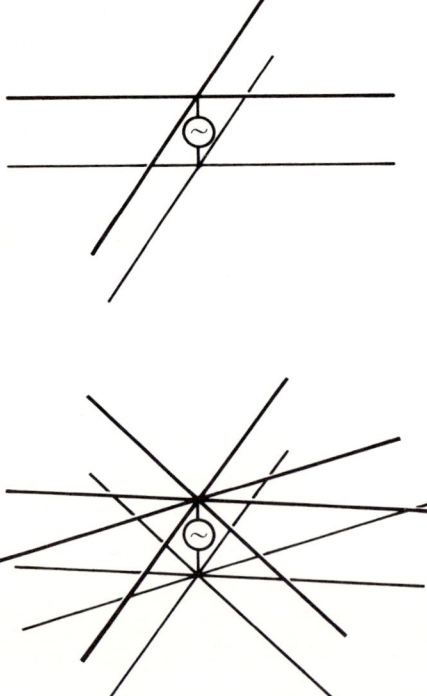

Fig. 2.5 Multiple transmission-line loadings.

SMALL ANTENNAS 39

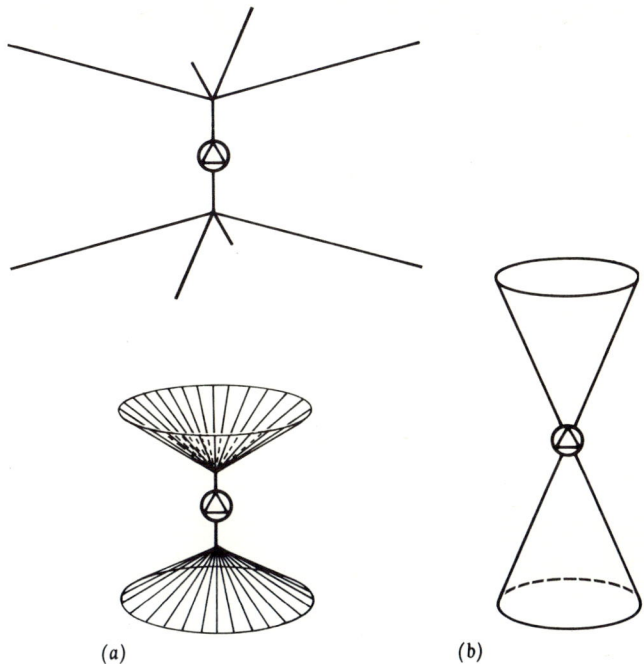

Fig. 2.6 (a) Divergent transmission-line loadings evolving into a biconical antenna; (b) biconical antenna.

In this limit, the structure evolves into a linear antenna. Consequently, we expect to find that linear antennas exhibit the same general features of input impedance behavior and current distribution (with more radiation and greater radiation resistance, however).

2.5 IMAGES

Unfortunately for the theory, antennas do not often act in free space. And sometimes the presence of nearby material bodies makes the theory dishearteningly complicated. However, with nearly perfectly conducting material in simple shapes, the idea of images can be borrowed from electrostatics and employed to give a reasonably simple account of the effects of such boundaries. The idea is to find an image system which together with the primary source generates a field that satisfies all the boundary conditions at the material boundaries.

As a preliminary, consider the fields generated by a pair of collinear point currents separated by a distance $2h$. The field everywhere is the superposition of the fields of each acting individually. In particular, consider the electric field components on the plane which is the perpen-

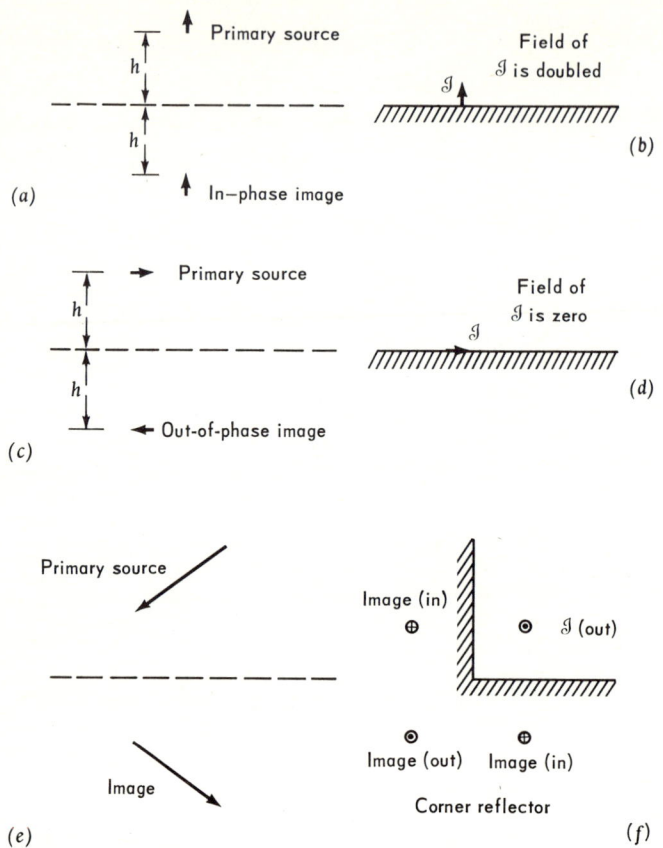

Fig. 2.7 Images of currents in conducting planes.

dicular bisector to the line joining the two currents. A simple sketch, together with the information on the components contained in Eqs. (15) and (16), Chap. 1, will show the reader that the electric field component tangent to this bisecting plane is zero.

Now, if a point current is aligned along a normal at a distance h above a conducting plane, as in Fig. 2.7a, the resulting fields must satisfy the condition that the tangential electric field is zero at the conductor. Thus, on the plane and in the region above, the boundary conditions and source condition are satisfied by the field produced by the first current together with another one, similarly directed and located a distance h below the surface of the conducting plane. The second source is referred to as the *image* current. The field is found by the superposition of the fields of the original source and its image, both acting in space without the conducting plane.

Next consider a pair of y-directed current elements, lying along the z axis and separated by a distance h. Again the total field is the superposition of the two. A simple sketch will show the reader that on the plane which is the perpendicular bisector of the line (z axis) joining the currents, the tangential components of $\mathbf{E}(E_x, E_y)$ are equal and in the same direction along the plane. But if the phase (or direction) of one of the currents is reversed, then the tangential components cancel, giving zero tangential $\mathbf{E}$ on the plane.

Next consider a current element parallel to and a distance h above a conducting plane (Fig. 2.7c). Again, the tangential component of $\mathbf{E}$ must vanish at the conducting plane. In view of the paragraph above, it is clear that this boundary condition may be satisfied by introducing, in place of the conductor, an out-of-phase image current located a distance h behind the surface of the plane.

Finally, consider an oblique current (Fig. 2.7e). The oblique current may be replaced by components that are parallel and perpendicular to a conducting plane. These parallel and perpendicular components are then individually imaged as just described.

The following simple special cases are often important. Note that if a perpendicular current is located at the surface of a conducting plane, its field is doubled (Fig. 2.7b). However, if a parallel current element lies at the surface of the plane (Fig. 2.7d), it coincides with its out-of-phase image and produces no external field.

More complicated current distributions also have images in conducting planes. To find the image system, simply regard the more complicated distribution as a superposition of current elements, and find the images for the current elements as described in the foregoing. A pictorial summary of image configurations is presented in Fig. 2.7.

If the conducting plane is bent so as to make a corner reflector, a current element in its vicinity produces a field that can be generated by the source element together with a multiple image system (Fig. 2.7f). The multiple images arise because each image has an image in the extensions of the planes making up the corner. If the angle between the conducting planes is $180/n$ where n is an integer, then the number of images is finite (some finally coincide). The images all lie on a circle whose center is at the bend.

2.6 SMALL ANTENNAS ABOVE GROUND

We may now reconsider each of the small-antenna types studied so far, in the light of the foregoing image theory. Suppose a current generator is connected between ground and one-half of the capacitor-plate antenna. From what has been said concerning images, it is clear that the effect of the ground plane is simply to provide the missing half of the capacitor-plate antenna by means of images. Thus, the fields above the ground plane are exactly the same as those produced by the complete capacitor-plate antenna with the same current. The capacitance of the half-structure over ground (sometimes called the monopole version) is twice that of the comp ete structure (called the dipole version). This is so because the currents and charges are the same in both, but the voltage from the single plate to ground is only half of that from one plate to the other. The radiation resistance of the monopole version is half that of the dipole version. The reason for this is that, since the fields are the same, the integral of **E** × **H*** over the hemisphere above ground is the same for both monopole and dipole; however, for the monopole above ground, the integral is zero on the other hemisphere. The total power radiated by the monopole above ground is then one-half of that of the complete structure in space. From the relationship $P_{\text{rad}} = \frac{1}{2}|I|^2 R_r$, it follows then that the radiation resistance must be half. Looking back to Eq. (2) for the dipole result, we find for the monopole $R_r \doteq 400(\Delta z/\lambda)^2$, or since $\Delta z = 2h$

$$R_r \doteq 1{,}600 \left(\frac{h}{\lambda}\right)^2 \tag{9}$$

where h is the height of the vertical section. This equation gives the

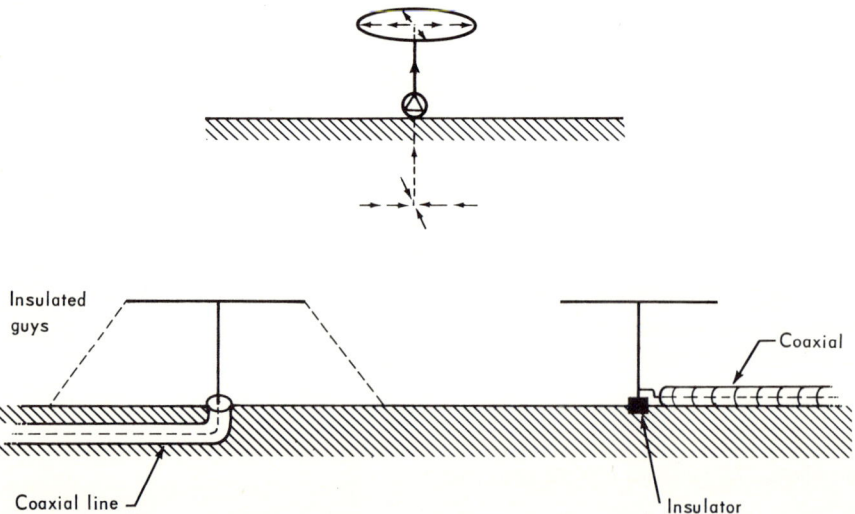

Fig. 2.8 Capacitively loaded antennas over ground.

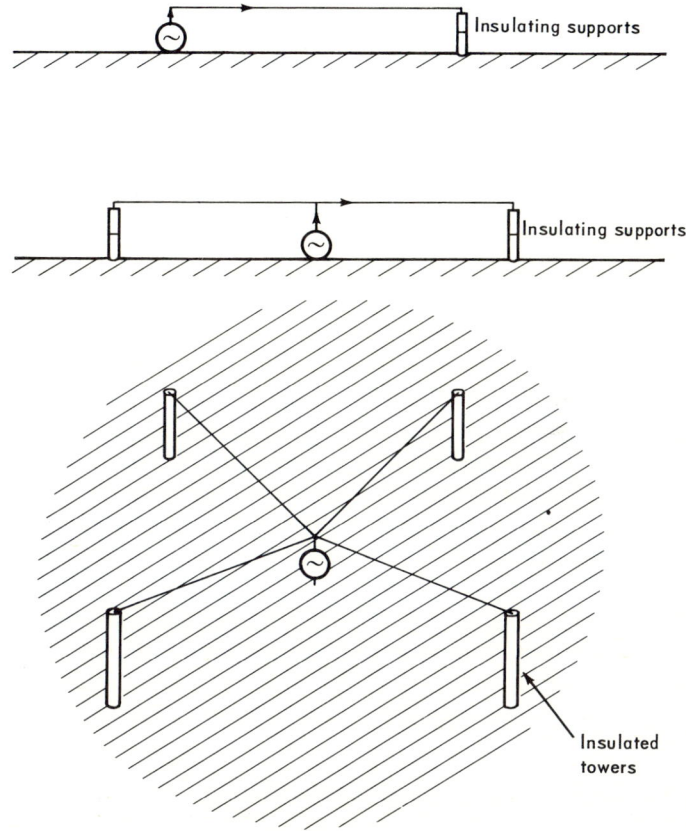

Fig. 2.9 Transmission-line-loaded antennas over ground.

radiation resistance of a monopole over a perfect ground. As with the corresponding dipole result, this equation is based on the assumption of a uniform current along the wire.

Although its desirability is open to question, the directivity of such small antennas over ground is sometimes stated relative to an isotropic radiator, $G_{di} = 3$.

The transmission-line-loaded structures considered earlier may, likewise, be split in half and mounted above a conducting ground plane. The currents are imaged in the obvious way, the fields are the same as those of the complete structure, and the input impedance is half, as before. The arrangement of a few monopole antennas above ground is indicated in Figs. 2.8 and 2.9.

2.6.1 PRACTICAL PROBLEMS

There are several practical problems associated with the implementation of the small antennas considered here. In the first place, at the lower frequencies at least, the structures must be supported with guys, and these will commonly be metal. If precautions are not taken, currents will flow in these guys, generally in such a direction as to cancel a portion of the radiation field of the radiating members. These effects of the guys may be minimized by insulating the guys from the tower and also by dividing the guys into short conducting lengths by the insertion of insulators.

Probably the most important practical limitation of the more efficient of the small antennas is associated with their extremely high input reactance. The high reactance means that in order to have appreciable input current and therefore significant or large amounts of power radiated, the voltage between the antenna and ground will be very high. The limitation on the power rating of a low-frequency transmitting installation is often determined mainly by the associated high voltages. Tuning may be employed to reduce the input impedance, but this does not relieve the voltage problem between the antenna and ground. Moreover, if the capacitive reactance of the antenna is high, the reactance of the tuning coil must be just as high. To provide this high inductive reactance, inevitably a coil resistance is introduced that will often swamp the small radiation resistance associated with the small antenna (as the reader is asked to demonstrate in the next problem). When an inductor is employed for tuning at low frequencies, great care is taken to obtain a low resistance winding. Construction with "litz" wire is common practice.

Problem 2.5 Consider, for operation at 100 kHz, a capacitively-loaded antenna which must be limited to about 30 ft (10 m) in height and to within a diameter of about 90 ft (30 m). Estimate the input reactance, and design a tuning coil to be used with this structure. Find the radiation efficiency of the antenna-and-coil combination. Consider the voltage to ground with various input powers or radiated powers up to megawatt levels.

Sometimes, for economic and structural reasons, the top loading and the tower guys are combined to give what is called an "umbrella"-loaded monopole. The umbrella struts provide capacitance to ground but are commonly too short to bring about a uniform current in the vertical. Also, since they slant downward, they are situated so that their currents partly oppose the currents in the main vertical. The general effect of umbrella loading is indicated in Figs. 2.10 and 2.11, which are based on data presented by Gangi, Sensiper and Dunn.[1] For the data presented,

[1] A. F. Gangi, S. Sensiper, and G. R. Dunn, *IEEE Trans.*, **AP-13**:864 (November, 1965).

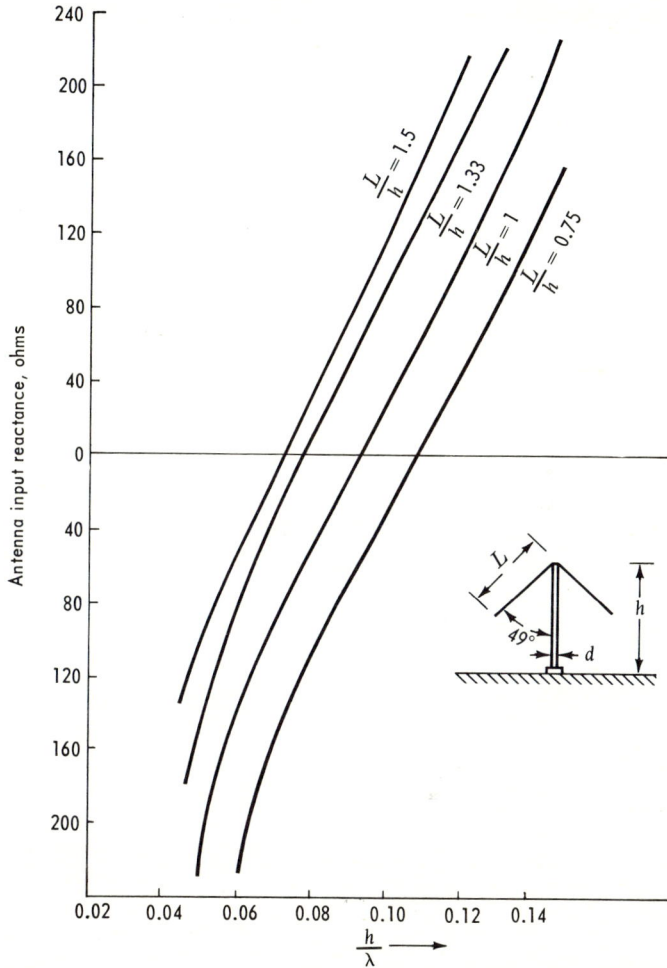

Fig. 2.10 Input reactance of monopole antennas with umbrella loading. Six struts. $h/d = 170$.

there were six umbrella struts, each making an angle of 49° from the vertical. The radiation resistance values are lower than those for a uniform current element for the two reasons mentioned—nonuniform current and lower effective current because of the slant in the loading wires.

Other practical problems of great importance are associated with the fact that many of the applications for small antennas require that they be installed over an imperfect ground plane, namely the actual earth. The imperfect conductivity of the earth brings about changes in both the

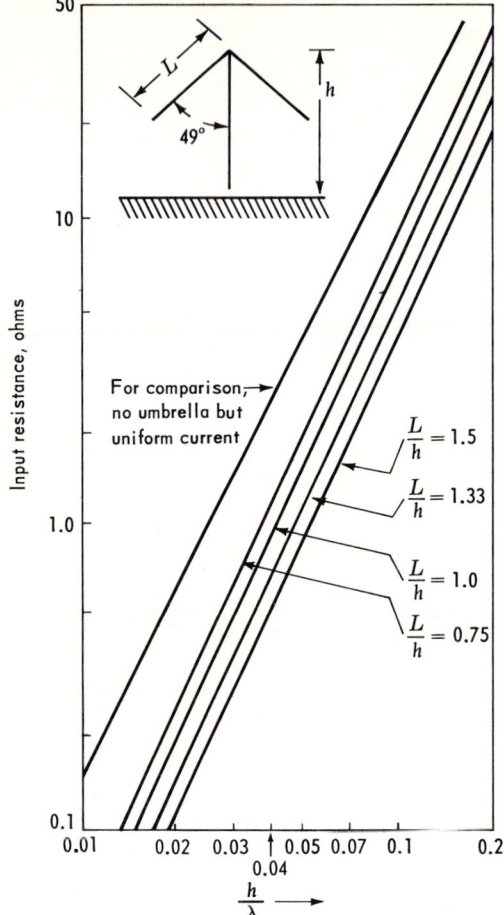

Fig. 2.11 Input resistance to monopole antennas with umbrella loading. Six struts.

radiation pattern and the input resistance. Let us consider first the increase in the input resistance.

The added input resistance introduced when the antenna is operated over an imperfect ground arises from the fact that currents are set up in the earth. These currents flow back to the input terminals and in the process they dissipate power in the earth. When a monopole antenna is operated above ground, the simplest type of connection to the earth (but by no means the most efficient) is probably a ground stake several feet long driven into the earth. The ground currents enter and leave through such a ground stake much as if they were flowing to and from a distant symmetrical electrode. The calculation of the effective resistance through this ground path is quite difficult for fields of arbitrary frequency. At dc, the problem is not as difficult, and we may obtain an optimistic estimate of the resistance at the lower frequencies from the dc value. Sunde[1] gives results for a ground stake as follows:

$$R = \frac{1}{2\pi\sigma L}\left[\ln\left(\frac{8L}{d}\right) - 1\right]$$

where L is the length and d is the diameter and $L \gg d$. For a typical earth conductivity (say about 0.01 mhos/m), this formula gives about 60 ohms for a ½-in. × 5-ft rod, a discouragingly high number. The total resistance may be cut down by employing larger stakes and/or several stakes in parallel. Sunde also gives a formula for n stakes arranged in a circle whose diameter D is very large compared with the length of the stakes:

$$R_n = \frac{1}{n 2\pi\sigma L}\left[\ln\left(\frac{8L}{d}\right) - 1 + \frac{L}{D}\sum_{m=1}^{n-1}\frac{1}{\sin\pi\frac{m}{n}}\right]$$

Probably the best way to decrease the ground resistance to the required low value is to install an artificial ground in the form of a copper screen and/or radial wires, on or just under the surface of the earth. Obviously, for economic reasons it is desirable to design ground systems that are as simple and small as possible. Although the determination of the resistance of various types of grounding arrangements is difficult, the subject is so important in practice that it has received considerable study. One of the classic papers is that of Brown, Lewis, and Epstein.[2] This paper has valuable experimental data on ground systems consisting of radial wires of various numbers and lengths. J. R. Wait,[3] and Maley and King,[4] more recently, have also published valuable data. Figures 2.12 to 2.15 are curves adapted from the data of Maley and King. Most of these data are theoretical, but the agreement with experimental results is satisfactory.

In making the calculation, the starting point is the known impedance and fields of a monopole antenna over a perfectly conducting, infinite ground plane. It is then assumed that the magnetic field over the actual ground will be almost the same as the field over the perfect ground. An average surface impedance for the parallel combination of the actual earth with the wire system imbedded in it is then calculated. Next, the electric field at the surface is determined from the product of the

---

[1] E. D. Sunde, "Earth Conduction Effects in Transmission Systems," D. Van Nostrand Company, Inc., Princeton, N.J., 1949.
[2] G. H. Brown, R. F. Lewis, and J. Epstein, *Proc. IRE*, **25**:753 (June, 1937).
[3] J. R. Wait, *Proc. IRE*, **46**:1539 (August, 1958).
[4] S. W. Maley, and R. B. King, *J. Res. Natl. Bur. Std.*, **66D**:175 (March, 1962); **68D**:157 (February, 1964); **68D**:297 (March, 1964).

48   ANTENNA ENGINEERING

tangential magnetic field and the surface impedance. With these field quantities, it is possible to relate the change in input impedance to an integral of the product of E and H over the plane. The actual calculations involve the evaluation of an integral which gives the difference between the input impedance of the antenna with the ground system in question and that of the same antenna over a perfect infinite ground plane (see Chap. 4, p. 152). It is this impedance increment that is plotted in the figures. To obtain the data presented, it was assumed that the antenna loading was sufficient to bring about a cosinusoidal

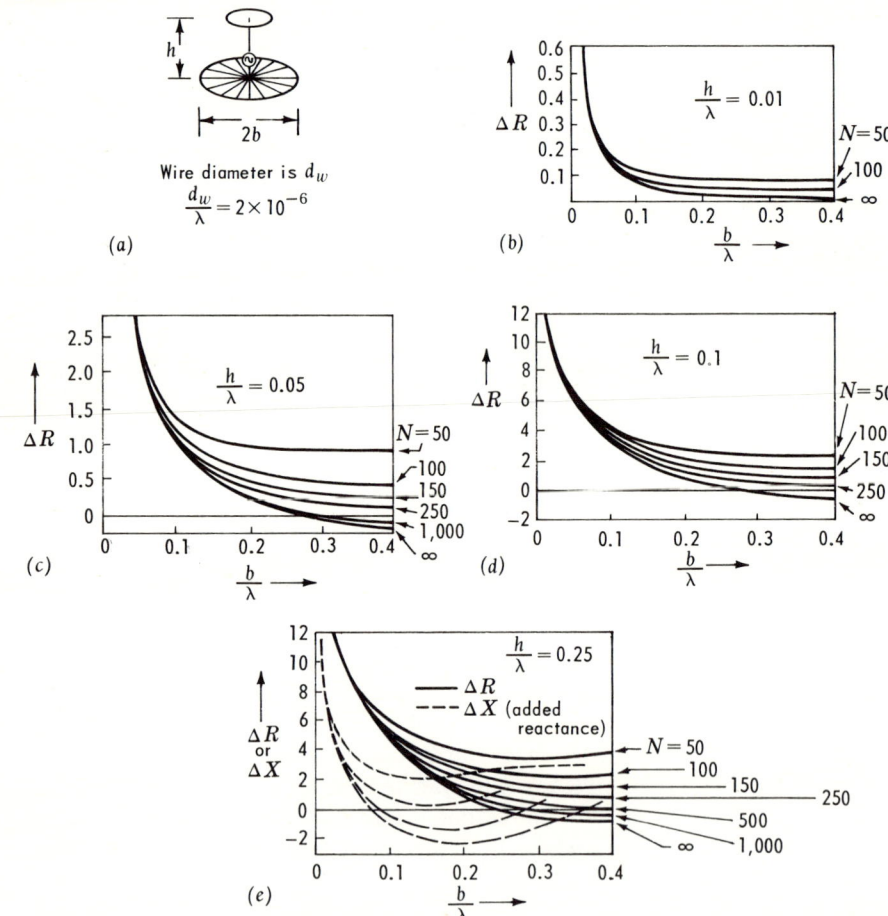

Fig. 2.12 Curves showing added resistance to resonant or loaded monopole antennas due to finite radial-wire ground screen on lossy earth, as a function of ground-screen radius, with number of wires as a parameter. Ground conductivity is such that $\omega\epsilon_0/\sigma = 0.01$ and $\omega\epsilon_0/\sigma \ll (\epsilon_r)^{-1}$.

current distribution along the wire (origin at the input). All the data presented apply to ground screens consisting of radial wires forming a "wheel" of radius b, as indicated in Fig. 2.12a. Figures 2.12b, c, and d show the resistance change for antennas of lengths 0.01, 0.05, and 0.1 wavelength, respectively. The resistance change is plotted versus the radius of the ground screen, with the number of radial wires as a parameter. Of course, the conductivity of the earth is a parameter[1] in these calculations, and the actual quantity of interest is $\sigma/\omega\epsilon$. The value of this quantity for Fig. 2.12 is $\sigma/\omega\epsilon = 100$. In all the data presented, it is assumed that the frequency, earth conductivity, and permittivity are such that $\epsilon \ll \sigma/\omega$ (that is, displacement currents in the earth are neglected). Figure 2.12e displays the corresponding data for a quarter-wavelength monopole. This figure also shows some reactance data, to show the trends. In all the calculations, the diameter of the radial wires has been assumed to be 2×10^{-6} wavelength. The actual wire diameter does not affect the results significantly, at least with wire diameters in the range from five to ten times larger or smaller than the stated value.

Figure 2.13 shows a typical variation of added resistance as a function of the number of radial wires, although these particular data apply for the specific case $h/\lambda = 0.05$, and $\sigma/\omega\epsilon = 100$. The shape of these curves indicates why the choice $N = 120$ is common in practice.

Figure 2.14 presents the same type of data as that in Fig. 2.12 except that the parameter $\sigma/\omega\epsilon$ is changed so that $(\omega\epsilon/\sigma)^{1/2} = 0.03$. Figure 2.15 presents more specific and detailed information on the effect of the earth conductivity-frequency ratio. These curves show the variation of resist-

[1] Earth conductivity varies widely from place to place but it is often in the range from 1 to 100 millimhos/m.

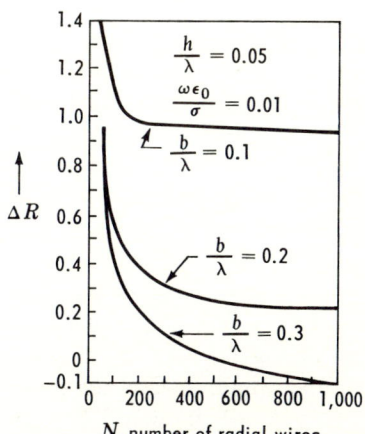

Fig. 2.13 Added resistance to top-loaded monopole antenna as a function of number of radial wires, with size of ground screen as parameter. $d_{\text{wire}}/\lambda = 2 \times 10^{-6}$.

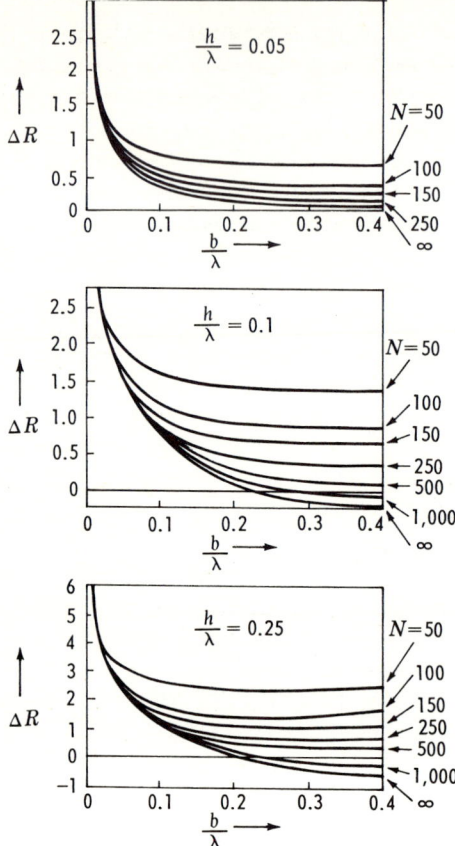

Fig. 2.14 Same as Fig. 2.12 except $\omega\epsilon/\sigma = 9 \times 10^{-4}$.

ance with the quantity $(\omega\epsilon/\sigma)^{1/2}$, with the radius of the ground screen as a parameter. The number of wires is fixed at 100 for these data.

As an example of the use of the curves, consider the design of a monopole antenna that is to be 0.1 wavelength in height. It is to be installed and operated in a location and at a frequency such that $\sigma/\omega\epsilon = 100$. Figure 2.12d is applicable to the design considerations. From that figure, it is clear that unless a ground-screen radius of at least 0.1 wavelength is employed, there is little advantage in employing more than 50 radial wires. The radiation resistance of the 0.1-wavelength monopole over a perfect ground would be 9 to 10 ohms; thus, even with a ground-screen radius as large as 0.1 wavelength, the radiation efficiency would be less than 75 percent. With 150 radials and a radius of 0.2 wavelength, the increment in resistance can be reduced to about 1.5 ohms, which means an increase in efficiency to about 85 percent. In many cases, it would be decided that the investment in larger ground screens to obtain the small possible increases in efficiency is not justified. As Fig. 2.14 shows, with locations and/or frequencies such that the quantity $\sigma/\omega\epsilon$ is greater (say 1,000), efficiencies of 85 percent or more can be obtained with ground screens of radius 0.1 wavelength, with only 100 radial wires. As a final point, note from Fig. 2.15 that with a ground screen of radius 0.2 wavelength or greater and at least 100 ground radials, the ground conductivity ratio $\sigma/\omega\epsilon$ can be considerably lower

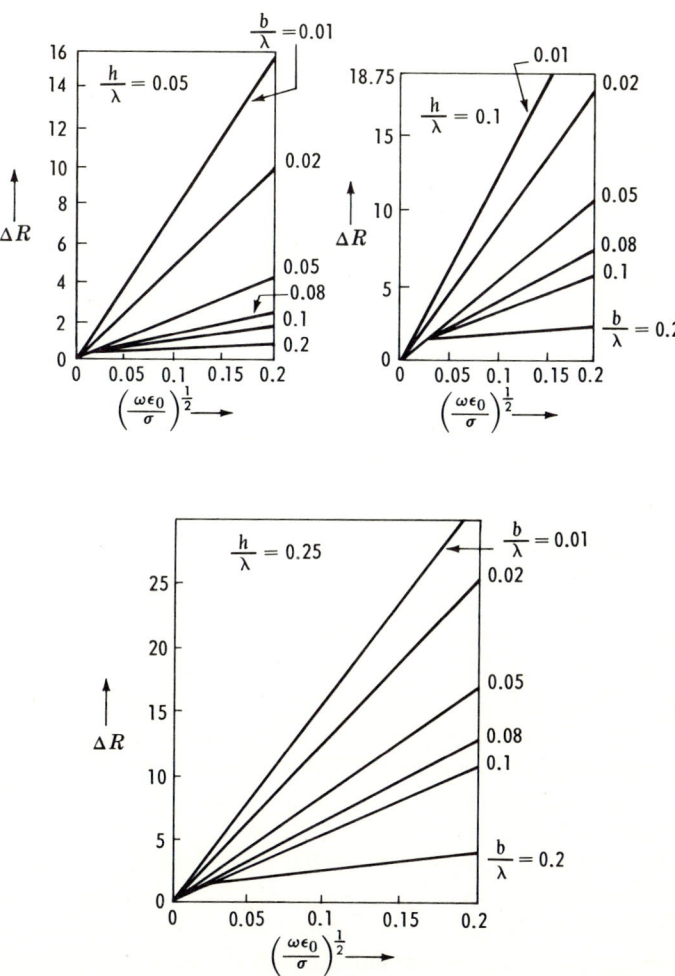

Fig. 2.15 Added resistance to top-loaded or resonant monopole antennas with ground screen consisting of 100 radial wires, as a function of ground conductivity or frequency, with ground-screen radius as a parameter. $d_w/\lambda = 2 \times 10^{-6}$.

52 ANTENNA ENGINEERING

than 100 without substantially increasing the resistance increment. This is not so if smaller ground screens are employed. The decisions concerning the extent of the ground screen at low frequencies will almost always be made on the basis of land availability and economics.

Grounding arrangements for small low-frequency antennas are also discussed in Jasik.[1] The resistance of some of the very large low-frequency installations has been brought down to a fraction of an ohm, but such installations have been extremely costly. It has been reported that the Navy VLF installation at Cutler, Maine, includes some 2,000 miles of #6 copper wire.

Another practical problem in the applications of small antennas stems from their limited bandwidth. That is, the input impedance of small antennas often changes very rapidly with frequency. Electrically small low-frequency antennas behave very much like circuit elements, and for the monopole types we have been considering, the input impedance looks like a small resistance in series with a capacitance. Such structures are often tuned with a series inductance, and the net result is then a series RLC circuit. If Δf is the 3-db bandwidth, a circuit analysis shows that

$$\frac{\Delta f}{f_0} = \frac{1}{Q} = 2\pi f_0 C R_{\text{total}}$$

where f_0 is the frequency at resonance and C is the antenna capacitance. This equation shows that the bandwidth may be increased by increasing either C or R_{total}. The radiation resistance is limited by the size of the structure; increasing the bandwidth by adding resistance of any other kind reduces the radiation efficiency. A useful figure of merit for small antennas is the product of the radiation efficiency times the bandwidth:

$$\eta \Delta f = \frac{R_r \Delta f}{R_{\text{total}}}$$
$$= 2\pi f_0^2 C R_r$$

This figure of merit turns out to be independent of the loss resistance, but it should be kept in mind that the loss resistance should be as low as economically feasible.

2.6.2 EFFECT OF THE EARTH ON RADIATION PATTERNS

The finite conductivity of the earth influences the radiation patterns of antennas as well as the impedance and radiation efficiency. That is, the field pattern over an actual earth is not given exactly by the sum of the field of the primary source and that of a simple image. Imperfect conductivity has its largest influence at angles near the horizon.

[1] H. Jasik, "Antenna Engineering Handbook," pp. 19-10 to 19-16, McGraw-Hill Book Company, New York, 1961.

SMALL ANTENNAS   53

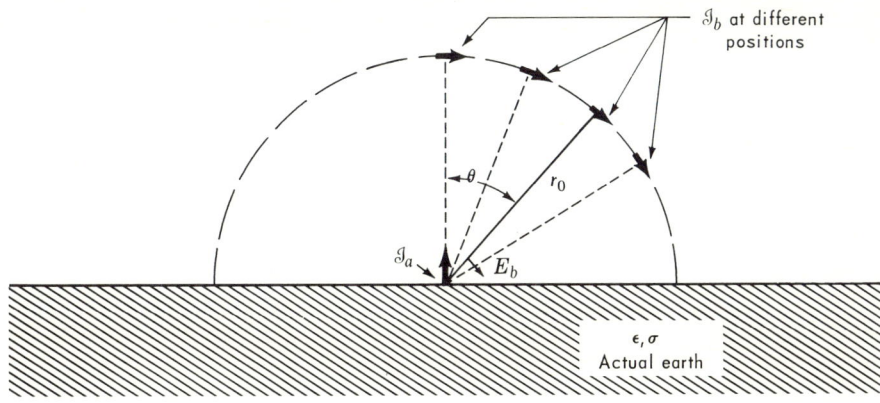

Fig. 2.16 Dipole near the surface of the earth "reacting" with point dipole at different distant-field positions.

As a specific example, let us calculate the radiation pattern of a small vertical antenna at or near the surface of the earth. In this situation, the reciprocity theorem [Eqs. (22) and (23), Chap. 1] is particularly useful, because by employing it we can avoid the rather difficult question of whether it is valid to employ plane-wave reflection coefficients to evaluate near-field effects. To apply the reciprocity theorem, imagine a unit point test source $\mathcal{J}_b$ taking up different positions on a large hemispherical surface (Fig. 2.16). Then according to Eq. (23), Chap. 1, the distant electric field produced by our antenna-current distribution $\mathcal{J}_a$ is

$$E_{a,\theta}(r,\theta) = \int \mathbf{E}_b \cdot \mathcal{J}_a \, dv$$

where $\mathbf{E}_b$ is the field produced at the position of $\mathcal{J}_a$ by the point test source $\mathcal{J}_b$. If for simplicity we take $\mathcal{J}_a$ to be a unit point source, perpendicular to the surface, then $\int \mathbf{E}_b \cdot \mathcal{J}_a \, dv = \mathbf{E}_b(0) \cdot \hat{\mathbf{z}}$. The problem then is simply to find the field produced by $\mathcal{J}_b$, at or just above the surface of the earth, as a function of its angular position θ. This is readily done, since by hypothesis the test source $\mathcal{J}_b$ is at a great distance, and therefore it produces essentially plane waves at the surface of the earth. Thus the reflected field may be found with the aid of well-known plane-wave reflection coefficients. The plane waves incident are of the type that have **H** perpendicular to the plane of incidence (**E** in the plane of incidence). Thus the solution may be written in terms of the reflection coefficients $\rho^{H\perp}$ peculiar to this polarization: $H_y = H_{\text{inc}}(1 + \rho^{H\perp})$. The E field follows from Maxwell's equation

$$E_{b,z} = \frac{\beta}{\omega \epsilon} \sin \theta \, H_{\text{inc}}(1 + \rho^{H\perp})$$

or

$$E_{a,\theta}(r,\theta) = \frac{j\omega\mu}{4\pi r} \sin\theta \, e^{-j\beta r}[1 + \rho^{H\perp}(\theta)]$$

Explicit expressions for $\rho^{H\perp}$ are well known.[1,2] For example,

$$\rho^{H\perp} = \frac{(\epsilon/\epsilon_0)^{1/2} \cos\theta - [1 - (\epsilon_0/\epsilon) \sin^2\theta]^{1/2}}{(\epsilon/\epsilon_0)^{1/2} \cos\theta + [1 - (\epsilon_0/\epsilon) \sin^2\theta]^{1/2}}$$

where the quantity $\epsilon = \epsilon' - j\sigma/\omega$ is the complex permittivity of the earth. The term $\sigma/\omega\epsilon$ is often very much larger than unity. When the quantity $n = (\epsilon/\epsilon_0)^{1/2} \doteq (-j\sigma/\omega\epsilon_0)^{1/2}$ is large, the expression for the reflection coefficient, to a good approximation, is

$$\rho^{H\perp} \doteq \frac{n \cos\theta - 1}{n \cos\theta + 1}$$

Thus the interesting term is

$$1 + \rho^{H\perp}(\theta) \doteq \frac{2n \cos\theta}{n \cos\theta + 1}$$

This factor clearly brings about a decrease in the field $E_{a,\theta}$ at low angles (horizon is $\theta = \pi/2$). The effect is indicated in Fig. 2.17. The falloff is more rapid for smaller values of n (smaller σ/ω). The equation is not strictly valid for very low angles, however. The field in the direction $\theta = \pi/2$ is not zero: it is the strength of the surface wave generated by the point-vertical dipole.

[1] E. C. Jordan, "Electromagnetic Waves and Radiating Systems," p. 141, Prentice-Hall, Inc., Englewood Cliffs, N.J., 1950.
[2] W. L. Weeks, "Electromagnetic Theory for Engineering Applications," p. 235 or p. 328, John Wiley & Sons, Inc., New York, 1964.

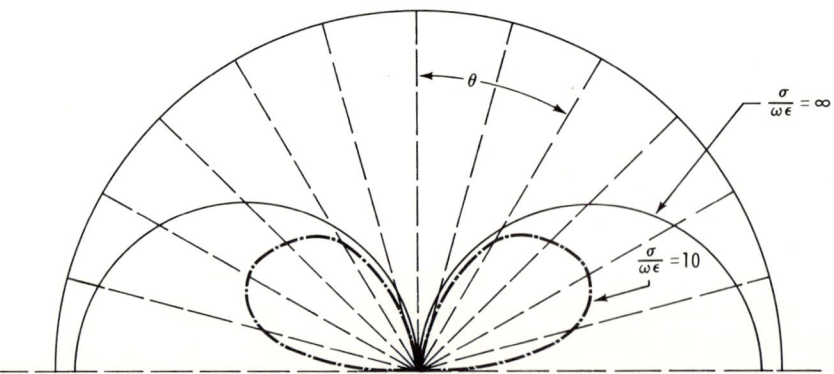

Fig. 2.17 Patterns of small vertical dipole over perfect and imperfect earth.

The exact calculation of the field of a small dipole at the surface of a flat imperfectly conducting earth is a moderately difficult problem and will not be discussed here. This problem was first solved by Sommerfeld about 1909. A good account of the solution is given in a book by Sommerfeld[1] which is now available. One difficulty is that even with the formal solution available, the evaluation of the integrals involved is rather intricate. Numerical values for the field strengths are available in the literature.[2]

A still more difficult but nonetheless practical question is: "What effect does a finite artificial ground screen have on the radiation patterns of antennas over imperfect ground?" This question is particularly relevant to antennas employed for ionospheric circuits: the reason is that long-distance ionospheric circuits usually depend on low-angle radiation, and the effect of the imperfect earth is to wipe out the low-angle radiation. Of course, if the ground screen is very small compared to a wavelength, its effect on the pattern is negligible. When the ground screen is made up of radial wires, if a good estimate of the current distribution in these wires can be made, the effect on the pattern may be calculated readily. The procedure is the same as that outlined in the beginning of this section: the term $\mathcal{J}_a$ may be defined so that it includes the known currents on the radial ground-screen wires as well as the currents on the actual antenna structure. Then as before the explicit formula for the field is

$$E_{a,\theta}(r,\theta) = \int \mathbf{E}_b \cdot \mathcal{J}_a \, dV$$

where again $\mathbf{E}_b$ is the field produced by a distant unit point dipole over the actual earth (no ground screen). Wait has considered the problem more generally; specific results are available in his papers.[3]

Problem 2.6 A Major Preliminary-Design Study. A transmitting-antenna installation is to be built for operation at a frequency of 100 kHz. An umbrella-loaded monopole is to be considered for the installation. For economic reasons, the height of the tower is to be limited to less than 500 ft (say 150 m). The whole installation must be contained in a rectangular plot of ground with dimensions 2,300 × 2,700 ft. The ground conductivity in the area is such that $\sigma/\omega\epsilon_0$ is 100. Present a preliminary design for the antenna, the tuning unit, if required, and the ground system, giving quantitative estimates of all the electrical quantities

---

[1] A. Sommerfeld, "Partial Differential Equations," chap. VI, Academic Press Inc., New York, 1949.
[2] K. A. Norton, *Proc. IRE*, **24**:1387 (1936); **25**:203 (1937).
[3] J. R. Wait, *IRE Trans.*, **AP-4**:179 (April, 1956); J. R. Wait and L. C. Walters, Influence of a Sector Ground Screen on the Field of a Vertical Antenna, *Natl. Bur. Std. Monograph* 60, Apr. 15, 1963.

of interest. Calculate the input power required to produce a field strength of 100 mv/m at a position which is 1,000 miles from the antenna and at an angle of 20° from the horizon.

2.7 SMALL LOOP ANTENNAS

Small loop antennas are frequently employed for low-frequency receiving applications. The radiation efficiency of such loops is usually very low, but this is not of first importance if the external noise levels are very high (as they usually are at low frequencies because of the electromagnetic noise generated by storms).

2.7.1 SQUARE OR RECTANGULAR LOOPS

Consider a square or rectangular loop lying in the $z = 0$ plane with its center at the origin. Let it be situated so that the currents are directed along x and y. If the loop is very small, the primary source will excite a current that is the same in all parts of the loop. Each leg then acts as an elemental current so the vector potentials are

$$A_x = \frac{IL_2}{4\pi}\left(\frac{e^{-j\beta R_1}}{R_1} - \frac{e^{-j\beta R_2}}{R_2}\right)$$

$$A_y = \frac{IL_1}{4\pi}\left(\frac{e^{-j\beta R_3}}{R_3} - \frac{e^{-j\beta R_4}}{R_4}\right)$$

where R_1, R_2, R_3, and R_4 are the distances from the observation point to each of the four legs (see Fig. 2.18). In the distant field, as far as the amplitudes are concerned, $R_1 \doteq R_2 \doteq R_3 \doteq R_4 \doteq r$, so that the potentials may be written

$$A_x = \frac{IL_2}{4\pi r}(e^{-j\beta R_1} - e^{-j\beta R_2}) \qquad A_y = \frac{IL_1}{4\pi r}(e^{-j\beta R_3} - e^{-j\beta R_4})$$

If we let the observation point be (x,y,z), then the geometry in the figure and the familiar relationships $x = r \sin\theta \cos\varphi$, $y = r \sin\theta \sin\varphi$, together imply the following relationships:

$$R_1 = [x^2 + (y + \tfrac{1}{2}L_1)^2 + z^2]^{\frac{1}{2}} \doteq r\left(1 + \frac{\tfrac{1}{2}L_1}{r^2}y\right) = r + \tfrac{1}{2}L_1 \sin\theta \sin\varphi$$

and similarly

$$R_2 = r - \tfrac{1}{2}L_1 \sin\theta \sin\varphi \qquad R_3 = r - \tfrac{1}{2}L_2 \sin\theta \cos\varphi$$
$$R_4 = r + \tfrac{1}{2}L_2 \sin\theta \cos\varphi$$

With the Euler identities, the components of the vector potential may be

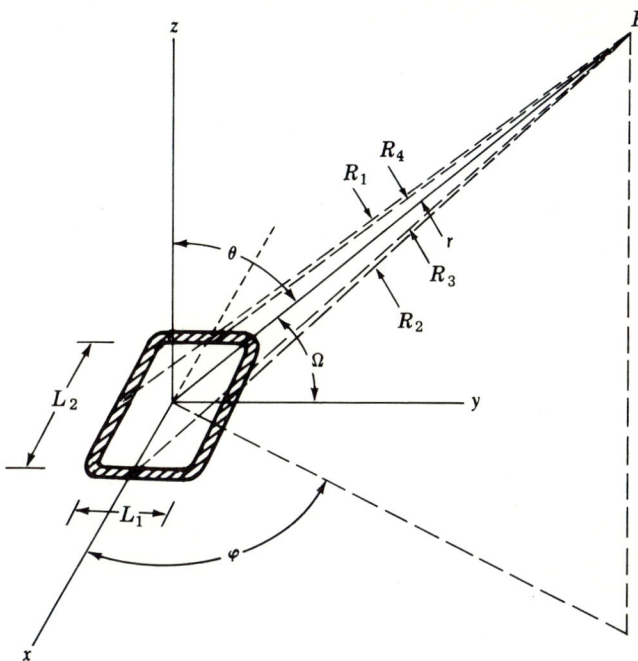

Fig. 2.18 Small loop antenna in xy plane.

written in the form

$$A_x = -\frac{IL_2}{4\pi r} e^{-j\beta r} 2j \sin(\tfrac{1}{2}\beta L_1 \sin\theta \sin\varphi)$$

$$A_y = \frac{IL_1}{4\pi r} e^{-j\beta r} 2j \sin(\tfrac{1}{2}\beta L_2 \sin\theta \cos\varphi)$$

Since the loop is small, the quantities βL_1 and βL_2 are very small compared to unity, and the sines may therefore be replaced by their arguments:

$$A_x = -jIL_2 \frac{e^{-j\beta r}}{4\pi r} \beta L_1 \sin\theta \sin\varphi$$

$$A_y = jIL_1 \frac{e^{-j\beta r}}{4\pi r} \beta L_2 \sin\theta \cos\varphi$$

The complete vector potential is then

$$\begin{aligned}\mathbf{A} &= A_x \hat{\mathbf{x}} + A_y \hat{\mathbf{y}} \\ &= j\beta IL_1 L_2 \frac{e^{-j\beta r}}{4\pi r} \sin\theta(-\hat{\mathbf{x}}\sin\varphi + \hat{\mathbf{y}}\cos\varphi)\end{aligned}$$

The quantity in parentheses may be recognized as the unit vector $\hat{\boldsymbol{\varphi}}$.

58 ANTENNA ENGINEERING

That is, the vector potential has only a φ component, $\mathbf{A} = A_\varphi \hat{\varphi}$. It therefore follows (either from a few thoughts about a vector field having a φ component only or by actually carrying out the operation $\nabla \cdot \mathbf{A}_\varphi \hat{\varphi}$ in spherical coordinates) that the divergence of $\mathbf{A}$ is zero. Since this is so, Eq. (7), Chap. 1 [$\mathbf{E} = -j\omega\mu\mathbf{A} + 1/j\omega\epsilon \, \nabla(\nabla \cdot \mathbf{A})$], provides a convenient way to find $\mathbf{E}$:

$$\mathbf{E} = \hat{\varphi} \omega\mu\beta I L_1 L_2 \frac{e^{-j\beta r}}{4\pi r} \sin\theta$$

The quantity $L_1 L_2$ is the area of the loop. The product of the loop current times the area is sometimes known as the *magnetic moment* of the loop. Multiplying and dividing by $(\epsilon)^{1/2}$, we may put the equation for the electric field in the form

$$\mathbf{E} = \hat{\varphi} \left(\frac{\mu}{\epsilon}\right)^{1/2} \frac{e^{-j\beta r}}{4\pi r} \beta^2 L_1 L_2 I \sin\theta$$

In the distant field, since the magnetic field and the electric field are related by $H_\theta = -(\epsilon/\mu)^{1/2} E_\varphi$, the magnetic field is given by the equation

$$H_\theta = -\frac{e^{-j\beta r}}{4\pi r} \beta^2 I L_1 L_2 \sin\theta$$

The total power radiated may be found by an integral of the type given in Eq. (27), Chap. 1. The process is similar to that given in Sec. 2.1.1:

$$P_{\text{rad,av}} = 10|I|^2 (\beta^2 L_1 L_2)^2$$

From this expression for the power radiated by the loop, we can find the radiation resistance of the loop, since $P_{\text{rad,av}} = \frac{1}{2}|I|^2 R_r$:

$$R_r = 20(\beta^2 L_1 L_2)^2$$

The other parts of the loop input impedance may be found in a variety of ways. One method is based on the impedance properties of transmission lines: the inductance per unit length of parallel-wire transmission line is

$$\mathcal{L} = \frac{\mu}{\pi} \cosh^{-1} \frac{L_1}{d}$$

where d is the wire diameter. The input reactance for short lengths is then

$$jX_{\text{in}} \doteq \frac{j\omega\mu}{\pi}\left[L_2 \cosh^{-1}\frac{L_1}{d} + L_1 \cosh^{-1}\frac{L_2}{d}\right]$$

The ohmic resistance per unit length is $\mathcal{R} = \frac{2}{\pi d} R_s \frac{L_1/d}{[(L_1/d)^2 - 1]^{1/2}}$ where

SMALL ANTENNAS

$R_s = (\pi f \mu / \sigma)^{1/2}$. The input resistance is then

$$R_{in} \doteq 20(\beta^2 L_1 L_2)^2 + \frac{2}{\pi d} R_s \frac{L_1 L_2}{d} \left\{ \frac{1}{[(L_1/d)^2 - 1]^{1/2}} + \frac{1}{[(L_2/d)^2 - 1]^{1/2}} \right\}$$

If the wire length-to-diameter ratio is large, the last term can be simplified

$$R_{in} = 20(\beta^2 L_1 L_2)^2 + \frac{2}{\pi d} R_s (L_1 + L_2)$$

Then according to Eq. (5) the radiation efficiency is

$$\eta = \frac{20(\beta^2 L_1 L_2)^2}{20(\beta^2 L_1 L_2)^2 + \dfrac{2}{\pi d} R_s (L_1 + L_2)}$$

As an example, to see orders of magnitude, consider the efficiency of a square loop 30 cm on a side (i.e., about 12 in.), made of copper wire 3 mm in diameter (i.e., about $\frac{1}{8}$-in. diameter). At 1 MHz (standard broadcast band), the numerical values of the parameters are as follows: $L/\lambda = 0.001$, $\beta^2 L^2 = 4 \times 10^{-5}$, $R_r = 3.2 \times 10^{-8}$ ohms, $R_s = \pi \times 10^{-4}$, $R_{wire} = 4 \times 10^{-2}$ ohms. For comparison, the reactance is about 25 ohms. Note that $R_{wire} \gg R_r$, and that the efficiency is about

$$\eta = 8 \times 10^{-7} \text{ or } 8 \times 10^{-5} \%$$

This level of efficiency is by no means lower than typical. In fact, it is sometimes said facetiously but candidly that the radiation efficiency of small loops is negligible. At 10 MHz, the figures for the same loop are $R_r = 3.2 \times 10^{-4}$ ohms, $R_{wire} = 0.126$ ohm, $\eta = 0.25$ percent. A useful formula for small square loops is

$$\eta = 8 \times 10^7 (f_{mc})^{-1/2} \left(\frac{L}{\lambda}\right)^3 \frac{d}{\lambda}$$

2.7.2 CIRCULAR LOOPS

Next, consider a circular loop of radius a lying in the xy plane. For variety we will employ the reciprocity theorem to find the radiation pattern of the loop. In accordance with the method reviewed in Sec. 1.10.2, we label the loop currents $\mathcal{J}_a$ and introduce a unit point dipole $\mathcal{J}_b$ at the observation point. Then according to the reciprocity theorem, Eq. (23), Chap. 1,

$$\mathbf{E}_a(r,\theta,\varphi) \cdot \hat{\mathbf{p}} = \int \mathbf{E}_b \cdot \mathcal{J}_a \, dv$$

where $\hat{\mathbf{p}}$ is a unit vector in the direction of $\mathcal{J}_b$. Clearly the only contribution to the integral on the right comes when **E** has a component in the xy plane. Let us obtain the radiation pattern in the xz plane by orienting the test dipole $\mathcal{J}_b$ along $\hat{\mathbf{y}}$ and considering it to be in different positions (θ), $\mathcal{J}_b = \hat{\mathbf{y}}\, \delta(z - z_0)\, \delta(x - x_0)\, \delta(y)$. In the plane being considered, this test source would produce a field $E_b = (-j\omega\mu/4\pi R)e^{-j\beta R}\hat{\mathbf{y}}$. We can represent the loop current as follows:

$$\mathcal{J}_a = \hat{\boldsymbol{\varphi}}\, \delta(\rho - a)\delta(z) I = (-\hat{\mathbf{x}} \sin\varphi + \hat{\mathbf{y}} \cos\varphi)\, \delta(\rho - a)\delta(z) I$$

The reciprocity integral then gives the electric field of the loop

$$E_\varphi(\theta)|_{\varphi=0} = -j\omega\mu I\, \frac{e^{-j\beta r}}{4\pi r}\, a \int_0^{2\pi} e^{j\beta a \sin\theta \cos\varphi} \cos\varphi\, d\varphi$$

$$= -j\omega\mu I\, \frac{e^{-j\beta r}}{4\pi r}\, a 2j\pi J_1(\beta a \sin\theta)$$

where the integral has been recognized as a standard form of a Bessel function. When βa is small, the following approximation may be employed

$$J_1(\beta a \sin\theta) \doteq \tfrac{1}{2}\beta a \sin\theta$$

$$E_\varphi(\theta)|_{\varphi=0} \doteq \left(\frac{\mu}{\epsilon}\right)^{1/2} \frac{e^{-j\beta r}}{4\pi r}\, \beta^2 a^2 \pi I \sin\theta$$

Note that, in terms of the magnetic moment (loop current times area), this result is the same as that obtained for a square loop.

Clearly the result for the electric field would be the same in any plane through the z axis, as long as the test dipole $\mathcal{J}_b$ is oriented along $\hat{\boldsymbol{\varphi}}$. If the test dipole is oriented along $\hat{\boldsymbol{\theta}}$, the integral around the loop is zero because the active component of the electric field (E_z for the pattern in the $\varphi = 0$ plane) is parallel to the current on one side of the loop and antiparallel to it on the other side. Thus, as with the square loop, the only component of the electric field is E_φ.

By comparison with the corresponding expressions for the square loop, we find the following formulas for the power radiated and the radiation resistance of small circular loops:

$$P_{\text{rad,av}} = 10|I|^2(\beta^2\pi a^2)^2$$

$$R_r = 20(\beta^2\pi a^2)^2$$

The inductance of a small circular loop $(a \gg d)$ is[1]

$$L = a\mu \left[\ln\left(\frac{16a}{d}\right) - 2\right]$$

where a is the loop radius and d is the wire diameter. As long as the wire diameter is at least ten skin depths, the wire resistance is $R_{\text{wire}} = 2R_s a/d$.

---

[1] S. Ramo, J. R. Whinnery, and T. Van Duzer, "Fields and Waves in Communication Electronics," p. 311, John Wiley & Sons, Inc., New York, 1965.

The radiation efficiency is then

$$\eta = \frac{10(\beta^2\pi a^2)^2}{10(\beta^2\pi a^2)^2 + a/dR_s}$$

For comparison, let us examine some typical numbers: if $a = 15$ cm, $d = 3$ mm, $f = 1$ MHz, then $R_r = 2 \times 10^{-8}$ ohms, $R_{\text{wire}} = 3.1 \times 10^{-2}$ ohms, $X = 6$ ohms, $\eta = 6.4 \times 10^{-7}$.

If either the rectangular loop or the circular loop has n turns of wire, where n is a small enough number that the current is uniform throughout all the turns, then the magnetic moment is nIA, the field strengths are increased by a factor of n, the power radiated is increased by a factor of n^2, and the radiation resistance is increased by a factor of n^2. The wire resistance also goes up by a factor of something greater than n, the factor depending on the geometrical arrangement (since the current distribution is disturbed by the proximity of other turns).

From the discussion presented, the reasons why small loops are not usually employed as transmitting antennas should be apparent. Their use as receiving antennas will be discussed in Chap. 8, wherein further use is made of the foregoing analysis.

In the material studied thus far, whenever we have dealt with an antenna current distribution that was more complicated than a simple point current, we were considering what might be called an array of sources. We have now reached the point at which it is appropriate to begin a formal study of such arrays. This aspect of antenna theory is the main subject of the next chapter.

3. ARRAYS

The field radiated by any small linear antenna is distributed uniformly in the plane perpendicular to the line of the antenna. This type of radiation pattern is desirable for many broadcast applications, but it is undesirable for point-to-point and preferred-coverage[1] applications. The field strength can be increased in selected directions by appropriately exciting several antennas simultaneously in what is called an antenna array. The purpose of this chapter is to describe the technology of such arrays. Much of the theory is applicable to any array of point sources whose distant field effects are governed by the wave equation (such as arrays of acoustic or seismic instruments).

In this chapter, we shall illustrate the calculation and synthesis of a quantity called the array factor, or space factor. This factor displays the features in the radiation pattern which are determined by the number and arrangement of the elements in the array. We shall also consider the novel features of active and adaptive arrays and time-modulated arrays. For the most part, we shall restrict our attention to linear arrays. Not only is this type of array very common in practice, but the principles for linear-array theory are essentially the same as those for the theory of two- and three-dimensional arrays and curved arrays. In much of the chapter, we shall consider the arrays as if they were transmitting; by reciprocity, the same structure will have an identical receiving pattern. Thus we can change points of view whenever it is convenient.

3.1 BASIC NOTIONS—THE ARRAY FACTOR

The basic idea of array theory is the principle of superposition of the fields. Consider the fields produced by a pair of z-directed current ele-

---

[1] That is, applications in which it is desired to radiate most of the energy in one direction, for example, toward population centers.

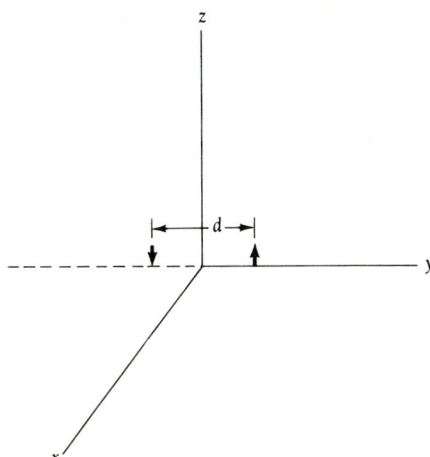

Fig. 3.1 A simple two-element array.

ments excited 180° out of phase. On a plane (the xz plane in Fig. 3.1) that is the perpendicular bisector to a line joining the centers of the elements, the net field adds to zero. However, along the y axis, the fields of the two will not in general cancel each other. In fact, if the separation d is in the vicinity of a half wavelength, the fields will add nearly in phase. The net field therefore depends on the relative phase and amplitude, the spacing, and the bearing of the observation point relative to the elements. The same simple factors are of importance when many elements act together, but most of us require the help of some formal mathematics to determine and visualize the field patterns.

The simplest array to consider is one in which all the element currents flow in the same direction and the elements are arranged in a line. In such an array, the elements may be oriented at right angles to the line of the array as in Fig. 3.2a, or they may be oriented collinearly as in Fig. 3.2b. We shall consider both of these possibilities, as well as a combination making up a two-dimensional array.

As a first example, suppose there are N elemental currents arranged collinearly along z as in Fig. 3.2b. This source distribution may be written as follows:

$$\mathcal{J}_z = \delta(x)\,\delta(y)\,[I_0\,\delta(z) + I_1\,\delta(z-d) + I_2\,\delta(z-2d) + \cdots]$$
$$= \delta(x)\,\delta(y)\,\sum_{n=0}^{N-1} I_n\,\delta(z-nd)$$

where I_n is the current in the nth element and d is the element spacing along z. As discussed in Sec. 1.8, Eq. (17), the vector potential function

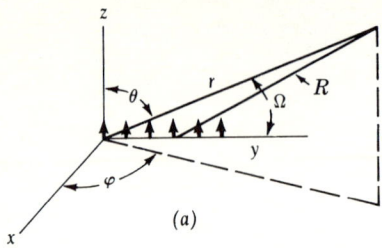

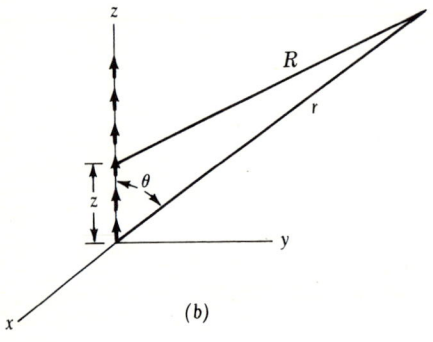

Fig. 3.2 (a) A linear array of current elements; (b) a collinear array of current elements.

for this array of elements is given by superposition:

$$A_z = \iiint \frac{e^{-j\beta R}}{4\pi R} \delta(x)\, \delta(y) \sum_{n=0}^{N-1} I_n\, \delta(z - nd)\, dx\, dy\, dz$$

In the distant field region, examination of Fig. 3.2b shows that the distance R may be approximated as follows:

$$R \doteq r - z \cos\theta$$

In the expression for the vector potential, this approximation is employed in the exponential phase factor, but in the denominator, the approximation $R \doteq r$ is sufficient:

$$A_z = \frac{e^{-j\beta r}}{4\pi r} \int e^{j\beta z \cos\theta} \sum_{n=0}^{N-1} I_n\, \delta(z - nd)\, dz$$

$$= \frac{e^{-j\beta r}}{4\pi r} \sum_{n=0}^{N-1} I_n e^{j\beta n d \cos\theta}$$

According to Eq. (19), Chap. 1, the electric field strength produced by the collinear array of elemental currents is then

$$E_\theta = j\omega\mu \frac{e^{-j\beta r}}{4\pi r} \sin\theta \sum_{n=0}^{N-1} I_n e^{j\beta n d \cos\theta} \tag{1}$$

The summation factor in the expressions for the fields is known as the *array factor*, or *space factor*, AF:

$$\text{AF} = \sum_{n=0}^{N-1} I_n e^{j\beta nd \cos \theta} \qquad (2)$$

(or sometimes the magnitude of AF or its square is defined as the space factor). Clearly the space factor contains all of the interesting pattern-directivity information except that inherent in the directional properties of the elements themselves. It does not usually contain the distance dependence.[1]

Next, let us consider an array of elemental currents arranged as indicated in Fig. 3.2a. Such a distribution of sources can be expressed as follows:

$$\mathcal{J}_z = \delta(x)\,\delta(z) \sum_{n=0}^{N-1} I_n\, \delta(y - nd)$$

where again d is the element spacing, this time along the y axis. The potential may be found from the superposition integral Eq. (17), Chap. 1, as before. The distance R between source points and observation point is simply expressed if we introduce an angle Ω which is a polar angle measured from the y axis (see Fig. 3.2a). (Ω is an angle with the y axis of the type that the usual angle θ is with the z axis.) In this case, $R \doteq r - y \cos \Omega$, so the result for the vector potential is

$$A_z = \frac{e^{-j\beta r}}{4\pi r} \int e^{j\beta y \cos \Omega} \sum_{n=0}^{N-1} I_n\, \delta(y-nd)\,dy$$

$$= \frac{e^{-j\beta r}}{4\pi r} \sum_{n=0}^{N-1} I_n e^{j\beta nd \cos \Omega}$$

---

[1] The idea of the array factor is more general than is suggested by the foregoing remarks. For example, consider a linear array of scalar sources of any kind that emits waves that superpose. If A_0 is the distant-field amplitude produced by source zero, A_1 that produced by source one, and so on, the total field is, by superposition,

$$F = \sum_{n=0}^{N-1} A_n e^{-j\beta r_n}$$

Let ϕ be the angle between the line of the array and the line drawn from the field point to the origin. Then, as before, the distance from the field point to the nth element is approximately $r_n = r_0 - nd \cos \phi$ and the field is

$$F = e^{-j\beta r_0} \sum_{n=0}^{N-1} A_n e^{j\beta nd \cos \phi}$$

and the electric field is

$$E_\theta = j\omega\mu \frac{e^{-j\beta r}}{4\pi r} \sin\theta \sum_{n=0}^{N-1} I_n e^{j\beta nd \cos\Omega} \qquad (3)$$

Note that this equation for the electric field of an array of noncollinear elements along the y axis is identical to Eq. (1) (collinear elements along z) except for the substitution of the angle Ω for the angle θ in the array factor. The expression of (3) in terms of the angle Ω is appropriate because the array factor is symmetrical with respect to the line of the array (in this case, the y axis). The factor, $\cos\Omega$, is readily expressed in terms of the usual spherical coordinate angles: $\cos\Omega = \sin\theta \sin\varphi$.

Note that in both Eqs. (1) and (3), the electric field of an array of elemental currents is the product of the element field pattern times the array factor. This is a general result, as we shall now demonstrate. The elements in the array may be more complicated than elemental currents. As a preliminary, let us examine the forms of the field expressions for a continuous collinear array along z, so that we may recognize these when they appear in more complicated solutions. Let the distribution along z be represented by $i(z)$, which for convenience is normalized so that the input current is unity. The representation of the continuous collinear array is then $\mathcal{J}_z = i(z)\,\delta(x)\,\delta(y)$. The vector potential is

$$A_z = \iiint \frac{e^{-j\beta R}}{4\pi R} \delta(x)\,\delta(y)\,i(z)\,dx\,dy\,dz$$

The distance R is approximated as before, $R \doteq r - z\cos\theta$; thus

$$A_z = \frac{e^{-j\beta r}}{4\pi r} \int e^{j\beta z \cos\theta} i(z)\,dz$$

and

$$E_\theta = j\omega\mu \frac{e^{-j\beta r}}{4\pi r} \sin\theta \int e^{j\beta z \cos\theta} i(z)\,dz \qquad (4)$$

These general forms apply to a linear current distribution along z; with these, we can permit the elements of an array to be more complicated than elemental currents.

With the experience gained by obtaining Eq. (4), let us now consider an array made up of linear elements each having the same distribution along z, $i(z)$. Let the elements lie along the y axis with a spacing d, as indicated in Fig. 3.3. Even though the individual elements all have the same distribution along z, they need not all have the same input current level. The variation of input currents may be included in a representa-

ARRAYS 67

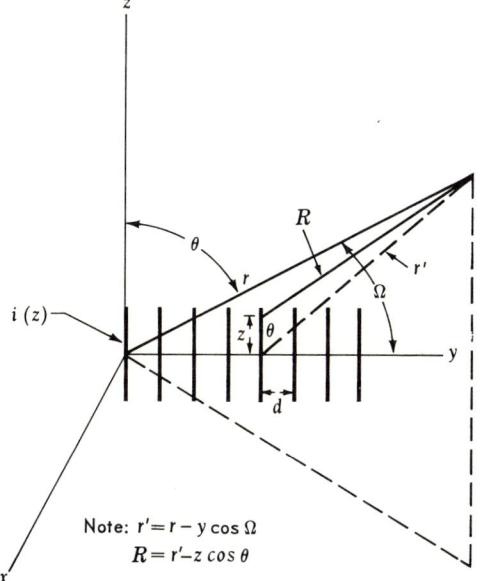

Note: $r' = r - y \cos \Omega$
$R = r' - z \cos \theta$

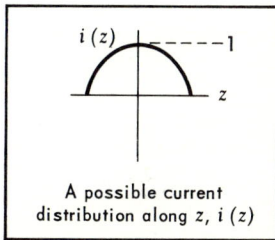

A possible current distribution along z, $i(z)$

Fig. 3.3 An array of z-directed current distributions.

tion of such a source distribution as follows:

$$\mathcal{J}_z = i(z)\, \delta(x) \sum_{n=0}^{N-1} I_n\, \delta(y - nd)$$

The procedure for the calculation of the fields produced by this distribution is the same as before, starting with Eq. (17), Chap. 1. The approximation for the distance R between source points and observation point is slightly more complicated, but is readily obtained (Fig. 3.3):

$$R = r - y \cos \Omega - z \cos \theta$$

68 ANTENNA ENGINEERING

The integral for the vector potential then reduces as follows:

$$A_z = \frac{e^{-j\beta r}}{4\pi r} \iint e^{j\beta y \cos \Omega + j\beta z \cos \theta} i(z) \sum_{n=0}^{N-1} I_n \, \delta(y - nd) \, dy \, dz$$

$$= \frac{e^{-j\beta r}}{4\pi r} \int e^{j\beta z \cos \theta} i(z) \, dz \int e^{j\beta y \cos \Omega} \sum_{n=0}^{N-1} I_n \, \delta(y - nd) \, dy$$

$$= \frac{e^{-j\beta r}}{4\pi r} \int e^{j\beta z \cos \theta} i(z) \, dz \sum_{n=0}^{N-1} I_n e^{j\beta nd \cos \Omega}$$

and the electric field is then

$$E_\theta = \left[j\omega\mu \frac{e^{-j\beta r}}{4\pi r} \sin \theta \int e^{j\beta z \cos \theta} i(z) \, dz \right] \sum_{n=0}^{N-1} I_n e^{j\beta nd \cos \Omega} \qquad (5)$$

Note that the factor in brackets is the field of a single element with a distribution $i(z)$ as given in Eq. (4); it multiplies an array factor of the type defined in Eq. (2). We have shown, then, that the field of the linear array is the product of the field of the elements in the array times the array factor.

If the elements in an array are arranged collinearly (Fig. 3.4) along z, the mathematical representation of the array requires a little more inge-

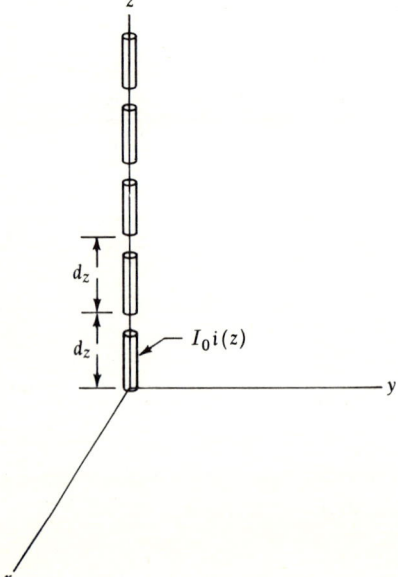

Fig. 3.4 Current distributions in a collinear array.

nuity, but the general procedures and results are the same. For example, let the distribution of currents be

$$\mathcal{J}_z = \delta(x)\,\delta(y)\,[I_0 i_0(z) + I_1 i_1(z) + I_2 i_2(z) + \cdots]$$

If all the elements in this array are identical but are simply displaced a distance d_z from each other along the z axis, then the individual distributions are related, for example, $i_1(z) = i_0(z - d_z)$, $i_2(z) = i_0(z - 2d_z)$, $i_n(z) = i_0(z - nd_z)$, and we can represent a collinear array of N such elements as follows:

$$\mathcal{J}_z = \delta(x)\,\delta(y)\sum_{n=0}^{N-1} I_n i_0(z - nd_z)$$

The distance R in this case involves only θ: $R \doteq r - z\cos\theta$. The potential integral is then

$$A_z = \frac{e^{-j\beta r}}{4\pi r}\sum_{n=0}^{N-1} I_n \int e^{j\beta z \cos\theta} i_0(z - nd_z)\,dz$$

To carry out the integral, we may introduce a change of variable, $z' = z - nd_z$:

$$A_z = \frac{e^{-j\beta r}}{4\pi r}\sum_{n=0}^{N-1} I_n e^{j\beta nd_z \cos\theta} \int e^{j\beta z' \cos\theta} i_0(z')\,dz'$$

or

$$E_\theta = j\omega\mu\,\frac{e^{-j\beta r}}{4\pi r}\sin\theta \int e^{j\beta z' \cos\theta} i_0(z')\,dz' \sum_{n=0}^{N-1} I_n e^{j\beta nd_z \cos\theta}$$

Again, the field of the collinear array of identical elements is the product of the element pattern times the array factor.

We can obtain similar results for two-dimensional arrays. A two-dimensional array may be represented by a combination of the representations of the two linear arrays considered previously:

$$\mathcal{J}_z = \delta(x)\sum_{n=0}^{N-1}\sum_{m=0}^{M-1} I_{nm}\,\delta(y - nd_y)i_0(z - md_z)$$

This source distribution represents an $n \times m$ two-dimensional array of the type indicated in Fig. 3.5. The distance R to points in the far field is

$$R \doteq r - y\cos\Omega - z\cos\theta$$

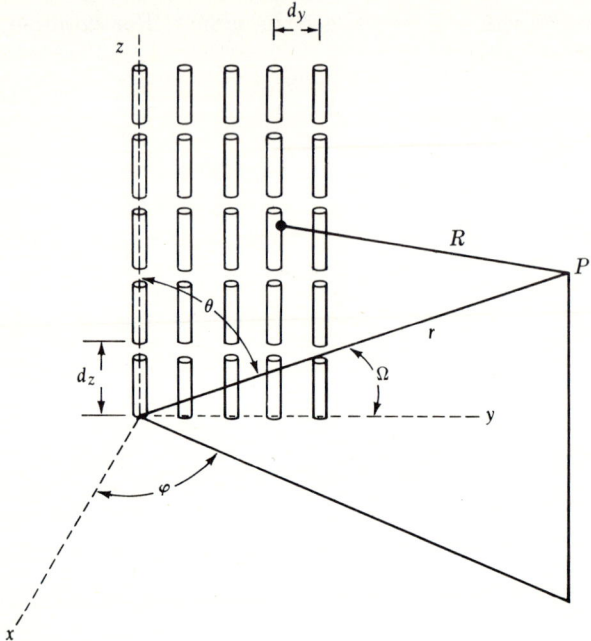

Fig. 3.5 A two-dimensional array.

The vector potential integral is

$$A_z = \frac{e^{-j\beta r}}{4\pi r} \iint e^{j\beta y \cos \Omega + j\beta z \cos \theta} \sum_{n=0}^{N-1} \sum_{m=0}^{M-1} I_{nm}\, \delta(y - nd_y) i_0(z - md_z)\, dy\, dz$$

$$= \frac{e^{-j\beta r}}{4\pi r} \sum_{n=0}^{N-1} \sum_{m=0}^{M-1} I_{nm} e^{j\beta n d_y \cos \Omega} \int e^{j\beta z \cos \theta} i_0(z - md_z)\, dz$$

With a change of variable, as before, $z' = z - md_z$, we obtain, finally, the electric field strength of a two-dimensional array

$$E = \left[j\omega\mu \frac{e^{-j\beta r}}{4\pi r} \sin \theta \int e^{j\beta z' \cos \theta} i_0(z')\, dz' \right] \cdot \sum_{n=0}^{N-1} e^{j\beta n d_y \cos \Omega} \sum_{m=0}^{M-1} I_{nm} e^{j\beta m d_z \cos \theta}$$

In the plane $\Omega = \pi/2$ ($\varphi = 0, \pi$), the field is thus basically the element pattern times the collinear array factor; in the plane $\theta = \pi/2$, it is the element pattern times the noncollinear array factor (as before, $\cos \Omega = \sin \theta \sin \varphi$).

Hold on. I agree that it is time to pause for a breath. Be assured that things will get simpler before they get more complicated.

In the past few pages, we have managed to obtain some useful general results concerning arrays and certainly have generated some complicated general formulas for the fields produced by arrays. Because they are general formulas, they do not tell us anything specific and are therefore hard to visualize. The basic result obtained is as follows: formulas for fields of arrays can be factored into the array factor times a formula for the field of the individual elements (or subarrays) making up the array. Discussions of the various types of element patterns will be presented in subsequent chapters. In this chapter we shall study the characteristics of array factors.

Our initial purpose will be to learn the main characteristics of common linear arrays. But we shall be able to do this better after some further preliminary discussions and changes in notation. We have seen that the array factor is symmetric about the line of the array, and that when expressed in terms of the polar angle with respect to the line of the array, it takes the form of Eq. (2):

$$\mathrm{AF} = \sum_{n=0}^{N-1} I_n e^{j\beta nd \cos\theta}$$

It has the same form as long as θ is imagined to be a polar angle with respect to the line of the array. But since θ is conventionally the polar angle with respect to the z axis, we have also used the angle Ω in the expressions for the array factor—in particular, in the case that the array was lined up along the y direction. The results which we shall obtain shortly for the array factor are independent of the coordinate system; therefore let us emphasize this by changing to a different symbol

$$\mathrm{AF} = \sum_{n=0}^{N-1} I_n e^{j\beta nd \cos\phi}$$

Here, ϕ is the polar angle with respect to the line of the array whatever the line of the array may be. It may be equal to θ, or Ω, but it is *not* equal to the spherical coordinate φ since the latter is not a polar angle; if the line of the array is the x axis, then ϕ and φ happen to be equal for points in the $\theta = \pi/2$ plane (xy plane). In the expression for the array factor, the quantities I_n represent the input currents at the different elements. We may choose to make these input currents differ in both amplitude and phase. Hence, in general we should regard the quantities I_n as complex numbers representing the different amplitudes and phases. In the applications we shall often be particularly interested in arrays in which the phase of the current differs by the same amount from each element to the next. For this reason, the expression for the array factor will now be written so as to display explicitly such a linear progressive phase variation. To do so, we introduce a variable A_n such that $A_n e^{jn\alpha} = I_n$:

$$\text{AF} = \sum_{n=0}^{N-1} A_n e^{jn(\beta d \cos \phi + \alpha)}$$

Here, α is the uniform phase-shift angle from one element to the next (a negative value of α represents a phase lag, i.e., a delay). In this expression for the array factor, the quantities A_n are commonly, but not necessarily, real. If they are complex, they represent the deviations from a uniform progressive phase shift. For example, the phase-shift constant α may be zero, and if it is, and the A_n's are real, then the currents are all in phase. The adoption of the symbol A_n in the array factor also helps us to remember that the array factor has more general application than simply to arrays of electric currents.

There is still one further change in variable and three general observations which are helpful to make before we turn our attention to particular examples. In the analysis, it turns out to be a great simplification to think of the array factor as a function of a single variable ψ

$$\text{AF}(\psi) = \sum_{n=0}^{N-1} A_n e^{jn\psi} \qquad (6)$$

where, by inspection, $\psi = \beta d \cos \phi + \alpha$.

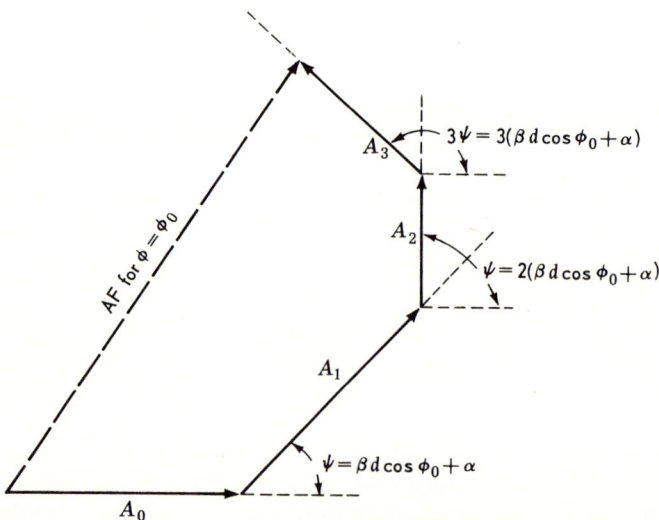

Fig. 3.6 Graphical calculation of the array factor of a four-element array, for the specific angle ϕ_0.

This form of the array factor suggests one method of computing values for an array factor. The method, a graphical one, is indicated in Fig. 3.6. In a particular angular direction ϕ_0, the variable ψ has the particular value $\psi = \beta d \cos \phi_0 + \alpha$. Then, with the element spacing and the phase-shift constant prescribed, the array factor is simply a sum of complex numbers. This sum may be found with the aid of a phasor diagram as indicated in Fig. 3.6.

Exercise A three-element array has a spacing $d = \lambda/4$. The phase shift from element to element is $\alpha = \pi/2$. Use the graphical method to find the array factor at the positions $\phi = 0$, $\phi = 90°$, $\phi = 120°$, and $\phi = 180°$, when the amplitudes are $A_0 = 1$, $A_1 = 3$, $A_2 = 2$. Repeat for the case $A_0 = 1$, $A_1 = 3/\!-\!5°$, $A_2 = 2/\!+\!10°$. In both cases, give a rough sketch of the amplitude of the array factor as a function of ϕ.

With the array factor expressed as a function of the variable ψ, another interesting property is easily observed; that is, the array factor is a *periodic* function of ψ, with period 2π. To see that this is so, consider $\mathrm{AF}(\psi + 2\pi)$, recalling that $e^{j2n\pi} = 1$:

$$\mathrm{AF}(\psi + 2\pi) = \sum_{n=0}^{N-1} A_n e^{jn(\psi + 2\pi)} = \sum_{n=0}^{N-1} A_n e^{jn\psi} \cdot e^{j2n\pi} = \mathrm{AF}(\psi)$$

Thus, since $\mathrm{AF}(\psi + 2\pi) = \mathrm{AF}(\psi)$ for any ψ, the array factor is periodic as stated.

Finally, let us note another general graphical aid of great value, particularly in the visualization and preliminary design processes. The technique is to superimpose the array factor, plotted as a function of ψ, and a graphical representation of the equation $\psi = \beta d \cos \phi + \alpha$, as indicated in Fig. 3.7. The latter sketch helps us to visualize radiation patterns by graphically showing the correspondence between the angular position of a field point ϕ and particular values of the variable ψ. In the sketch, the quantity βd is plotted as a radius vector. As this radius vector rotates from the ψ direction by the angle ϕ, its projection on the ψ axis is $\beta d \cos \phi$. The tip of the radius vector traces a circle whose center is displaced from $\psi = 0$ by an amount α.

Since the angle ϕ is a polar angle (or conical angle), its range of values extends only from 0 to π; i.e., all points in space are covered by values of ϕ between 0 and π. Thus, although the array factor as a function of ψ is defined for all values of ψ, only the portion of the array factor $\mathrm{AF}(\psi)$ that lies in the range of ψ from $\alpha - \beta d$ to $\alpha + \beta d$ is associated with actual physical angles ϕ. Only this portion of the array factor shows up in an actual radiation pattern. This portion or range of the function $\mathrm{AF}(\psi)$ is usually called the *visible* range.

We are now ready to examine some particular types of arrays.

74 ANTENNA ENGINEERING

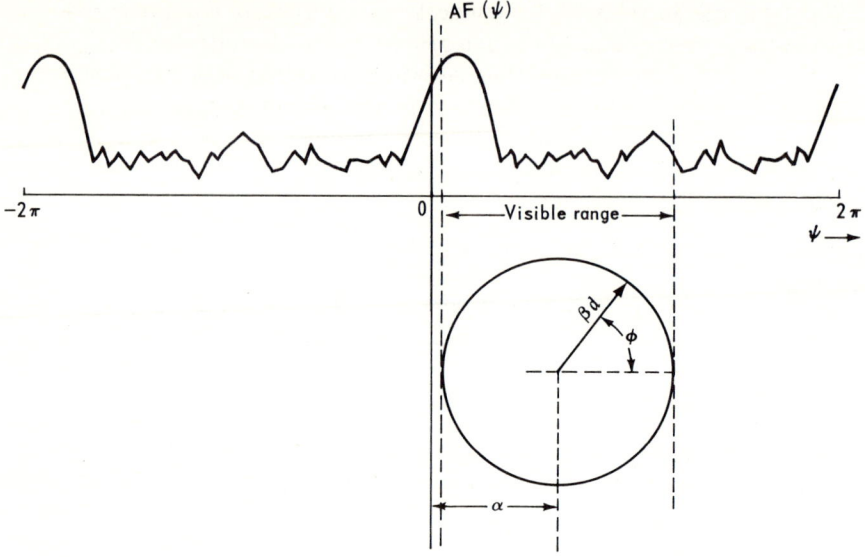

Fig. 3.7 Array factor as a function of ψ, showing a construction that relates ψ to the field angle ϕ.

3.1.1 UNIFORM LINEAR ARRAY

One of the more common array types is one in which the element currents have equal amplitudes. In this case, the array factor

$$\text{AF} = A_0 \sum_{n=0}^{N-1} e^{jn\psi}$$

is determined by a sum which can be identified as a geometrical series with the common ratio $e^{j\psi}$. The sum of N terms of such a geometric series is well known to be

$$\sum_{n=0}^{N-1} e^{jn\psi} = \frac{e^{Nj\psi} - 1}{e^{j\psi} - 1}$$

so that the array factor is

$$\text{AF} = A_0 \frac{e^{j\frac{1}{2}N\psi}(e^{j\frac{1}{2}N\psi} - e^{-j\frac{1}{2}N\psi})}{e^{j\frac{1}{2}\psi}(e^{j\frac{1}{2}\psi} - e^{-j\frac{1}{2}\psi})} = A_0 e^{j\frac{1}{2}(N-1)\psi} \frac{\sin \frac{1}{2}N\psi}{\sin \frac{1}{2}\psi} \quad (7)$$

Alternatively, the origin may be placed at the center of the array. If the number of elements in the array is odd, then the center element may be called the zeroth and the array factor written in the form

$$\text{AF}(\psi) = A_0 \sum_{n=-\frac{1}{2}(N-1)}^{\frac{1}{2}(N-1)} e^{jn\psi}$$

This is also a geometric series with the first term equal to $A_0 e^{-\frac{1}{2}(N-1)j\psi}$. The common ratio is still $e^{j\psi}$ so the sum is

$$\mathrm{AF}(\psi) = A_0 e^{-\frac{1}{2}(N-1)j\psi} \frac{e^{jN\psi} - 1}{e^{j\psi} - 1} = A_0 \frac{\sin \frac{1}{2}N\psi}{\sin \frac{1}{2}\psi}$$

This equation, which physically represents the same array as Eq. (7), shows that the phase factor $e^{j\frac{1}{2}(N-1)\psi}$, in Eq. (7), arises because of the initial choice of the origin at the end of the array. This phase factor disappears when the origin is taken at the center and hence is not usually of fundamental concern to us[1] (except for arrays of arrays).

About the only satisfactory way to visualize the $\sin \frac{1}{2}N\psi/\sin \frac{1}{2}\psi$ function for $\mathrm{AF}(\psi)$ is to examine a graph. Figures 3.8 and 3.9 show the uniform linear array factors as a function of ψ for two different values of N, together with the radiation patterns for two different values of βd. Examination of the graphs quickly reveals that if the elements are all in phase ($\alpha = 0$), then the maximum in the radiation pattern is at $\phi = 90°$ (broadside). If the spacing is restricted in size so that βd is less than $(2\pi - \pi/N)$, there is a single lobe in the radiation pattern. However, if βd becomes large, then two, three, or more "major" lobes may appear in the pattern, as indicated in Fig. 3.8b. The extra major lobes

---

[1] If the number of elements in the array is even, the placement of the origin at the center of the array requires some additional manipulation since the distance from the origin to the nearest element is then $d/2$. If the elements are numbered in both directions from the origin from 1 to $N/2$, positive and negative, the array factor may be written in the form

$$\mathrm{AF} = \sum_{n=1}^{\frac{1}{2}N} A_n' e^{j\frac{1}{2}(2n-1)\beta d \cos\phi} + \sum_{n=1}^{\frac{1}{2}N} A_n' e^{-j\frac{1}{2}(2n-1)\beta d \cos\phi}$$

Then, if the phase shift from element to element is represented by $A_n' = A_n e^{j\frac{1}{2}(2n-1)\alpha}$, along with the restriction $A_n' = A_n'^*$, the usual ψ variable may be introduced to give an array factor as follows:

$$\mathrm{AF} = \sum_{n=1}^{\frac{1}{2}N} A_n [e^{j\frac{1}{2}(2n-1)\psi} + e^{-j\frac{1}{2}(2n-1)\psi}]$$

$$= \sum_{n=1}^{\frac{1}{2}N} 2 A_n \cos \frac{1}{2}(2n-1)\psi$$

For the case of the uniform array, this simplifies to

$$\mathrm{AF} = 2A_1 \sum_{n=1}^{\frac{1}{2}N} \cos \frac{1}{2}(2n-1)\psi$$

In this alternative representation, the array factor is clearly a real number without the phase dependence on ψ as was found in Eq. (7). The finite sum of cosines is equivalent to the $\sin \frac{1}{2}N\psi/\sin \frac{1}{2}\psi$ factor in Eq. (7).

are often referred to as "grating" lobes, since they arise in a manner similar to the multiple lobes of a diffraction grating. The submaxima are commonly called *side lobes*.

Note that when there is an element-to-element phase shift ($\alpha \neq 0$), the direction of the main beam is shifted to a position ϕ_m. The angular position of the maximum and the phase shift α are related through the

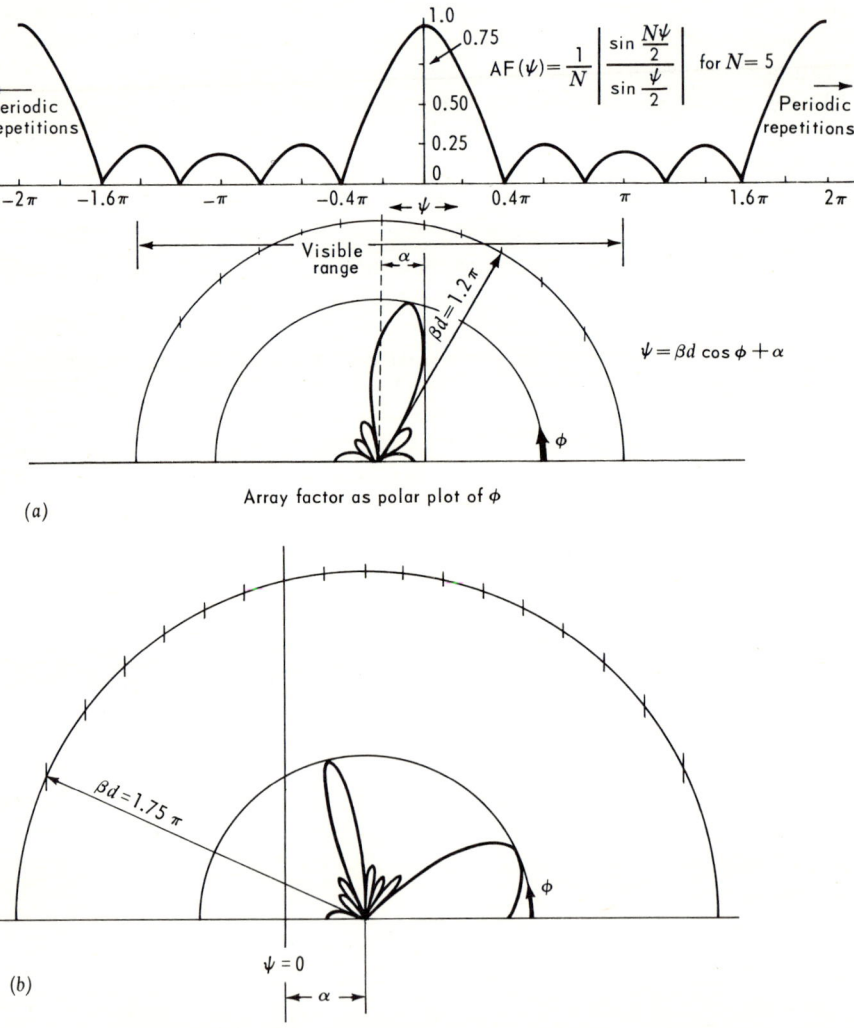

Fig. 3.8 (a) Array factor for a uniform five-element array. (b) Array factor as a polar plot of ϕ (five elements, uniform excitation). Note grating lobe at about $\phi = 120°$ brought in by large spacing.

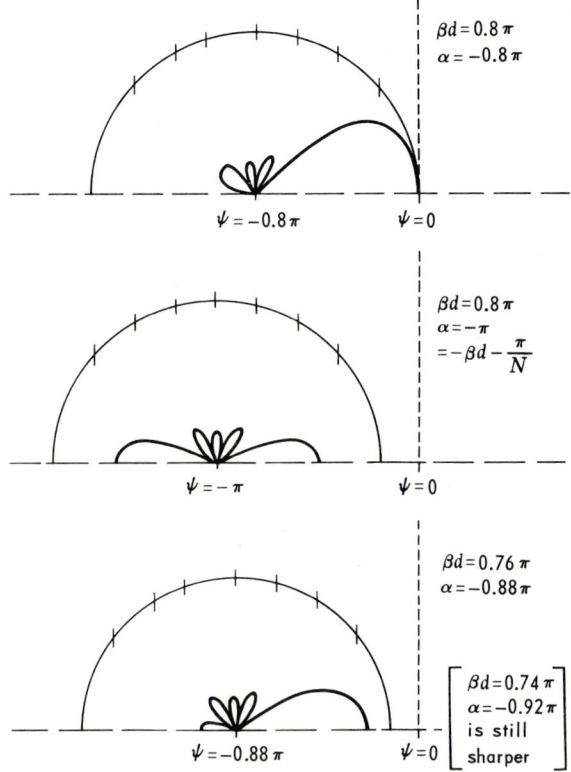

Fig. 3.8c Array factors: five-element end-fire arrays.

definition of ψ, $\psi_m = 0 = \beta d \cos \phi_m + \alpha$ or $\alpha = -\beta d \cos \phi_m$. An alternative expression for the variable ψ is therefore

$$\psi = \beta d(\cos \phi - \cos \phi_m)$$

One of the practical implications of the direct connection between the position of the beam maximum and the element-to-element phase shift is that if the latter can be varied (say by phase shifters), then the position of the main beam can be varied (i.e., "scanned") without changing the mechanical orientation of the array.

When the phase shift α is equal to $\pm \beta d$, the radiation pattern has a maximum at $\phi = 0°$ or $180°$ ("end fire"). To obtain a single end-fire lobe (i.e., to avoid the grating lobes mentioned above), the element spacing should be small enough that $\beta d \leq \pi(1 - 1/2N)$.

One of the most significant points to be learned here is that, given the number of elements in a uniform linear array, all radiation patterns are

78 ANTENNA ENGINEERING

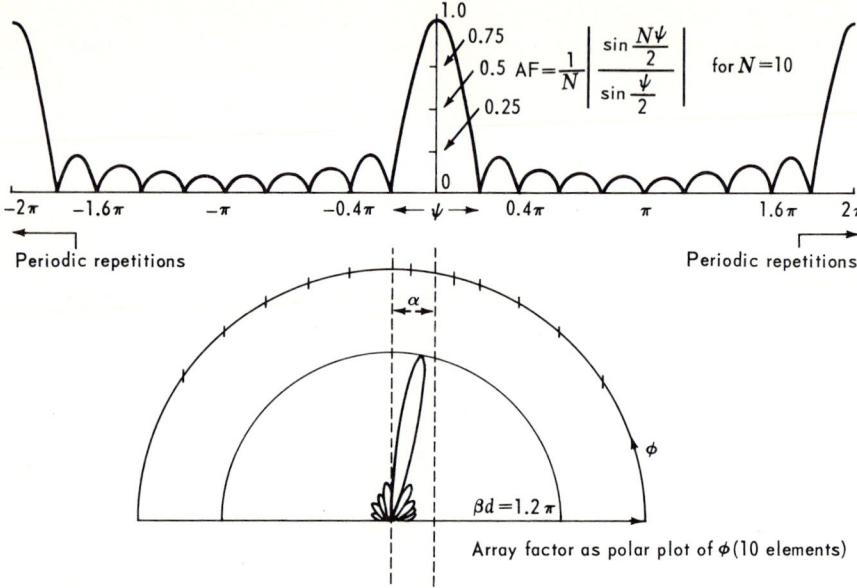

Fig. 3.9 Array factor for a ten-element uniform array.

derived from the same basic curve. This fact not only simplifies and unifies preliminary design considerations, but also means that all uniform arrays have certain features in common. For example, the ratio of the main-lobe level to the side-lobe level is the same in all uniform linear arrays with the same number of elements, N. The only exception to this is the end-fire array with $\alpha > \beta d$. With end-fire arrays, if α is made slightly greater than βd, the main beam can be sharpened up (greater directivity) at the expense of an increase in the relative side-lobe level (see Fig. 3.8c).

The reader should study Figs. 3.8 and 3.9 with compass or dividers in hand in order to fully appreciate the foregoing remarks.

Problem 3.1 Design a five-element uniform array of small antennas so as to produce a maximum in the radiated field strength at 45° to the line of the array. Sketch the principal plane radiation pattern.

3.1.2 CONTINUOUS LINEAR ARRAYS

In some instances, the individual elements of an array are so close together that they form an essentially continuous array. The equivalent sources for "aperture" antennas (for example, the magnetic currents equivalent to an electric field in an aperture, or the Huygens sources) frequently comprise an exactly continuous array. Hence, continuous arrays are of practical interest and importance.

In a continuous array of the type indicated in Fig. 3.10, the current distribution may be represented in the form

$$\mathcal{J} = \hat{\mathbf{z}}\,\delta(y)\,i(z)I(x)$$

The resulting radiation pattern may be calculated either by superposition integrals or by the reciprocity theorem. The superposition integral, Eq. (17), Chap. 1, gives for the vector potential

$$A_z = \frac{e^{-j\beta r}}{4\pi r} \int e^{j\beta z \cos\theta} i(z)\,dz \int e^{j\beta x \cos\phi} I(x)\,dx$$

wherein the approximation $R \doteq r - x\cos\phi - z\cos\theta$ has been employed (and here $\cos\phi = \sin\theta\cos\varphi$); as before

$$E_\theta = j\omega\mu A_z \sin\theta$$

By comparison with Eq. (5), we see that we may identify the array factor

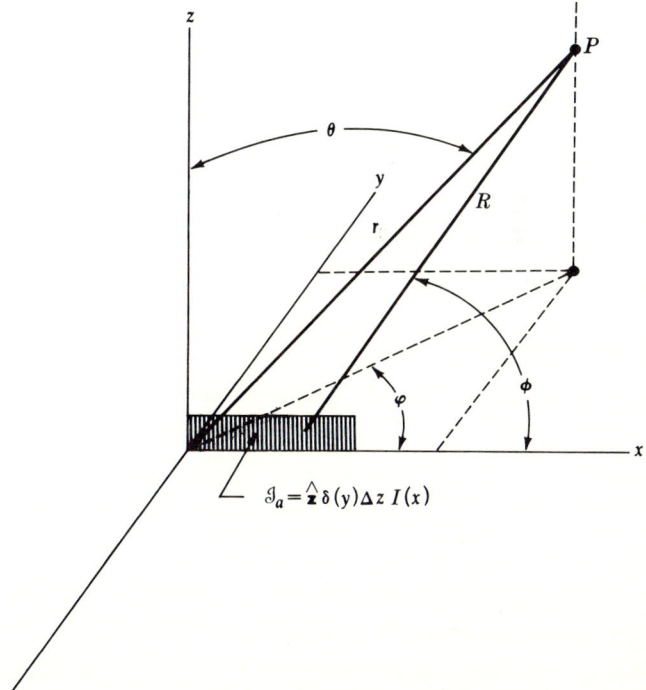

Fig. 3.10 Geometry for continuous array along x.

for continuous linear arrays as

$$AF = \int I(x) e^{j\beta x \cos \phi} \, dx$$

The distribution $I(x)$ may contain a linear progressive phase shift. It will, for example, if the sources are excited by a matched transmission line along the x direction. As with discrete arrays, we can explicitly display a linear progressive phase shift; for example, let $A(x)e^{jkx} = I(x)$, so that the array factor can be written in the form

$$AF = \int A(x) e^{j(\beta \cos \phi + k)x} \, dx \tag{8}$$

Of course, the same result can be obtained by starting with the array factor for discrete arrays and letting d decrease indefinitely and N increase indefinitely in such a way that the product $(N - 1)d$ remains a constant, L, the length of the array. The quantity nd then becomes the continuous variable x, the quantity n becomes the phase shift per unit length, and A_n becomes $A(nd) = A(x)$.

Exercise Approximate the array factor for the continuous linear array with a finite sum by breaking the array length L into N parts of length d. Show the relationship of this sum to the array factor of a discrete array having N elements spaced d apart and with element currents $I_n = I(nd + \tfrac{1}{2}d)d$.

The first and most important example of a continuous linear array is one of finite length having a uniform current distribution along the length. The reader may show readily (as an exercise) that for this uniform linear continuous array, viz.,

$$A(x) = \begin{cases} C_0 & 0 < x < L \\ 0 & \text{otherwise} \end{cases}$$

the array factor is

$$AF = C_0 L e^{j\psi_1} \frac{\sin \psi_1}{\psi_1} \tag{9}$$

where $\psi_1 = \tfrac{1}{2}(\beta \cos \phi + k)L$.

The $\sin \psi_1 / \psi_1$ factor is plotted in Fig. 3.11. A geometrical construction of the same type as that shown in Figs. 3.7 to 3.9 may be employed to associate the angle variable ϕ with ψ_1. The maximum in the radiation pattern occurs at the angle corresponding to $\psi_1 = 0$, if this value is within the visible range. The angle of the maximum in the pattern is given by $\cos \phi_{\max} = -k/\beta$. The direction of the main beam may therefore be controlled by controlling the phase constant k. The maximum in the pattern is at $\phi = \pi/2$ (broadside) for $k = 0$ (no phase shift) and at $\phi = 0$ (end fire) for $k = -\beta$. The condition $k = -\beta$ implies a phase delay corresponding to that of a plane electromagnetic wave traveling down the line of the array. Practice with the construction of Fig. 3.11 makes it possible for an engineer to determine by quick inspection the effects of other linear progressive phase shifts (i.e., those corresponding to "fast" waves in which $k < \beta$ so that the phase velocity v is greater than the velocity of light c and "slow" waves in which $k > \beta$ and $v < c$).

A few points concerning uniform continuous arrays are noteworthy. First of all, there is no possibility of multiple grating lobes since, unlike the case of arrays having discrete elements, the array factor is not periodic and has only one principal lobe. Next, provided the maximum point $\psi_1 = 0$ is included within the visible range, every uniform current distribution having $L/\lambda > 1.5$ has the same relative side-lobe level, -13.3 db.

Finally, note (from the graphical construction in Fig. 3.11) that if the phase-shift constant k is slightly greater in magnitude than β, then the pattern is end fire with a narrower main-lobe and higher relative side-lobe levels than when the value $\psi_1 = 0$ is within the visible range. Since this is so, it is to be expected that the directivity of such an end-fire array might be increased by a judicious selection of the phase constant k. Hansen and Woodyard[1] considered this question many years ago. They found by computation that the function

$$\frac{\{\sin [\tfrac{1}{2}(\beta - k)L]/\tfrac{1}{2}(\beta - k)L\}^2}{\int_0^\pi \{\sin [\tfrac{1}{2}(\beta \cos \phi - k)L]/\tfrac{1}{2}(\beta \cos \phi - k)L\}^2 \sin \phi \, d\phi}$$

has a broad maximum centered about the value $k = \beta + 2.94/L$. In the analysis, they assumed $L \gg \lambda$. Thus, provided the directivity is not influenced by the pattern of the individual elements, a value of the phase constant k which is larger than β by the amount indicated gives maximum directivity in the end-fire linear array (see Figs. 3.11b and 3.11c).

Another important special case of a continuous array is one whose amplitude varies sinusoidally along the array. Such a distribution with half-period L is an example of a class of amplitude distributions known as tapered apertures. The general effect of a tapered amplitude distribution is to broaden the main beam and to decrease the side-lobe levels. This general effect may be illustrated by considering a sine-tapered amplitude distribution

$$A(x) = \begin{cases} C_0 \sin \dfrac{\pi x}{L} & 0 < x < L \\ 0 & \text{otherwise} \end{cases}$$

According to Eq. (8), the array factor is

$$\mathrm{AF} = C_0 \int \sin \frac{\pi x}{L} e^{j(\beta \cos \phi + k)x} \, dx$$

[1] W. W. Hansen, and J. R. Woodyard, *Proc. IRE*, **26:**333 (March, 1938).

82 ANTENNA ENGINEERING

This integral may be readily evaluated by writing the sine function in exponential form, or by consulting standard integral tables (e.g., Peirce, No. 414). The result, in terms of the variable $\psi_1 = \frac{1}{2}(\beta \cos \phi + k)L$, is

$$\text{AF} = C_0 e^{j\psi_1} \frac{2L}{\pi} \frac{\cos \psi_1}{1 - (2\psi_1/\pi)^2} \tag{10}$$

Note that the first null in the array factor for the sine-tapered aperture distribution appears at $\psi_1 = 3\pi/2$, compared with $\psi_1 = \pi$ for the case of the uniform distribution. In fact, at the value $\psi_1 = \pi$, the sine-tapered array factor has a value of $\frac{1}{3}$ relative to its maximum value. This shows that the beam produced by the sine-tapered amplitude distribution is significantly broader than that of the uniform distribution (see Fig. 3.12). Note also that the relative level of the first side lobe is -23 db, as compared to -13.3 db with the uniform distribution. Such results are typical of the general effects of tapered amplitude distributions.

We note in conclusion that the current distribution on which our discussion in this section began, $\mathcal{J}_z = \delta(y) i(z) I(x)$, is actually a two-dimen-

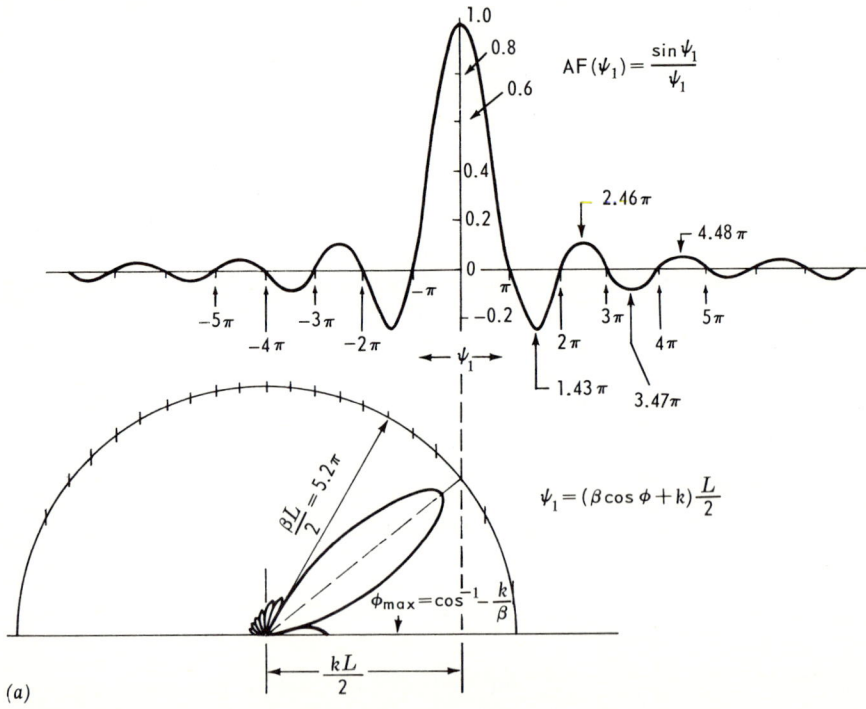

(a)

Fig. 3.11a Array factor for uniform continuous linear array.

sional array. The vector potential and electric field for such an array are given in the general equations on page 79. In most of the discussion in this section, we have singled out the integral on x for detailed discussion and called it the continuous linear array factor. But note that the integral on z has the same form, so if $i(z)$ is a continuous function in a finite interval, the integral on z is also a continuous linear array factor; the electric field is the product of these two array factors times the field of an elemental current.

Problem 3.2 Design a sine-tapered continuous array of elemental currents so as to produce a maximum in the field strength at 30° to the line of the array, and to have a beamwidth between half-power points of 10°. Sketch the principal-plane radiation pattern.

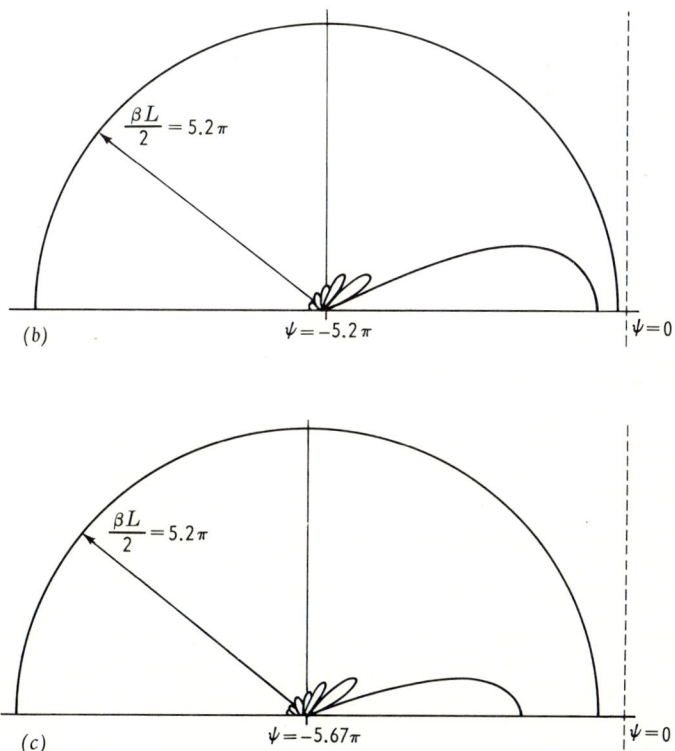

Fig. 3.11 (b) end-fire pattern, $k = -\beta$, $\beta L/2 = 5.2\pi$, $(L/\lambda = 5.2)$, $KL/2 = -5.2\pi$; (c) end-fire pattern, $|k| > \beta$. In this example, k is adjusted according to the Hansen-Woodyard condition so that $|kL/2| = \beta L/2 + 1.47$; in this example, $kL/2 = -5.67\pi$.

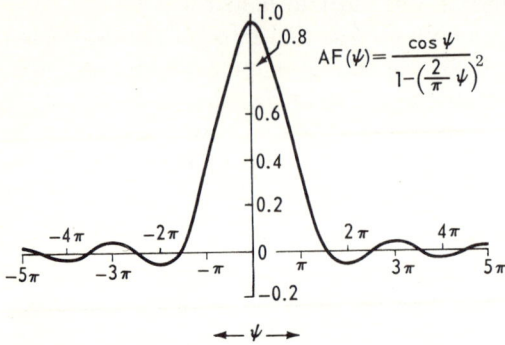

Fig. 3.12 Array factor for sine-tapered amplitude distribution.

3.2 POLYNOMIAL REPRESENTATIONS FOR LINEAR ARRAYS

S. A. Schelkunoff[1] recognized that the array factor for discrete arrays [Eq. (6)] could be thought of as a polynomial in the complex variable $z = e^{j\psi}$ as follows:

$$\text{AF} = A_0 + A_1 z + A_2 z^2 + A_3 z^3 + A_{N-1} z^{N-1} \tag{11}$$

which is of course the same as Eq. (6). Schelkunoff's observations are very important because they make possible the application of the large body of knowledge concerning the algebra of polynomials to the design of arrays. Some of his observations are stated as theorems, for example: Every linear array with commensurable[2] spacing between the elements can be represented by a polynomial; and every polynomial can be interpreted as the array factor of some linear array. An array of N elements is represented by a polynomial of degree $N - 1$ [Eq. (11)]. It follows then from the fundamental theorem of algebra that this polynomial has exactly $N - 1$ roots and can be factored into the form

$$\text{AF} = A_{N-1}(z - z_1)(z - z_2)(z - z_3) \cdots (z - z_{N-1}) \tag{12}$$

where the z_n's are the zeros (roots) of the polynomials. This form is interesting for several reasons. In the first place, it displays explicitly the zeros in the array factor, since the zeros are those ψ values which correspond to z_1, z_2, and so on. Of course, zeros in the array factor give zeros in the radiation pattern provided they are within the visible range. Also, the form shows that an array factor may be interpreted as the product of the array factors of $N - 1$ two-element arrays, each of which has a zero at one of the zeros of the larger array factor.

[1] S. A. Schelkunoff, *Bell System Tech. J.*, **22**:80 (1943).
[2] Some of the amplitudes can be zero to make the spacings commensurable.

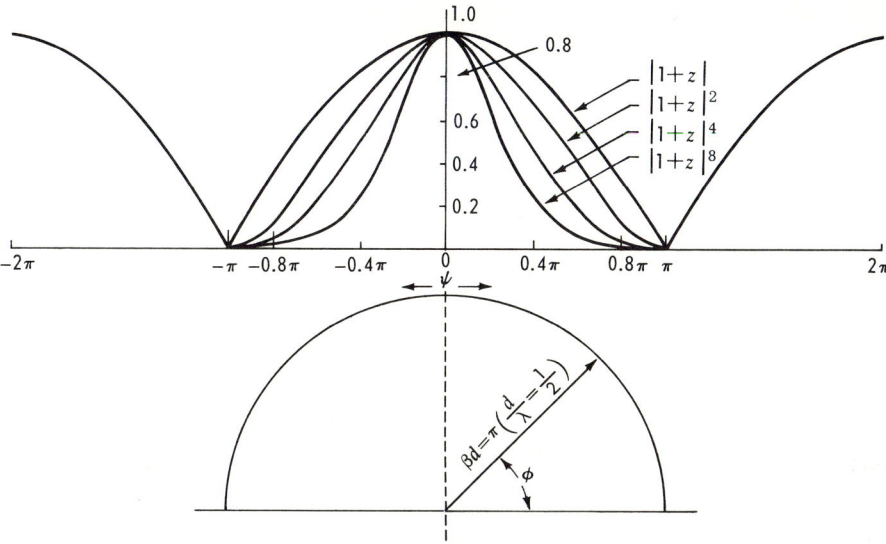

Fig. 3.13 Array factors for binomial arrays.

As an example of the type of thinking suggested by these representations, consider a simple two-element array

$$AF = 1 + e^{j\psi} = 1 + z$$

If the element spacing is less than one-half wavelength, this array factor has the simple appearance given in Fig. 3.13 and of course has no side lobes. Consider next an array formed by taking the product (square) of two array factors of this simple type

$$AF = (1 + z)(1 + z) = 1 + 2z + z^2$$

This is a three-element array with current amplitudes in the ratio $1:2:1$. Since it is the square of the simple pattern above, it has a sharper beam, but the same zeros, and no side lobes. This multiplicative process may be continued to generate what are known as binomial array factors: $AF = (1 + z)^n$. The current amplitudes in these array factors are in the ratios of the binomial coefficients; that is, the coefficient of the term z^r in this binomial array is $_nC_r = n!/r!(n-r)!$. An array constructed according to this array factor has $n - 1$ elements and (theoretically) no side lobes. The array factors of several of these binomial arrays are sketched in Fig. 3.13.

In certain applications, arrays must be constructed so as to produce zeros (nulls) in certain directions, for example, to prevent radio interference. The factored form of the array factor [Eq. (12)] is useful in such problems, since it is so easy to insert the specified zeros. For example, consider the design of a broadcast antenna consisting of a three-element array with quarter-wavelength spacing lying along a north-south line. A population center lies at about $\phi = 90°$ relative to the array, but a null must be produced at $\phi = 26°$ for interference suppression. Suppose the elements are excited in phase; then, since $\psi = \beta d \cos \phi$, the ψ value corresponding to the specified zero is $\psi_1 = 2\pi/\lambda \times \lambda/4 \times 0.9 = 0.45\pi$. Thus one of the two roots of the polynomial representing the three-element array is $z_1 = e^{j0.45\pi}$. Since no further information about the pattern is specified, the second zero may be selected arbitrarily. If there were little population in the general direction of $\phi = 154°$, we could decide to produce a pattern that is symmetric about the broadside direction. Then the second zero could be placed at $\phi = \pm(180° - 26°)$ which corresponds to a ψ value of -0.45π so that the second zero is at $z_2 = e^{-j0.45\pi}$. The factored form of this array factor is

$$\begin{aligned}
\text{AF} &= (z - e^{j0.45\pi})(z - e^{-j0.45\pi}) \\
&= z^2 - z(e^{j0.45\pi} + e^{-j0.45\pi}) + 1 \\
&= z^2 - (2 \cos 0.45\pi)z + 1 \\
&= z^2 - 0.313z + 1
\end{aligned}$$

Thus, an array constructed with amplitudes of element currents in the ratio $1:0.313:1$ and with element spacing of a quarter wavelength will give the desired null positions. The actual array factor for this array is plotted in Fig. 3.15.

The radiation pattern of this and other arrays may be estimated readily from a geometrical interpretation of the array-factor polynomial written

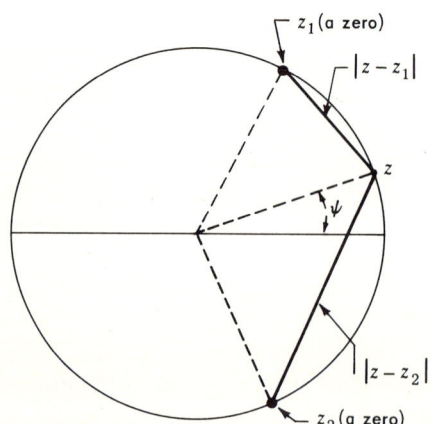

Fig. 3.14 Complex plane of z. All values of the variable z pertaining to the array-factor polynomial lie on the unit circle. $|\text{AF}(z)| = |z - z_1| \cdot |z - z_2|$

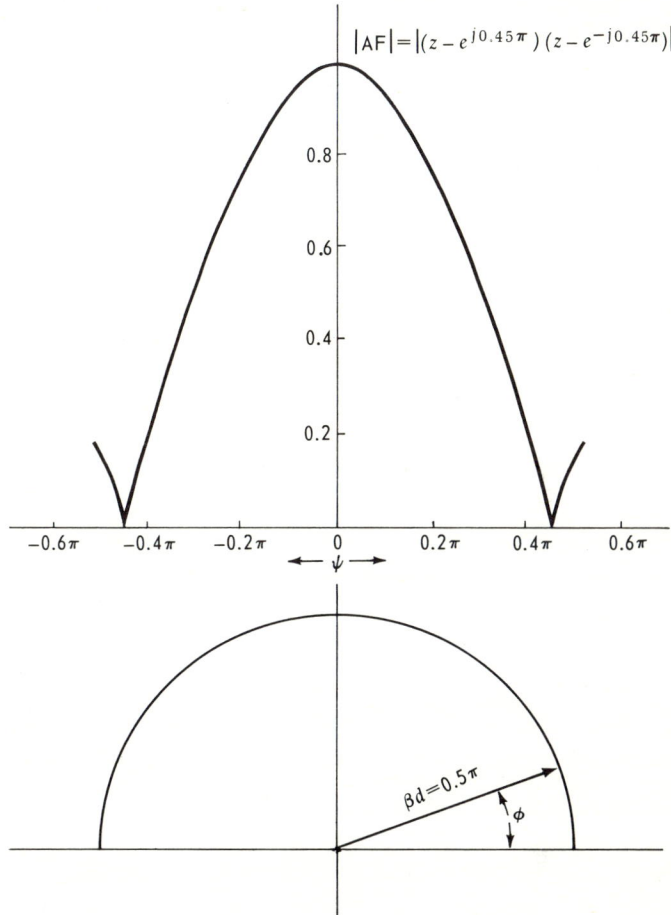

Fig. 3.15 An array factor with specified zeros (the first example).

in factored form. Note in the first place that all values of the variable z lie on the unit circle in the complex plane. Consequently, all the zeros lie on the unit circle. Let the point z in Fig. 3.14 represent an arbitrary angular position with respect to a particular array whose array factor has the zeros z_1 and z_2 as indicated. The magnitude of the array factor can be written

$$|AF| = |z - z_1| \cdot |z - z_2|$$

The magnitudes $|z - z_1|$ and $|z - z_2|$ are indicated in the figure. By inspection, the magnitude of the array factor at any position z is given by the product of the distances from that point to each of the zeros of

88   ANTENNA ENGINEERING

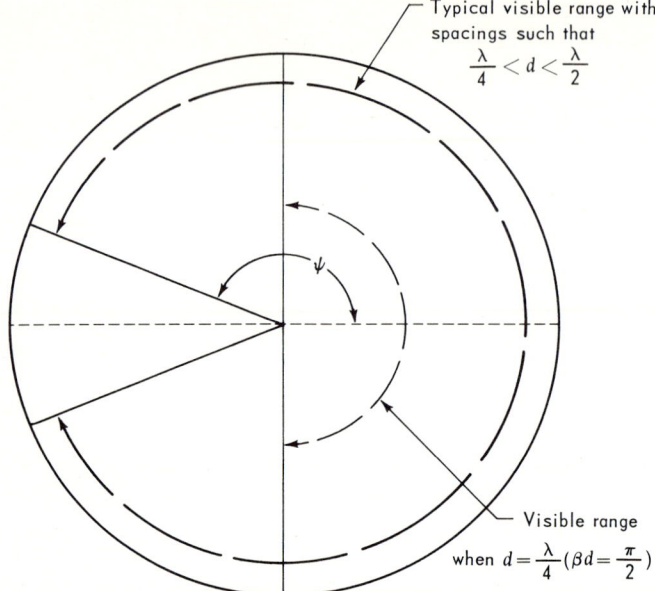

Fig. 3.16 Complex z plane showing the range of z that is visible when $\alpha = 0$. The visible range depends on βd and extends from $\psi = -\beta d$ to $\psi = \beta d$.

the polynomial. In the complex plane, the "visible" range of the ψ and z values (that is, $0 \leq \phi \leq \pi$) depends of course upon βd and the phase-shift constant α. If α is zero, the range of ψ is from $-2\pi d/\lambda$ to $2\pi d/\lambda$ (see Fig. 3.16). With this information, a rough calculation of the pattern in the visible range may be made quickly.

In the example just considered, if the specification had been to produce a zero at $\phi = 26°$, with the radiation directed generally away from this null, the array might have been constructed differently. Here, the position of the second zero in the three-element array can be placed at the discretion of the designer. If he so desires, the designer may place the zero outside the visible range. For example, to place the maximum in the radiation pattern in the direction $\phi = 180°$, the second zero could be located at $\psi = 0.55\pi$ (see Fig. 3.17). The array factor of this array is then

$$\text{AF} = (z - e^{j0.45\pi})(z - e^{j0.55\pi})$$
$$= z^2 - z(e^{j0.45\pi} + e^{j0.55\pi}) + e^{j\pi}$$

or

$$\text{AF} = z^2 - z(2j \sin 81°) - 1$$

so the current amplitudes are in the ratio $-1:-j1.97:1$. The array factor is sketched in Fig. 3.18. In these examples, the beginnings of a limited synthesis technique may be seen emerging.

Exercise Find the current excitation coefficients in a two-element array with spacing of a quarter wavelength which must have a null at (1) $\phi = 0°$, (2) $\phi = 60°$. Sketch the patterns.

As a final example, consider Schelkunoff's design for an end-fire array. It will be recalled that in order to achieve maximum radiation in the end-fire direction, the element-to-element phase shift along the array should be the same as the phase shift in a uniform plane wave traveling along the array; that is, the phase-shift constant α is taken equal to $-\beta d$. The ψ variable is then $\psi = \beta d(\cos \phi - 1)$ and the visible range is from $\psi = 0$ to $\psi = -2\beta d$. Schelkunoff made the astute observation that the spacing of the zeros of the array polynomial equally within the visible range of ψ will result in a pattern with lower side lobes and a narrower main beam than a uniformly excited array having the same length and number of elements. Recalling that if an array has N elements, it can have $N - 1$ distinct zeros, then it is clear that a set of equally spaced

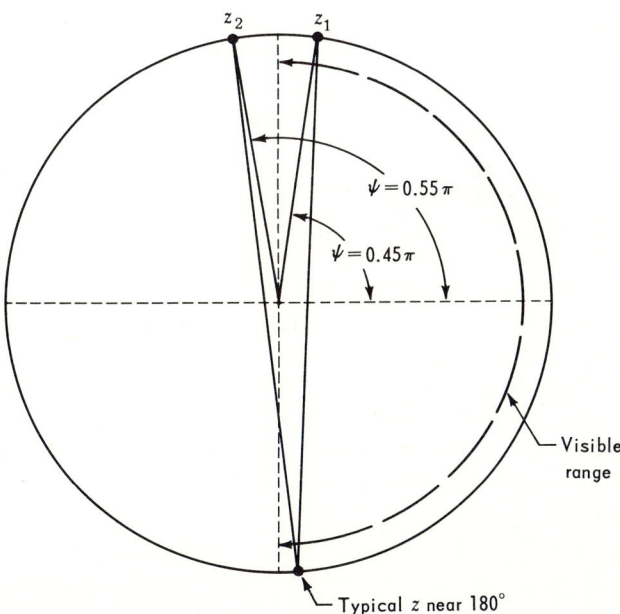

Fig. 3.17 Zeros and visible range of array factor in second example.

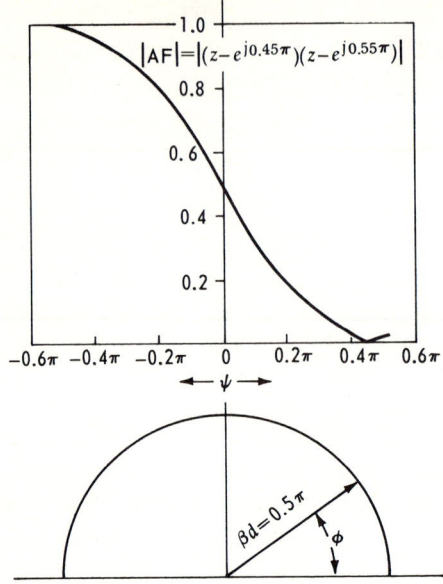

Fig. 3.18 Array factor with specified zeros (the second example).

zeros in the visible range are located at the z values $\exp(-2j\beta d/N - 1)$, $\exp(-4j\beta d/N - 1)$, $\exp(-6j\beta d/N - 1) \cdots \exp(-2j\beta d)$. Then the array factor may be written in the compact form (serial product)

$$\mathrm{AF} = \prod_{n=1}^{N-1}\left(z - \exp\frac{-nj2\beta d}{N-1}\right) \tag{13}$$

As a specific example, consider a seven-element array with a spacing of a quarter wavelength. The quantity $2\beta d$ is then π, so there are to be six roots with angular separation $\pi/6$. The visible range and the positions of the zeros are shown in Fig. 3.19. The array factor is then

$$\mathrm{AF} = (z - e^{-j\pi/6})(z - e^{-2j\pi/6})(z + j)(z - e^{-j2\pi/3}) \cdot (z - e^{-j5\pi/6})(z + 1)$$

The multiplication of the factors to give the familiar series form for the array factor displays the required current excitation coefficients. The process is straightforward although a little tedious. The pattern for this seven-element array is shown in Fig. 3.20, taken from a paper by DuHamel.[1]

Figure 3.20 also shows the pattern produced by a uniformly excited linear array with the same spacing and the same number of elements. In terms of the polynomial representations, the analysis for the uniformly excited array proceeds as follows: We have already shown that the

[1] R. H. DuHamel, *Proc. IRE*, **41**:652 (May, 1953).

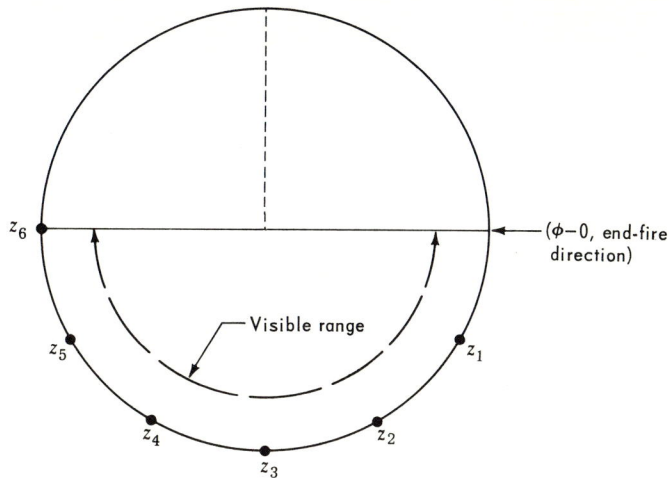

Fig. 3.19 Complex z plane for Schelkunoff's seven-element end-fire array, showing the equally spaced zeros within the visible range. ($\alpha = -\beta d$.)

array factor may be recognized as the sum of a finite number of terms in a geometric progression $\mathrm{AF} = (z^N - 1)/(z - 1)$. The zeros are then those values of z for which $z^N = 1$ but $z \neq 1$. That is, they are the Nth roots of unity $z^N = e^{j2\pi n}$ or $z = \exp(j2\pi n/N)$, where n is an integer less than N. These six zeros are spread uniformly through the range $0 < \psi < 2\pi$, consequently some of the zeros are outside the visible range. The changes in the radiation pattern brought about by Schelkunoff's

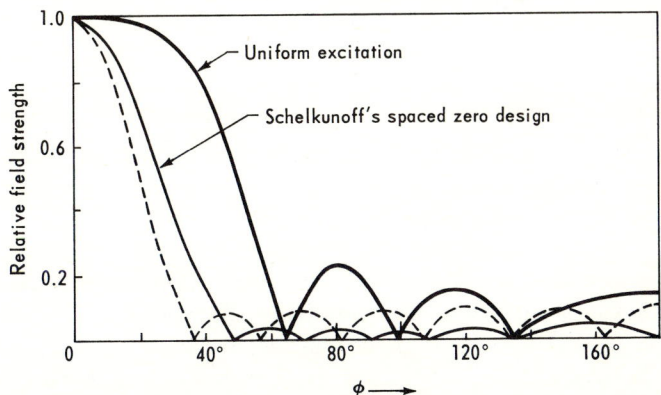

Fig. 3.20 Comparison of end-fire array designs: seven elements. [Dashed curve is DuHamel's optimum which will be considered later (see Sec. 3.4 and Prob. 3.7).]

92   ANTENNA ENGINEERING

positioning of all the zeros within the visible range are vividly displayed in the figure.

The visible range of ψ may be decreased arbitrarily by decreasing the element spacing βd. If at the same time the number of elements in the array is large, then a large number of zeros may be placed within this small visible range, and consequently the main beam may be made almost arbitrarily sharp. This type of array is commonly referred to as a *super-gain* array (i.e., it has theoretically much more directive gain than a uniformly excited array of the same size). Although a moderate amount of "supergaining" may certainly be accomplished, there is a basic practical difficulty associated with supergain arrays. The difficulty is that when the current amplitudes are calculated, it is found that adjacent elements (closely spaced, of course) must have large currents, oppositely directed. As a result, the net current over the length of the array is very small compared to the element currents. This means that the amount of radiation per unit of input current is very small, which, from a practical standpoint, means poor efficiency. Moreover, the required currents must be established with high precision.[1,2]

Problem 3.3   Design a broadside array of five elements with quarter-wavelength spacing. Employ the technique of spacing the zeros in the visible range. Compare the array factor of your design with that of the conventional uniformly excited array of the same length and spacing. Note the differences from Schelkunoff's end-fire design.

3.3 FOURIER REPRESENTATIONS AND ANTENNA PATTERN SYNTHESIS

It has already been shown, Eq. (8), that the array factor for a continuous array is given by an integral

$$\text{AF} = \int I_0(x) e^{j(\beta \cos \phi + k)x} \, dx$$

which closely resembles one of the members of a Fourier transform pair. The resemblance is made more apparent by the introduction of a variable $\xi = \beta \cos \phi + k$:

$$\text{AF}(\xi) = \int I_0(x) e^{j\xi x} \, dx$$

It will also be recalled that the derivation of this equation for the array factor is subject to the condition that the extent of the source $I_0(x)$ be very small in comparison to the distance to the point of observation. To make this explicit, let us now require that $I_0(x)$ be zero except at points relatively close to the origin; in this case, the integral can just as

---

[1] E. C. Jordan, "Electromagnetic Waves and Radiating Systems," pp. 449–450, Prentice-Hall, Inc., Englewood Cliffs, N.J., 1950.
[2] N. Yaru, *Proc. IRE*, **39**:1081 (September, 1951).

well be written with infinite limits

$$\mathrm{AF}(\xi) = \int_{-\infty}^{\infty} I_0(x) e^{j\xi x}\, dx \tag{14}$$

This form is indeed a member of a Fourier transform pair; it displays the array factor as the inverse transform of the current distribution $I_0(x)$. The associated direct transformation is then

$$I_0(x) = \frac{1}{2\pi} \int_{-\infty}^{\infty} \mathrm{AF}(\xi) e^{-j\xi x}\, d\xi \tag{15}$$

Whether Eq. (14) is called the inverse transform or whether Eq. (15) is so designated is of course arbitrary—a change of variable $\xi' = -\xi$ or $j = -i$ changes sign to the desired conventional form.

Equations (14) and (15) tell us that the array factor and the current distribution in the array are Fourier transforms of one another. This observation is important for two reasons: (1) It forcefully points out that all the known results relating Fourier transform pairs (often used to find the frequency spectra of specified time functions) can be applied immediately to the task of finding the array factors produced by specified current distributions; and (2) it shows us how to find the current distribution that is required to produce a specified array factor—that is, it suggests a technique for radiation-pattern synthesis.

The synthesis problem requires some additional discussion both because it is an important practical problem and because the technique is not quite straightforward. The difficulty is that in practice only a limited range of ξ values are correlated with real observation angles (visible range). As far as the pattern is concerned, the values of $\mathrm{AF}(\xi)$ outside the visible range $(-\beta + k < \xi < \beta + k)$ are quite arbitrary. In practice, however, the choice of values is important since, for example, if the value zero were selected for $\mathrm{AF}(\xi)$ outside the visible range, then the required current distribution would extend over an infinite range of x. Besides being impossible to achieve in practice, such a current distribution of infinite size violates the condition under which the original equation was derived. The task of the designer is then (1) to specify the array factor within the visible range so as to obtain the desired pattern—this is relatively straightforward; and (2) to specify the array factor outside the visible range in such a way that the required current distribution is confined to a finite interval (or better, to some previously specified length L along x).

Collin[1] has shown that if the array factor is expressible in the form

$$\mathrm{AF}(\xi) = \frac{\sin \frac{1}{2}\pi\xi L}{\frac{1}{2}\pi\xi L} \frac{P(\frac{1}{2}\xi L)}{\prod_{n=1}^{N} [(\frac{1}{2}\xi L)^2 - n^2]}$$

[1] R. E. Collin, *IRE Trans.*, **MTT-10**:393 (September, 1962).

where $P(\tfrac{1}{2}\xi L)$ is an arbitrary polynomial of degree $2N$, then the required $I_0(x)$ will always be zero outside the range 0 to L.

3.3.1 DISCRETE ARRAY SYNTHESIS

The determination of the linear array of discrete elements required to produce a prescribed array factor is particularly simple and interesting. Since, as was pointed out on page 73, the array factor for discrete arrays is a periodic function of ψ, it follows that a desired array factor may always be approximated by a finite number of terms from a Fourier series. To proceed, the desired array factor is specified within a range of 2π in ψ; then the balance of $\mathrm{AF}(\psi)$ consists of periodic repetitions of the specified function.

Suppose initially, for simplicity, that the desired array factor has the required symmetry so that the Fourier series approximating it may consist of cosine terms only. Thus

$$\mathrm{AF}(\psi) = \tfrac{1}{2}a_0 + \sum_{n=1}^{N} a_n \cos n\psi$$

or in terms of the ξ variable,

$$\mathrm{AF} = \tfrac{1}{2}a_0 + \sum_{n=1}^{N} a_n \cos n\xi d \qquad (16)$$

Then, according to Eq. (15) (the inverse transform of the specified array factor), the required current distribution is

$$I_0(x) = \frac{1}{2\pi} \int_{-\infty}^{\infty} (\tfrac{1}{2}a_0 + \sum_{n=1}^{N} a_n \cos n\xi d)\, e^{-j\xi x}\, d\xi$$

The problem then is to find the transform of each term in the series approximation for the array factor. It will be recalled that the transform of the delta, or impulse, function $\delta(x - x_0)$ is an exponential $\int_{-\infty}^{\infty} \delta(x - x_0) e^{j\xi x}\, dx = e^{j\xi x_0}$, and consequently the transforms of cosines are

$$\frac{1}{2\pi} \int_{-\infty}^{\infty} (a_n \cos n\xi d) e^{-j\xi x}\, d\xi = \tfrac{1}{2}a_n[\delta(x + nd) + \delta(x - nd)]$$

Thus the associated current distribution is

$$I_0(x) = \tfrac{1}{2}a_0\, \delta(x) + \sum_{n=1}^{N} \tfrac{1}{2}a_n[\delta(x + nd) + \delta(x - nd)]$$

This current distribution can be recognized as a discrete linear array with $2N + 1$ elements, excited so that the amplitudes of the currents are symmetric about the center (a_0) element. The element spacing is d. The required excitation coefficients are given by the usual Fourier formula

$$a_n = \frac{1}{\pi} \int_{-\pi}^{\pi} \mathrm{AF}(\psi) \cos n\psi \, d\psi \qquad \psi = \xi d$$

The same conclusions may be reached in a less formal but perhaps more illuminating fashion by a manipulation of the basic equation of the array factor

$$\mathrm{AF}(\psi) = \sum_{n=0}^{N-1} A_n e^{jn\psi}$$

If the array has an odd number of elements with amplitudes that are symmetric about the center element, the array factor can be written in the form

$$\begin{aligned}
\mathrm{AF}(\psi) &= A_0 + \sum_{n=1}^{\frac{1}{2}(N-1)} A_n e^{jn\psi} + A_{-n} e^{-jn\psi} \\
&= A_0 + 2 \sum_{n=1}^{\frac{1}{2}(N-1)} A_n \cos n\psi \qquad A_n = A_{-n}
\end{aligned}$$

This is essentially the postulated form, Eq. (16). As before, we conclude that if the array factor can be satisfactorily represented by a series of this form, then it can be produced by a symmetrical array of N elements (N odd) with the current in the nth element equal to A_n. In this case, the desired pattern must be such as to make the array factor an even function of ψ.

There is one further difficulty that may arise, even though the desired pattern has the desired symmetry. The trouble is that for a given element spacing the visible range in ψ may be either larger or smaller than 2π. In fact, since the visible range in ψ is $2\beta d$, the visible range will correspond to 2π only for the particular spacing $d = \lambda/2$. The consequence of this is that if $d > \lambda/2$, the possible patterns that may be represented in this fashion are further restricted in that they must be consistent with the periodicity restriction on $\psi(2\pi)$. Or, on the other hand if d is less than $\lambda/2$, then the specification of the pattern in the visible range does not prescribe the array factor in a complete period in ψ. For example, consider the pattern

$$\mathrm{AF}(\phi) = \begin{cases} \sin^2 5\phi & 0.4\pi < \phi < 0.6\pi \\ 0 & \text{otherwise} \end{cases}$$

This pattern has the required symmetry. However, if we try to produce this pattern with an array that has an element spacing $d = \lambda/4$, the visible range in ψ is from $-\pi/2$ to $+\pi/2$; thus the array factor is not speci-

96 ANTENNA ENGINEERING

fied in the ranges between $-\pi/2$ and $-\pi$ or between $\pi/2$ and π. The lack of specification does not invalidate the method; the missing range may simply be filled in at the discretion of the designer. The "fill-in" function should be selected to provide a more accurate realizable pattern within the visible range.

Equation (16) may be generalized to have the form of a complete Fourier series representation:

$$\text{AF}(\psi) = \tfrac{1}{2} a_0 + \sum_{n=1}^{N-1} a_n \cos n\xi d + b_n \sin n\xi d \tag{17}$$

The symmetry restriction on the pattern is removed under this representation. The corresponding current distribution is

$$I(x) = \tfrac{1}{2} a_0 \delta(x) + \sum_{n=1}^{N-1} \tfrac{1}{2}(a_n - jb_n)\delta(x+nd) + \tfrac{1}{2}(a_n + jb_n)\delta(x-nd)$$

From an alternate point of view, suppose we write the array factor in the standard form, assuming an odd number of elements and numbering the center element zero,

$$\text{AF}(\psi) = \sum_{n=-N}^{N} A_n e^{jn\psi}$$

This expression is a finite number of terms from an exponential form of Fourier's series. Consequently, any array factor that may be represented by such a series may be synthesized by this method. The current excitation coefficients are calculated by the usual Fourier formula

$$A_n = \frac{1}{2\pi} \int_{-\pi}^{\pi} \text{AF}(\psi) e^{-jn\psi} \, d\psi$$

Problem 3.4 You are asked to design an array so as to give, in the range $0 < \phi < \pi$, the array factor

$$\text{AF}(\phi) = \begin{cases} C_0 & \dfrac{\pi}{4} < \phi < \dfrac{3\pi}{4} \\ 0 & \text{otherwise} \end{cases}$$

Decide on an element spacing (in wavelengths), find the current excitation coefficients, and present an argument for your choice of the number of elements in the array.

3.3.2 CONTINUOUS ARRAY SYNTHESIS

A pattern-synthesis procedure may be set up in another way. The mathematics is similar to that used in communication theory to derive the sampling theorem. Essentially, the desired array factor is sampled at an appropriate point and a corresponding space harmonic is introduced into the current distribution in such a way as to produce the correct relative field strength at the sample point.

To make best use of the Fourier transform relationship between the array factor and the current, we shall first represent the required current distribution by a complex Fourier series. However, the current distribution must of course be confined to a finite length L, so we shall multiply the series by a pulse-function multiplier defined to be zero outside the range $-\tfrac{1}{2}L < x < \tfrac{1}{2}L$ and unity inside that range:

$$I(x) = \sum_{-\infty}^{\infty} a_n e^{j2\pi nx/L}[U(x + \tfrac{1}{2}L) - U(x - \tfrac{1}{2}L)] \tag{18}$$

(This current distribution plays the role of the band-limited frequency response.) We may now make use of the Fourier transform relationship, Eq. (14), to display the associated array factor:

$$\mathrm{AF}(\xi) = \int_{-\infty}^{\infty} I(x)e^{jx\xi}\, dx = \int_{-\infty}^{\infty} \sum_{-\infty}^{\infty} a_n e^{j2\pi nx/L}[U(x + \tfrac{1}{2}L) - U(x - \tfrac{1}{2})L]e^{j\xi x}\, dx$$

$$= \sum_{-\infty}^{\infty} a_n \int_{-\tfrac{1}{2}L}^{\tfrac{1}{2}L} e^{j(2n\pi/L+\xi)x}\, dx$$

The integral is readily evaluated:

$$\int_{-\tfrac{1}{2}L}^{\tfrac{1}{2}L} e^{j(2n\pi/L+\xi)x}\, dx = L\,\frac{\sin(2n\pi/L + \xi)\tfrac{1}{2}L}{(2n\pi/L + \xi)\tfrac{1}{2}L}$$

The array factor is then, in terms of a variable $\psi_1 = \tfrac{1}{2}\xi L$,

$$\mathrm{AF}(\psi_1) = L \sum_{n=-\infty}^{\infty} a_n \frac{\sin(\psi_1 + n\pi)}{(\psi_1 + n\pi)} \tag{19}$$

(ψ_1 is the same variable employed earlier in the analysis of continuous arrays since $\xi = \beta \cos \phi + k$, although here we imagine k to be zero.) Note that the array factor is a series of $\sin \psi_1/\psi_1$-type functions each of which is displaced along the ψ_1 axis by an amount π from the next lower order member of the series. This fact is indicated in Fig. 3.21. Note also the particular value of each member of the series at $\psi_1 = m\pi$, where m is an integer; since the argument of the sine functions is then $(m + n)\pi$, all terms in the series are zero except the one term for which $n = -m$. Thus we have the equation

$$\mathrm{AF}(m\pi) = La_{-m} + 0$$

or

$$\mathrm{AF}(-n\pi) = La_n + 0 \tag{20}$$

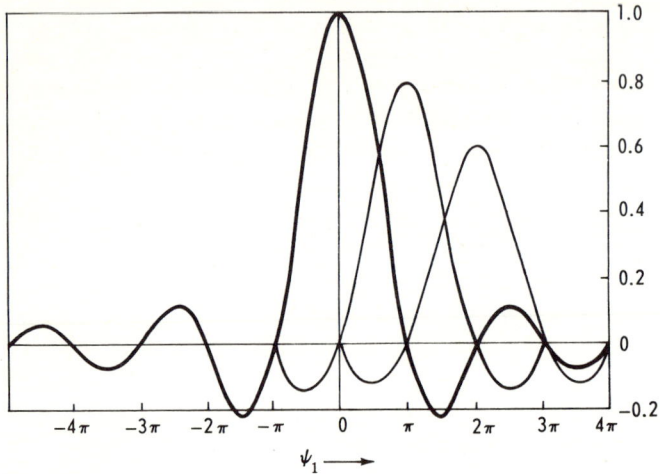

Fig. 3.21 Three terms in the series, Eq. (19), showing the $\sin \psi_1/\psi_1$ curves displaced along the ψ_1 axis by amounts $n\pi$.

According to this equation, we may find the amplitude of the nth term in the series expression for the current distribution [Eq. (18)] very simply because it is proportional to the value of the required array factor at the particular point $\psi_1 = -n\pi$:

$$a_n = \frac{1}{L} \text{AF}(-n\pi) \tag{21}$$

It is interesting to interpret these results physically. Looking back to Eq. (18), it is seen that each term in the series represents a uniform current distribution over a length L with a phase shift per unit length of $2n\pi/L$. We learned earlier [Eq. (9)] that such a current distribution gives a $\sin \psi_1/\psi_1$-type pattern, with the direction of the maximum in the pattern determined by the phase-shift constant, in this case $\cos \phi_{\max} = -2n\pi/\beta L = -n\lambda/L$. The individual terms in the series expression for the current distribution may therefore be interpreted as individually contributing $\sin \psi_1/\psi_1$ patterns having main beams at the angles $\phi_{\max} = \cos^{-1}(-2n\pi/\beta L)$. As the length becomes quite large, these maxima are essentially the rays corresponding to the plane waves heading in the direction $\phi_{\max}$. This interpretation is similar to that given originally by Woodward.[1]

We may now summarize the procedure for synthesis of an array factor. In practice, the desired array factor is given as a function of ϕ in the visible range $0 < \phi < \pi$. The array factor is sampled at points $\psi_{1s} =$

---

[1] P. M. Woodward, *Proc. IEEE (London)*, pt. IIIA, **93**:1554 (March–May, 1946).

$-n\pi$. It follows from the definition $\psi_{1s} = \frac{1}{2}\beta L \cos \phi_s$ that the sample points in ϕ are given by $\cos \phi_s = -2n\pi/\beta L = -n\lambda/L$. Since the visible range is confined to values $\cos \phi$ between 1 and -1, the largest value of $\cos \phi$ that is sampled is such that $n_{\max}\lambda/L \leqq 1$ or $n_{\max} \leqq L/\lambda$. This means that the array factor is sampled at $2n_{\max} + 1$ points in the visible range, where $n_{\max}$ is the largest integer less than or equal to L/λ. The values of the array factor at these points determine the coefficients:

$$\mathrm{AF}(\phi_s) = \mathrm{AF}\left(\cos^{-1}\frac{-n\lambda}{L}\right) = a_n L$$

The synthesized array factor is then a sum of $2n_{\max} + 1$ functions and furthermore has exactly the prescribed values at the $2n_{\max} + 1$ sample points. If this number of terms gives an acceptable approximation at positions in between the sampled points and gives an acceptable $\mathrm{AF}(\frac{1}{2}\xi L)$ outside the visible range, then the task is complete. Generally, there will be some "ripple" in between the sampled points. If this ripple is too great, it may be minimized by the assignment of a phase angle to the coefficients a_n. This is possible because at the sampled points we determined (and generally require) only the *amplitudes* of the a_n. At the sample points, only one of these coefficients has a value, so the phase of the a_n's does not affect the magnitude of the array factor at the sample points. At points in between the sample points, the individual phases do play a part. The values of the array factor at points between the sample points is in the main determined by two terms—the "beams" on either side of the sample points. Consequently, a systematic adjustment of the relative phases of adjacent coefficients a_n and a_{n+1} can often force the synthesized array factor to have the correct value in between the sample points.

As in the analogous problem in communication or signal theory, the accuracy of the synthesis depends on the "sampling rate." The more rapidly the array factor varies as a function of angle (i.e., the higher the angular harmonic content), the more "frequently" the function must be sampled for an accurate representation. In array theory, the "sampling rate" depends on the length of the array measured in wavelengths. To sample at something like the Nyquist rate may require a relatively long array. If it is helpful in improving the accuracy of the approximation, the array factor may be smoothly continued into the invisible range of ψ, and a few more terms may be included in the series expansions. The synthesis procedure is perhaps best illustrated by means of an example.

Example *Pattern synthesis by the sampling theorem.* The problem is to produce a field amplitude which is specified in the visible range to be

$$\mathrm{AF}(\phi) = e^{-3(\phi-0.4\pi)^2} \qquad 0 < \phi < \pi$$

100 ANTENNA ENGINEERING

The antenna may not occupy more than some finite number of wavelengths; as an example we take $L/\lambda = 6$. The required current distribution is

$$I(x) = \sum_{n=-\infty}^{\infty} a_n e^{j2n\pi x/L}[U(x + \tfrac{1}{2}L) - U(x - \tfrac{1}{2}L)]$$

in which the coefficients a_n are determined from $a_n = \mathrm{AF}(-n\pi)/L$ or

$$\mathrm{AF}(\phi_s) = \mathrm{AF}\left(\cos^{-1}\frac{-n\lambda}{L}\right) = \mathrm{AF}\left(\cos^{-1}\frac{n}{6}\right)$$

In the visible range, the sample points are angles such that $\cos\phi_s = \pm n/6$, $n = 0, 1, 2, \ldots 6$. The values of the array factor at the sampling points are given in the following table:

n	ψ_1	$\cos\phi_s$	$\mathrm{AF}(\phi_s) = a_n L$
6	-6π	-1	4×10^{-5}
5	-5π	$-5/6$	6×10^{-3}
4	-4π	$-4/6$	0.04
3	-3π	$-1/2$	0.12
2	-2π	$-1/3$	0.287
1	-1π	$-1/6$	0.493
0	0	0	0.742
-1	π	$1/6$	0.941
-2	2π	$1/3$	1.0
-3	3π	$1/2$	0.840
-4	4π	$2/3$	0.593
-5	5π	$5/6$	0.260
-6	6π	1	0.01

Since the higher- and lower-order coefficients fall off rapidly in amplitude, it is not necessary to determine more of these by specifying the array factor outside the visible range. The extent to which the thirteen terms determined above approximate the array factor is indicated in Fig. 3.22. A better approximation implies more sample points and hence a longer array. Note the ripple in the synthesized pattern A at points between the sample points. The character of this ripple may be changed markedly by adding a phase angle to the a_n's. For example, the curve B results if the phases of alternate coefficients are shifted by 180°. Note that the sample points are still correct in amplitude but the ripple is greatly increased. The phase angles of the coefficients may sometimes be adjusted to smooth the ripples. This is equivalent to specifying the phase of the array factor at the sample points.

Fig. 3.22 Example of pattern synthesis by the sampling theorem. Solid curve is desired pattern. Dashed curve [A is synthesized pattern of example (very slight ripple)]. Dotted curve B shows drastic result of alternating signs (phase) of the a_n's.

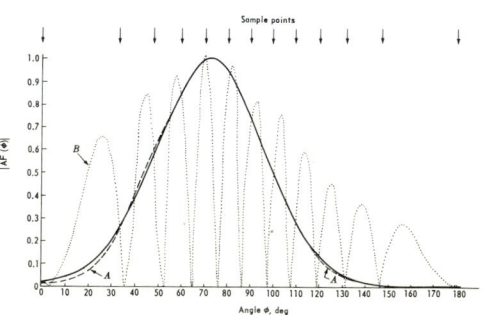

An interesting array factor for academic purposes is one of the type

$$AF(\tfrac{1}{2}\xi L) = e^{-C(\xi/\beta)^2}$$

corresponding to a pattern in the visible range

$$AF(\phi) = e^{-C \cos^2 \phi}$$

This is a pattern, symmetric about $\phi = \pi/2$, that has a single lobe, but has as narrow a beamwidth as is desired according to the constant C. The synthesis of this array factor is suggested as a problem for the interested reader.

One of the principal difficulties with the practical application of the synthesis theory is that the theory does not tell us how to obtain the required current distributions. Very little work has been done to show how, in connection with the foregoing synthesis procedure, the required beams may be generated. There has been a suggestion[1] that in certain aperture antennas (in particular, traveling-wave slot antennas), the desired phase velocities might be obtained by the excitation of the normal modes of a system of coupled waveguides. The experimental work reported was for a single pair of coupled waveguides, however, and more complex systems were not systematically investigated.

Problem 3.5 A continuous linear array is to provide an array factor $AF(\phi) = \sin^2 \phi$ in the range $0 < \phi < \pi$. Find the required current distribution, specify the length, and compare your synthesized array factor with the specified array factor.

3.4 CHEBYSHEV ARRAYS

In antenna system design, it is frequently desirable to achieve both a narrow beamwidth and a low side-lobe level. However, these features

[1] W. L. Weeks, *IRE WESCON Conv. Record*, pt. **1**:236 (August, 1957).

are normally tied together in such a way that an improvement of one is associated with a deterioration of the other. Consequently, Dolph[1] made an important contribution when he demonstrated that array factors derived from Chebyshev polynomials optimize the relationship between beamwidth and side-lobe level. We shall now describe a design procedure for Dolph-Chebyshev arrays.

Let us first digress slightly in order to review the main mathematical facts concerning the Chebyshev polynomials. The polynomials may be defined by the equations (the symbol T is a vestige of an older common spelling "Tchebysheff"):

$$T_m(x) = \cos(m \cos^{-1} x) \qquad -1 < x < 1 \qquad (22)$$

$$T_m(x) = \cosh(m \cosh^{-1} x) \qquad |x| \geq 1 \qquad (23)$$

From these equations, the specific forms for the lower-order polynomials are

$$T_0(x) = 1$$

$$T_1(x) = x$$

The higher-order polynomials may be found by a recursion relationship

$$T_{m+1}(x) = 2x T_m(x) - T_{m-1}(x)$$

or by setting $\delta = \cos^{-1} x$ and expanding $\cos m\delta$ in powers of $\cos \delta$ (for example, $\cos 4\delta = 8 \cos^4 \delta - 8 \cos^2 \delta + 1$) so that

$$T_4(x) = 8x^4 - 8x^2 + 1$$

The following general form for even-order Chebyshev polynomials may be established:[2]

$$T_{2m}(x) = \sum_{i=0}^{m} A_{2i}{}^{2m} x^{2i}$$

where

$$A_{2i}{}^{2m} = \frac{(-1)^{m-i} 2m(m+i-1)! \, 2^{2i-1}}{(2i)!(m-i)!}$$

The functions $T_2(x)$, $T_3(x)$, and $T_4(x)$ are shown in Fig. 3.23. Note that in the range $-1 < x < 1$, the polynomials all oscillate in value between $+1$ and -1. The mth-order polynomial $T_m(x)$ crosses the axis m times in the range $|x| < 1$. In the range $|x| > 1$, the polynomials eventually increase without limit at a rate proportional to x^m.

---

[1] C. L. Dolph, *Proc. IRE*, **34**:335–348 (June, 1946).
[2] H. B. Dwight, "Tables of Integrals and Other Mathematical Data," The Macmillan Company, New York, 1947.

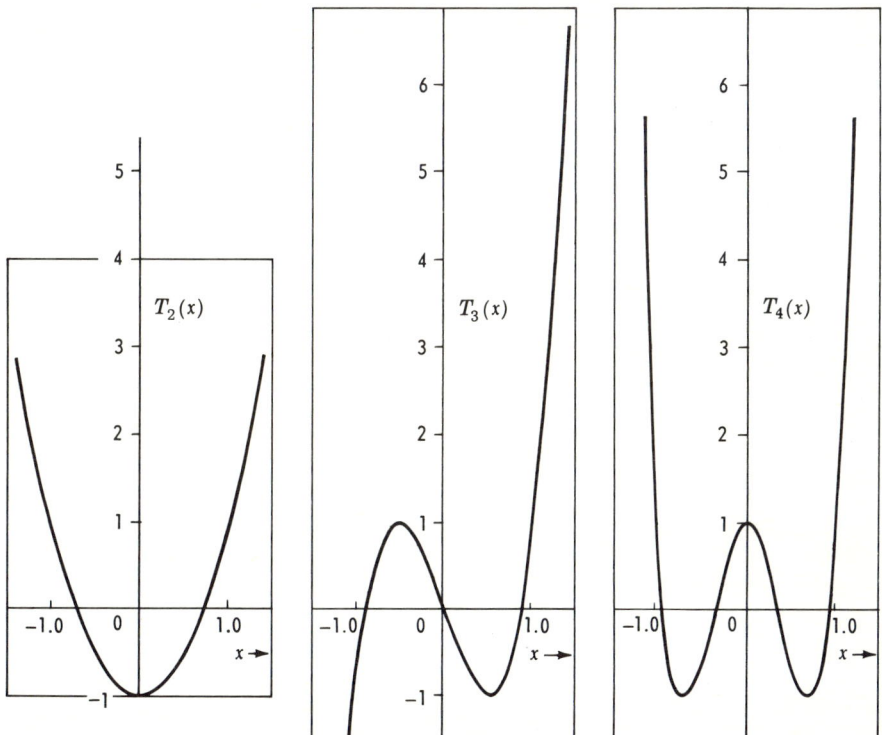

Fig. 3.23 Chebyshev polynomials.

The object of identifying an array factor with a Chebyshev polynomial is to arrange that the side-lobe levels are all equal and that (1) the main beam is a specified number of decibels above the side-lobe level or (2) the main beam has a specified angular width. These features are accomplished by positioning the visible range with respect to the variable x of the Chebyshev function so that the side lobes arise from the region $|x| < 1$ and the main beam extends as far into the range $|x| > 1$ as is required to obtain the specified ratio of main beam to side-lobe level, or the specified beamwidth. A proof that the Chebyshev array is optimum (in the sense of giving the narrowest beamwidth for a given side-lobe level) may be found in the papers by Dolph[1] and Riblet.[2]

To get into the details, the problem is to find the excitation coefficients in an array such that over a restricted range of x a Chebyshev polynomial

[1] Dolph, *op. cit.*
[2] H. J. Riblet, *Proc. IRE*, **35:**489 (May, 1947).

is equal to the array factor:

$$T_M(x) = \sum_{n=0}^{N-1} A_n e^{jn\psi} \qquad x_1 \leq x \leq x_2$$

where of course x must be some function of ψ and $\psi = \beta d \cos \phi + \alpha$, as before. We are thinking of x as a real variable so it is appropriate to change the form of the array factor so that it is a series of real functions. This may be done by imposing on the current distribution a symmetry condition about the center of the array. In this text, we shall consider only arrays with odd numbers of elements. The analysis for even numbers of elements may be done similarly (or equivalent procedures may be found in the literature). For organizational purposes, we shall renumber the elements so that the center element is the zeroth. With the symmetry condition about the center element, $A_n = A_{-n}$, the array factor may be written

$$\begin{aligned}
\text{AF}(\psi) &= \sum_{n=-\frac{1}{2}(N-1)}^{\frac{1}{2}(N-1)} A_n e^{jn\psi} \\
&= A_0 + \sum_{n=1}^{\frac{1}{2}(N-1)} (A_n e^{jn\psi} + A_{-n} e^{-jn\psi}) \\
&= A_0 + 2 \sum_{n=1}^{\frac{1}{2}(N-1)} A_n \cos n\psi
\end{aligned}$$

To identify the Chebyshev polynomial with the array factor, we now impose the condition

$$T_M(x) = A_0 + 2 \sum_{n=1}^{\frac{1}{2}(N-1)} A_n \cos n\psi \qquad x_1 \leq x \leq x_2 \qquad (24)$$

where again x is some function of ψ. In order that Eq. (24) be satisfied exactly, it is necessary that the Chebyshev polynomial $T_M(x)$ be exactly representable by a finite cosine series. The problem is to choose the functional relationship between x and ψ so that (1) this exact representation is possible, (2) the ratio of main beam amplitude to side-lobe level is as specified (or that the main beamwidth is as specified), and (3) the main beam has the required direction in space. Finally, the appropriate current excitation coefficients must be identified.

The functional relationship between x and ψ is not unique. However, there is one general form of it which guarantees that the Chebyshev polynomial can be expressed in a finite cosine series. In particular, if x is simply a linear function of the cosine of ψ, then $T_M(x)$ will have a representation in the form of a finite cosine series. To see this, recall that $T_M(x)$ is a sum of powers of x (i.e., it is a polynomial in x). Therefore, if x has the form $x = a \cos \psi + b$, then the Chebyshev polynomial will consist of a sum of terms of the form $x^n = (a \cos \psi + b)^n$, which of course is a sum of terms in powers of $\cos \psi$. But it follows from standard trigonometric identities that powers of $\cos \psi$ may be expressed as a sum of terms of the type $\cos n\psi$, where if M is the highest power of x and $\cos \psi$, then n takes on values $n = 0, 1, 2, 3, \ldots M$.

These latter statements and the process of finding the current amplitudes are best illustrated by means of an example. Consider an array factor which is to be equal to the Chebyshev polynomial $T_2(x) = 2x^2 - 1$. Taking $x = a \cos \psi + b$, we have

$$T_2(x) = 2(a^2 \cos^2 \psi + 2ab \cos \psi + b^2) - 1$$

But since $2 \cos^2 \psi = \cos 2\psi + 1$, then

$$T_2(x) = a^2 + 2b^2 - 1 + 4ab \cos \psi + a^2 \cos 2\psi$$

In this latter form, the Chebyshev polynomial is clearly expressed in the form of a finite cosine series. By comparison to Eq. (24) and inspection, the current excitation coefficients may be identified as follows:

$$A_0 = a^2 + 2b^2 - 1$$
$$2A_1 = 4ab \tag{25}$$
$$2A_2 = a^2$$

The formulas for the current excitation coefficients required to produce Chebyshev array factors of other orders may be found similarly; however, this is not usually necessary since by now the values for the more common types are tabulated in the literature.[1,2,3]

To sum up, the transformation $x = a \cos \psi + b$ guarantees that an Mth-order Chebyshev polynomial may be expressed as a finite cosine series having $M + 1$ terms. This implies that in the series for the array factor [Eq. (24)], the quantity $\frac{1}{2}(N - 1)$ must equal M; that is, the number of elements in the array is $N = 2M + 1$. The procedure being described is somewhat more general than that given in Dolph's original paper. It has the advantage of being applicable to arrays with arbitrary element spacing and to end-fire array calculations.

We have yet to find the numbers a and b. These are determined from antenna pattern specifications. Suppose that the ratio of main beam amplitude to side-lobe level is specified to be R and of course $R > 1$. Since in the range $|x| < 1$ all the Chebyshev polynomials have a maxi-

---

[1] DuHamel, *op. cit.*
[2] H. Jasik, "Antenna Engineering Handbook," McGraw-Hill Book Company, New York, 1961.
[3] J. L. Brown, *IRE Trans.*, **AP-10**:215 (March, 1962).

mum value of unity, we can establish the desired side-lobe level by associating the main beam position with a value of x, call it x_0, that is in the range $|x| > 1$; that is, $T_M(x_0) = R$. Our first condition on a and b is then

$$x_0 = a \cos (\beta d \cos \phi_{\max} + \alpha) + b \qquad (26)$$

Let us restrict our treatment first to broadside beams, so that $\alpha = 0$. Then we need only one additional equation in order to determine both of the constants a and b. The determination of this additional condition is the subject of the following paragraph.

To see what the additional equation should be, it is helpful to employ a graphical representation of the transformation $x = a \cos \psi + b$. This can be regarded as the projection of a rotating radius vector similar to that employed earlier to relate ψ and the visible angle ϕ. When completed, the diagram will resemble that in Fig. 3.24. Therein, the visible

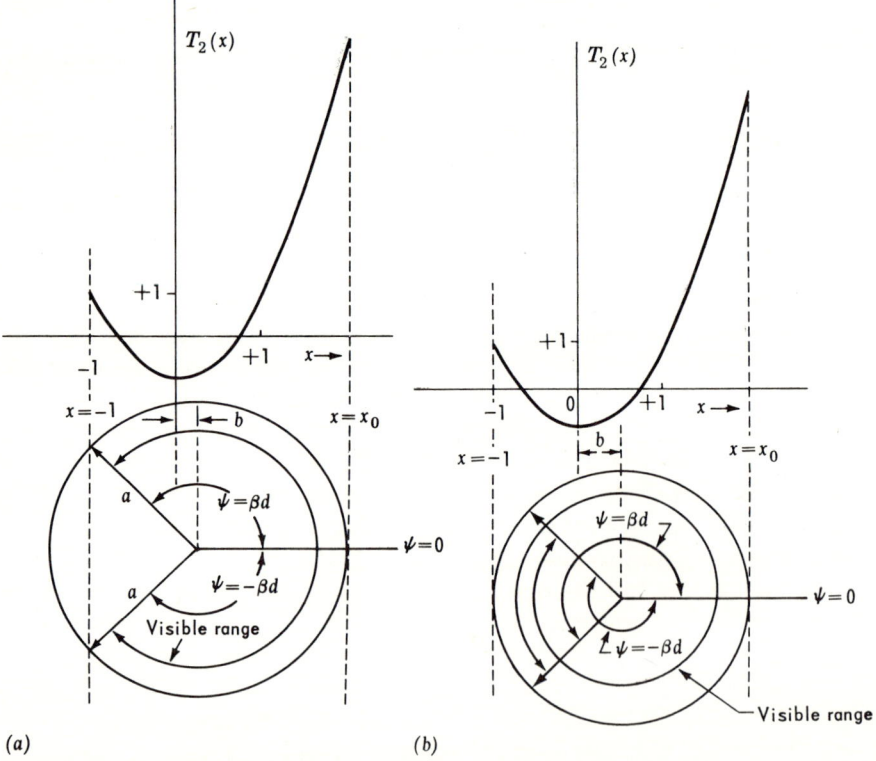

Fig. 3.24 A graphical representation showing the correspondence between ψ and $x = a \cos \psi + b$. (a) Here βd is less than π; (b) here βd is greater than π.

range of ψ is shown in correspondence with a visible range in x which extends from $x = -1$ to $x = x_0$. The problem of identifying the constants a and b may be recognized as the problem of finding the radius a and the location of the center $x_c = b$ of the locus circle. In order to have the beam maximum at broadside, we want the position $\psi = 0$ to correspond to $x = x_0$ (recall $\psi_{max} = \beta d \cos \phi_{max} = \beta d \cos \frac{1}{2}\pi = 0$), which is just the condition imposed by Eq. (26). The second condition to be imposed is that the limit of the visible range of ψ is to correspond to the point $x = -1$.[1] The quantity βd is of course known (specified), and since the visible range of ϕ is from 0 to π, the visible range of ψ is from βd to $-\beta d$. The equation representing the second condition is then

$$-1 = a \cos \beta d + b \tag{27}$$

Solving this equation simultaneously with Eq. (26), evaluated at $\phi = \pi/2$ ($\alpha = 0$) gives for the case that $\beta d < \pi$

$$a = \frac{x_0 + 1}{1 - \cos \beta d} \tag{28}$$

$$b = \frac{x_0 \cos \beta d + 1}{\cos \beta d - 1} \tag{29}$$

As the visible range of ψ is increased by the increase of the quantity βd, changes are brought about which we may visualize by a study of Fig. 3.24. The circle therein is always tangent to the line $x = x_0$. Thus, as βd increases, the radius a must decrease until βd reaches the value $\beta d = \pi$; at that point, the parameter a is determined by $a = (x_0 + 1)/2$, since the circle is then tangent to the line $x = -1$ as well as $x = x_0$. But note further that if the quantity βd is greater than π, the visible range still includes the value $\psi = \pi$; so again we should make the position $\psi = \pi$ correspond to the point $x = -1$ and again the circle should be tangent to the lines $x = -1$ and $x = x_0$; also, the center of the circle is just halfway between the points $x = -1$ and $x = x_0$. In summary then, if $\beta d \geq \pi$, the constants are

$$a = \frac{x_0 + 1}{2} \tag{30}$$

$$b = \frac{x_0 - 1}{2} \tag{31}$$

We see therefore that the specification of x_0 and βd (the element spacing) determines the constants a and b. These constants in turn determine the required element currents through equations of the type of Eq. (25).

[1] The choice $x = -1$ rather than some larger value gives the narrowest beam for a given βd, still giving equal side lobes.

The number x_0 is determined from the specified side-lobe level R together with the defining equation, (23), for the polynomials as follows:

$$R = \cosh(M \cosh^{-1} x_0)$$

so $\cosh^{-1} R = M \cosh^{-1} x_0$ and

$$x_0 = \cosh\left(\frac{1}{M} \cosh^{-1} R\right) \tag{32}$$

Sometimes, the design problem may be met in another way. The task may be to generate a beam with a stated beamwidth between nulls. Again the problem is to determine the current amplitudes and spacing such that the pattern is optimum in the sense of having the lowest side-lobe level for the required beamwidth. The zeros in the Chebyshev polynomial and therefore the zeros in the pattern may be determined from Eq. (22). Clearly the function is zero when x takes on values determined by the equation

$$M \cos^{-1} x_r = \pm \frac{(2k-1)\pi}{2} \qquad k = 0, 1, 2, 3, \ldots$$

The roots of the polynomials x_r are then given by the formula

$$x_r = \cos \frac{2k-1}{M} \frac{\pi}{2}$$

The corresponding values of ψ are

$$\psi_r = \cos^{-1} \frac{\cos \frac{2k-1}{M} \frac{\pi}{2} - b}{a}$$

If the beamwidth is specified by giving the angle of the first null ϕ_1, we note that the first null correlates with the root given by placing $k = 1$ in the expression above; thus we can equate two expressions for ψ_r:

$$\beta d \cos \phi_1 = \cos^{-1}\left[\frac{\cos(\pi/2M) - b}{a}\right]$$

or

$$a \cos(\beta d \cos \phi_1) = \cos \frac{\pi}{2M} - b$$

The constants a and b may be determined from this latter equation together with Eq. (27) or by the equation $-1 = -a + b$ (depending on the value of βd); then x_0 may be found since $x_0 = a + b$. The final result for the case $\beta d < \pi$ is

$$x_0 = \frac{1 + (1 - \cos \beta d) \cos(\pi/2M) - \cos(\beta d \cos \phi_1)}{\cos(\beta d \cos \phi_1) - \cos \beta d} \tag{33a}$$

In the event that $\beta d > \pi$, the formula is

$$x_0 = \frac{1 + 2\cos(\pi/2M) - \cos(\beta d \cos \phi_1)}{1 + \cos(\beta d \cos \phi_1)} \tag{33b}$$

Graphical constructions indicative of the patterns and processes are displayed in Fig. 3.25. On the left is a construction similar to that indicated in Fig. 3.24. The array factor as a function of ψ may be determined from this construction. Having determined $AF(\psi)$, the construction described earlier (Figs. 3.7 and 3.8) may be employed to obtain the pattern. The latter construction is displayed on the right of Fig. 3.25. It is clear from these constructions that (1) if $\beta d \ll \pi$, there are large lobes just outside the visible range; (2) if $\beta d \gg 2\pi$, there will be secondary maxima (grating lobes) in the pattern. (Experience has shown that large lobes just outside the visible range are usually associated with large reactive energy in the array and poor bandwidth and efficiency.)

Example We shall now illustrate the design procedure by applying it to obtain a design for a seven-element broadside array that has the optimum pattern for a side-lobe level of -20 db. The element spacing is to be one-half wavelength.

A seven-element array gives an optimum pattern if it is based on a Chebyshev polynomial of order $M = (N-1)/2 = 3$. For a side-lobe level of -20 db, the ratio R of the field strength at the main beam maximum to that of the side lobes is 10. The beam maximum is then to be associated with a value of the polynomial x_0 where, by Eq. (32),

$$x_0 = \cosh(\tfrac{1}{3} \cosh^{-1} 10) = 1.541$$

Thus, by Eqs. (30) and (31), $a = 2.541/2 = 1.2705$, $b = 0.541/2 = 0.2705$. The Chebyshev polynomial must equal the array factor

$$T_3(x) = 4x^3 - 3x = 4(a\cos\psi + b)^3 - 3(a\cos\psi + b)$$
$$= A_0 + \sum_{n=1}^{3} 2A_n \cos n\psi$$

Expanding $(a\cos\psi + b)^3$, collecting terms, and identifying coefficients (or consulting existing tables) gives, for the current excitation coefficients,

$A_0 = 6a^2b + 4b^3 - 3b = 1.90$
$2A_1 = 3a^3 + 12ab^2 - 3a = 3.46$
$2A_2 = 6a^2b = 2.63$
$2A_3 = a^3 = 2.05$

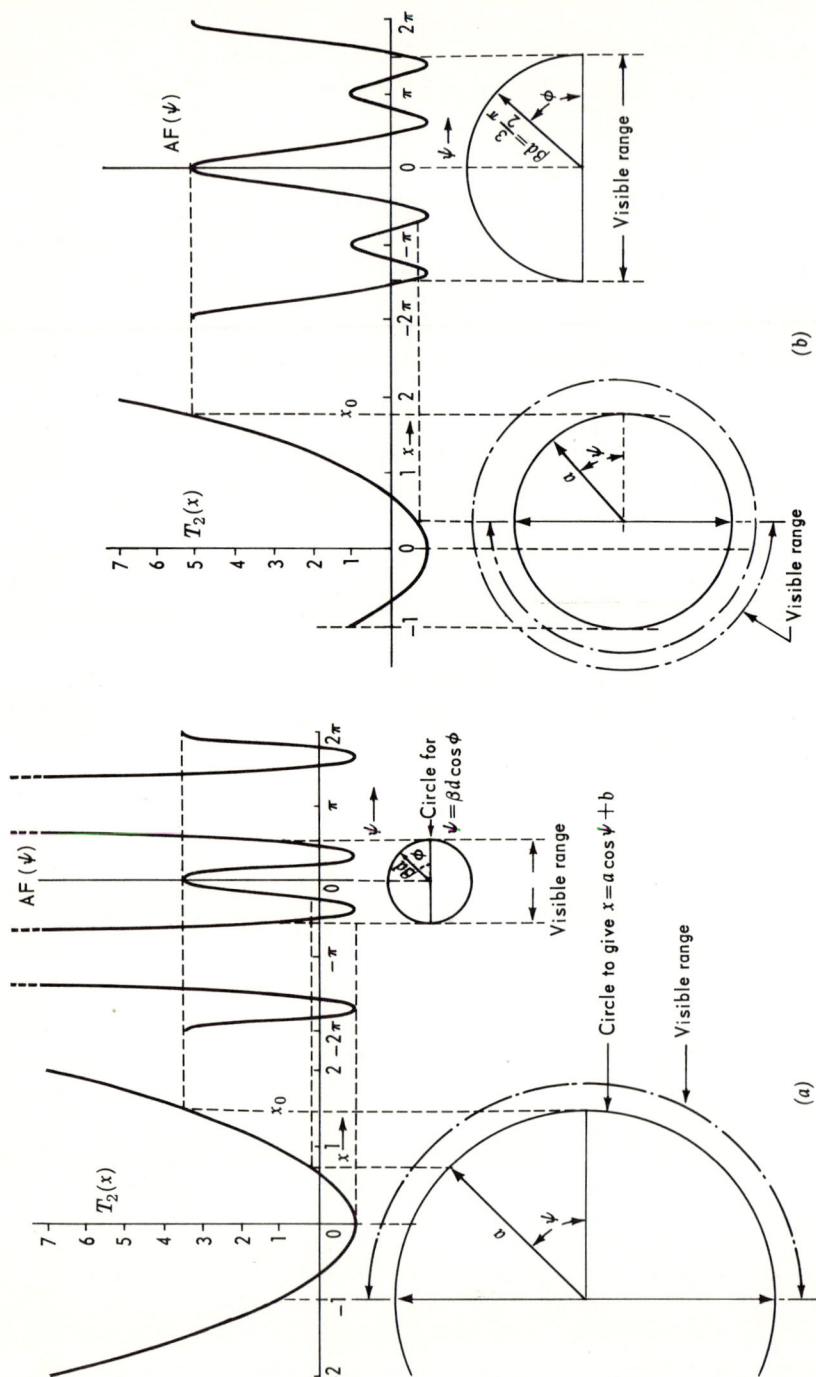

Fig. 3.25 (a) Graphical construction showing optimum pattern from five-element array based on T_2 polynomial, in the case that $d < \lambda/2$ (actually here $\beta d = \pi/2$, so $d = \lambda/4$); (b) similar construction but with $d > \lambda/2$ (actually here $\beta d = 3\pi/2$, so $d = 3\lambda/4$).

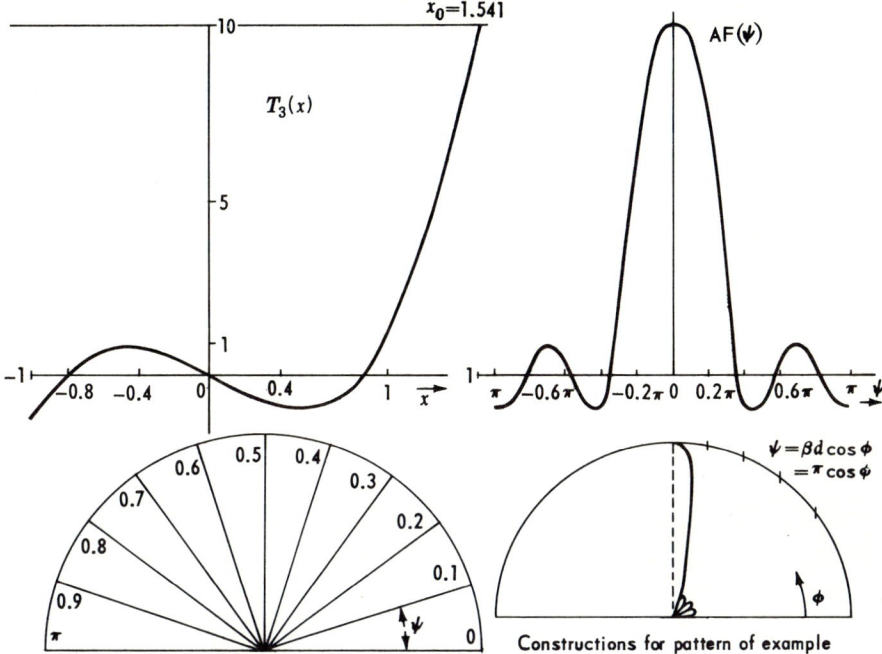

Fig. 3.26 Constructions for the example—a seven-element Chebyshev pattern with −20-db side lobes.

This completes the design. The patterns may be sketched from graphical constructions as indicated in Fig. 3.26.

Problem 3.6 Design a seven-element array to radiate an optimum broadside pattern with a beamwidth between nulls of 30°. Determine the side-lobe level and sketch the pattern. Also, determine the patterns at frequencies 10 percent above and 10 percent below the design frequency, assuming that the element currents are not changed by the frequency change.

Finally, let us consider the design of Chebyshev arrays for other than broadside beams. For all except end-fire antennas, the general procedure is the same as for broadside beams and the result is but a slight modification of Eqs. (26) to (29). The beam direction is controlled by including an element-to-element phase shift. The variable $\psi = \beta d \cos \phi + \alpha$ includes the phase shift and is most commonly written in the form $\psi = \beta d (\cos \phi - \cos \phi_m)$, where ϕ_m is the angle at which the pattern is to be a maximum. The visible range of ψ is then from $-\beta d(1 + \cos \phi_m)$ to $\beta d(1 - \cos \phi_m)$. The ψ value for the beam maximum $\psi = 0$ must corre-

112   ANTENNA ENGINEERING

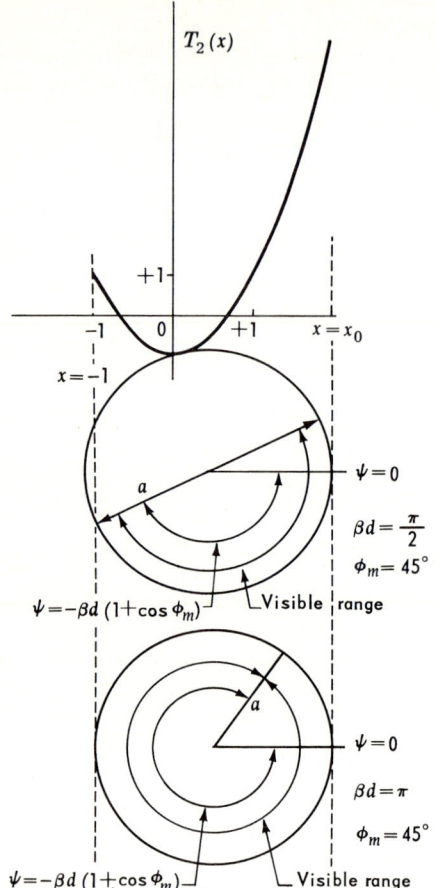

Fig. 3.27 Construction showing correspondence between ψ variable and Chebyshev polynomial for non-broadside beams.

spond to the largest value x_0 of the Chebyshev polynomial

$$x_0 = a + b$$

This is the same as Eq. (26). No part of the visible range should extend beyond $x = -1$. Hence, as before, we must consider two cases, those in which the ψ value is always less than π in the visible range, and those for which the visible range includes $\psi = \pi$. We note the following: if $0 < \phi_m < \pi/2$, then $\cos \phi_m$ is positive; therefore if $\beta d(1 + \cos \phi_m) < \pi$ then the largest value of $|\psi|$ is $|-\beta d(1 + \cos \phi_m)|$. Thus we make this value of ψ correspond to $x = -1$ and so obtain a slightly modified form of Eq. (27) (see Fig. 3.27):

$$-1 = a \cos[-\beta d(1 + \cos \phi_m)] + b$$

(Note: $\cos(-x) = \cos x$; thus the equation is the same for the case $\pi/2 < \phi_m < \pi$.) Now we have two equations that may be solved for a and b. The result is a slightly modified form of Eqs. (28) and (29):

$$a = \frac{x_0 + 1}{1 - \cos[\beta d(1 + \cos\phi_m)]}$$

$$b = \frac{x_0 \cos[\beta d(1 + \cos\phi_m)] + 1}{\cos[\beta d(1 + \cos\phi_m)] - 1}$$

When the value $\psi = \pi$ is included in the visible range (that is, $\beta d(1 + |\cos\phi_m|) > \pi$), the value $\psi = \pi$ must be brought into correspondence with $x = -1$, or

$$-1 = -a + b$$

and the equations for a and b in terms of x_0 are the same as for the broadside case, Eqs. (30) and (31).

These results hold for beam directions up to the end-fire direction. For end-fire arrays, if the same procedure is followed, the resulting pattern is based on a Chebyshev polynomial, and has equal-level side lobes, but for spacings such that $\beta d < \pi/2$, the pattern is not optimum. The reason is related to the fact that the element-to-element phase shift may be made greater than that of a plane wave traveling along the array (i.e., greater than β) and this results in a narrower end-fire beam (review pages 77–78 and 81–83).

Thus for end-fire Chebyshev arrays, the phase-shift constant α must be regarded as an unknown. This makes some difference in the analysis. In particular, the value of ψ corresponding to the beam maximum is not zero as it was for the non-end-fire beams; rather, it is $\psi_m = \beta d + \alpha$ since the beam maximum is in the direction $\phi = 0$. This value of ψ_m must correspond to the position $\pm x_0$ in the Chebyshev polynomial

$$\pm x_0 = a\cos(\beta d + \alpha) + b$$

This is the first of the three equations required to find the unknowns a, b, and α. The sign ambiguity will be resolved by examination of the visible range. The visible range of ψ is $\beta d + \alpha$ to $-\beta d + \alpha$, and the value $\psi = 0$ corresponds to an angle ϕ_0 such that $0 = \beta d \cos\phi_0 + \alpha$. We expect this angle to be in the visible range. Examination of Fig. 3.28 shows that in order to have the maximum position at end fire with $\psi_m = \beta d + \alpha$, it is necessary that $\psi = 0$ correspond to $x = +1$, that is,

$$1 = a + b \qquad (34)$$

114 ANTENNA ENGINEERING

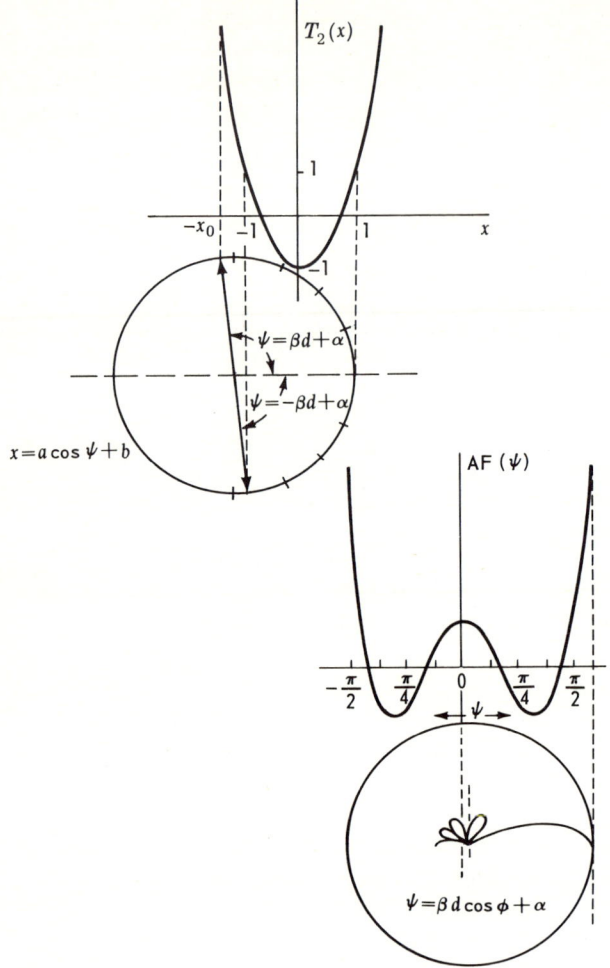

Fig. 3.28 Constructions showing the transformations taking a Chebyshev polynomial into an end-fire antenna pattern.

and ψ_m correspond to $-x_0$

$$-x_0 = a \cos(\beta d + \alpha) + b \tag{35}$$

The last equation of the required set of three is obtained by bringing the other end of the visible range $\psi = -\beta d + \alpha$ into correspondence with $x = -1$:

$$-1 = a \cos(\beta d - \alpha) + b \tag{36}$$

Equations (34) to (36) may be solved for the three unknowns a, b, and α.

The result is

$$a = \frac{x_0 + 3 + (2 \cos \beta d)(2x_0 + 2)^{1/2}}{2 \sin^2 \beta d} \tag{37}$$

$$b = 1 - a \tag{38}$$

$$\sin \alpha = \frac{x_0 - 1}{2a \sin \beta d} \tag{39}$$

If the visible range includes both $\psi = 0$ and $\psi = \pi$, then Eqs. (37) to (39) are not applicable. Generally, when the values $\psi = 0$ and π are both included, the spacing will be of the order of a half wavelength or greater, as a result of which the grating lobes will begin to appear, and the unidirectional end-fire beam will be lost.

Problem 3.7 Design a seven-element array, with element spacing of a quarter wavelength, to give an optimum end-fire pattern with a side-lobe level of -20 db. Sketch the pattern and compare it with DuHamel's optimum pattern displayed in Fig. 3.20.

3.5 SUMMATION PROCEDURES FOR ARRAY FACTORS

As was pointed out in Eq. (5), the array factor of a linear array of N equally spaced elements may be written

$$\text{AF}(\psi) = \sum_{n=0}^{N-1} A_n e^{jn\psi} \qquad \psi = \beta d \cos \phi + \alpha$$

One of the problems in array analysis is to carry out this finite summation. The process should be applicable to array factors having current distributions of arbitrary form and preferably should yield the final result in closed form. The latter is not possible in general, but a few of the special techniques that are known will be discussed in this section.

It will be recalled that the array factor of a uniformly excited array turns out to be the sum of N terms in a geometric series. The series is also geometric if the current amplitudes, although not uniform, are arranged so that there is a constant ratio between the amplitudes from one element to the next, $A_{n+1} = KA_n$, where K is a constant independent of n. In this case the common ratio is $Ke^{j\psi}$; thus the summation may be written in the closed form

$$\text{AF}(\psi) = \frac{A_0(K^N e^{jN\psi} - 1)}{Ke^{j\psi} - 1}$$

The use of Z transforms to help sum array factors has been discussed by Cheng,[1] Christiansen,[2] and others. The point of view taken is that the element currents in a discrete array represent sample points of a continuous current distribution $I(x)$ in which the samples are taken at intervals $x = nd$, where d is the spacing. The Z transform of a function $f(x)$ is $Z\{f(x)\} \equiv \sum_{n=0}^{\infty} f(x)z^{-n}$. If, in the expression for the array factor, we put $z = e^{-j\psi}$, the array factor may be written

$$\text{AF} = \sum_{n=0}^{N-1} I(nd)z^{-n}$$

As $N \to \infty$, the array factor tends toward the Z transform of $I(nd)$. In any case, we may write the array factor as a difference between two infinite sums and so obtain the array factor as a difference of Z transforms:

$$\text{AF} = \sum_{n=0}^{\infty} I(nd)z^{-n} - \sum_{N}^{\infty} I(nd)z^{-n}$$
$$= Z\{I(nd)\} - z^{-N}Z\{I(Nd + d)\}$$

Alternatively, the array factor may be written as the infinite sum times a gate function $g_N(x)$:

$$\text{AF} = \sum_{n=0}^{\infty} I(nd)z^{-n} \cdot g_N(x)$$

where

$$g_N(x) = \begin{cases} 0 & x < 0 \text{ and } x > (N-1)d \\ 1 & 0 \leq x \leq (N-1)d \end{cases}$$

It follows that the array factor is a Z transform as follows:

$$\text{AF} = Z\{I(nd) \cdot g_N(x)\}$$

Existing tables of Z transforms may thus be employed to sum array factors in closed form.

Laplace transforms may also be employed for the purpose.[3] Regarding n temporarily as a continuous variable x, we note that the Laplace

[1] D. K. Cheng and M. T. Ma, *IRE Trans.*, **AP-8**:255 (May, 1960); D. K. Cheng, *IEEE Trans.*, **AP-11**:593 (September, 1963).
[2] P. L. Christiansen, *IEEE Trans.*, **AP-11**:198 (March, 1963); **AP-12**:647 (September, 1964).
[3] R. E. Collin, *IEEE Trans.*, **AP-12**:368 (May, 1964).

transform of $I(nd)e^{jn\psi}$ is

$$i(s) = \int_0^\infty I(xd)e^{(j\psi-s)x}\,dx$$

and the inverse transform may be expressed as an integral in the complex plane

$$I(xd)e^{j\psi x} = \frac{1}{2\pi j}\int_c e^{sx}i(s)\,ds$$

The forms hold for all x, in particular $x = n$; thus

$$\sum_{n=0}^{N-1} I(nd)e^{jn\psi} = \frac{1}{2\pi j}\sum_{n=0}^{N-1}\int_c e^{sn}i(s)\,ds$$

Provided the integrand is uniformly convergent, the order of summation and integration may be interchanged. Then a geometric series may be recognized and summed, and since the left-hand side of the above equation is the array factor for an N-element array, the equation becomes

$$\mathrm{AF}(\psi) = \frac{1}{2\pi j}\int_c i(s)\frac{e^{Ns}-1}{e^s-1}\,ds$$

The integral contour may be closed in the left half-plane, and residue theory applied to give the result

$$\mathrm{AF}(\psi) = \sum_{\substack{\text{residues at}\\ \text{poles in LHP}}} i(s)\frac{e^{sN}-1}{e^s-1}$$

It should be mentioned that the array factors for specific arrays having large numbers of elements can be obtained by straightforward computational procedures on a high-speed digital computer.

Exercise Consider an array of N elements with element currents given by $I(nd) = (knd)^2 e^{-an}$, where k and a are constants. Evaluate the array factor by means of the (1) Z-transform technique, (2) Laplace transform technique.

3.6 TOLERANCE PROBLEMS AND THE EFFECTS OF UNEQUAL ELEMENT-TO-ELEMENT SPACING

Certain radiation patterns, especially those having more gain than the pattern produced by a uniform array of the same size, require the establishment of the prescribed element current amplitudes and phases to a high accuracy. Also, in large high-gain uniform arrays, the performance in general deteriorates if the current amplitudes, phases, and/or element positions depart from the specified values. Studies have been made of systematic phase errors and of random errors of all types. Some results are given in Jasik[1] and a short bibliography[2] is presented here for reference. Expressions are available for the loss in gain and the effect on the side-lobe level of such errors. As an example of the types of results that are available, the gain G of an array having errors in the current excitations of the elements is approximately

$$G = G_0 \left(1 + \frac{3\pi}{4} \left(\frac{d}{\lambda}\right)^2 \bar{\epsilon}^2 \right)^{-1}$$

where
G_0 = gain of same array with perfect excitation
d = spacing between elements
$\bar{\epsilon}^2$ = total mean square error in current excitation

In large arrays, having hundreds of elements or more, the problem of establishing the proper current values and even the proper positioning of the elements is very severe. Structures flex, and the ground and supports heave and move about. The associated problems are so great that alternatives have been earnestly sought. The idea of decreasing the number of elements by "thinning," either systematically or randomly, has been studied. The analysis of thinned arrays leads into the subject of unequally spaced arrays. Likewise, analysis of the effects of inaccurate positioning involves unequal spacings in arrays.

The investigation of the subject of unequally spaced arrays for its own sake has revealed some interesting properties, for example: (1) Unequally spaced arrays having large element-to-element spacings need not necessarily exhibit grating lobes, as do arrays having equal spacing.[3] (2) The selection of appropriate unequal element spacings can reduce side-lobe levels.[4] In addition, it has been demonstrated that preliminary designs of large arrays may be obtained by a statistical process, in which the element positions are described by a probability density function.[5]

The first of the interesting properties listed above may be demonstrated by examining the condition for periodicity in the array factor. As will be recalled, the grating lobes arise when the visible range is enlarged (by increasing the element spacing in wavelengths βd) to the extent that it

---

[1] Jasik, *op. cit.*, pp. 2-30.
[2] R. N. Bracewell, *IRE Trans.*, **AP-9**:49 (January, 1961); L. A. Rondinelli, *IRE Natl. Conv. Record*, **pt. 1**:174 (1959); B. Y. Mills et al., *Proc. IRE*, **46**:67 (January, 1958); R. S. Elliott, *IRE Trans.*, **AP-6**:114 (January, 1958); E. N. Gilbert and S. P. Morgan, *BSTJ*, **34**:637 (May, 1955); H. F. O'Neill and L. V. Baillin, *IRE Trans.*, **AP-4**:93 (December, 1952); L. L. Bueller and M. J. Ehrlich, *Proc. IRE*, **41**:235 (February, 1953); A. S. Ashmead, *IRE Trans.*, **AP-1**:81 (December, 1952); J. Ruze, *Nuovo Cimento, Suppl.*, **9**:364 (1952).
[3] D. D. King et al., *IRE Trans.*, **AP-8**:380 (July, 1960).
[4] R. F. Harrington, *IRE Trans.*, **AP-9**:380 (July, 1960).
[5] Y. T. Lo, *IEEE Trans.*, **AP-12**:257 (May, 1964).

includes more than one period of a periodic array factor (Fig. 3.8). With an array having a general spacing, let x_n be the distance between the nth element and the zeroth; the array factor is then

$$\text{AF}(\xi) = \sum_{n=0}^{N-1} A_n e^{j\xi x_n} \qquad \xi = \beta \cos \phi + \alpha_1$$

The periodicity condition, $\text{AF}(\xi) = \text{AF}(\xi + \xi_0)$, where ξ_0 is the period, therefore implies the equation

$$\sum_{n=0}^{N-1} A_n e^{j\xi x_n} = \sum_{n=0}^{N-1} A_n e^{j(\xi+\xi_0)x_n}$$

For this equation to be true for all ξ, the coefficients of the corresponding terms in the series of exponentials $e^{j\xi x_n}$ should be equal. This implies the following condition for periodicity in the array factor: $e^{j\xi_0 x_n} = 1$, or $\xi_0 x_n = \pm 2n\pi$, or $x_n = \pm 2n\pi/\xi_0$. The latter equation says that each distance x_n is an integral multiple of some constant factor. Thus with unequally spaced arrays, the grating lobes need not necessarily appear even though the element separation is made large.

Next, let us form a quantitative estimate of the effects of element-positioning errors. With the same analysis, we can at the same time demonstrate the second of the interesting properties listed above (sidelobe reduction by the adjustment of element-to-element spacing). Consider an array composed of N elements, excited equally in amplitude and phase, but with arbitrary positions x_n. The array factor is then

$$\text{AF}(\xi) = \sum_{n=0}^{N-1} e^{j\xi x_n}$$

To study the effects of positioning errors in a uniformly spaced array, we can regard a position x_n as being not too different from the corresponding position in the equally spaced array and write $x_n = nd + \Delta_n$, where Δ_n measures the departure of the nth element from its position in an equally spaced array with spacing d. The array factor is then

$$\text{AF}(\xi) = \sum_{n=0}^{N-1} e^{j\xi nd} e^{j\xi \Delta_n}$$

This equation may be written in terms of the more familiar ψ variable as follows:

$$\text{AF}(\psi) = \sum_{n=0}^{N-1} e^{jn\psi} e^{j\psi \Delta_n/d}$$

The array with accurately uniform spacing would have an array factor

$$\mathrm{AF}_e(\psi) = \sum_{n=0}^{N-1} e^{jn\psi}$$

which of course is the familiar $\sin \tfrac{1}{2}N\psi / \sin \tfrac{1}{2}\psi$ function. The error in the pattern arising because of element-positioning error may be described in terms of some appropriate function of the difference between the actual array factor $\mathrm{AF}(\psi)$ and that $\mathrm{AF}_e(\psi)$ in which the elements are accurately positioned with equal spacing. This difference in the array factors may be expressed as follows

$$\mathrm{AF}(\psi) - \mathrm{AF}_e(\psi) = \sum_{n=0}^{N-1} e^{jn\psi} e^{j\psi \Delta_n/d} - \sum_{n=0}^{N-1} e^{jn\psi}$$

$$= \sum_{n=0}^{N-1} e^{jn\psi} (e^{j\psi \Delta_n/d} - 1)$$

Presumably the departure from equal spacing is small compared to the wavelength; thus $\psi \Delta_n/d \ll 1$. Consequently, to a good approximation, the exponential may be written

$$e^{j\psi \Delta_n/d} \approx 1 + \frac{j\psi \Delta_n}{d}$$

thus

$$\mathrm{AF}(\psi) - \mathrm{AF}_e(\psi) = \sum_{n=0}^{N-1} e^{jn\psi} \frac{j\psi \Delta_n}{d}$$

To save writing, define the left-hand side of this equation to be $D(\psi)$; then we have

$$\frac{D(\psi)}{\psi} = \frac{j}{d} \sum_{n=0}^{N-1} \Delta_n e^{jn\psi}$$

The quantity on the right is essentially a finite Fourier series which gives the difference function in terms of the positioning errors. But note particularly the fact that $D(0) = 0$; that is, the field strength in the direction of the maximum in the main lobe is not affected by the small positioning errors. At any position ψ, the difference function $D(\psi)$ may be explicitly evaluated in terms of the finite series, given the positioning errors Δ_n. If the quantities Δ_n are regarded as tolerances on position, the worst possible and most probable pattern errors may be obtained in a straightforward fashion. Other measures of the pattern error, such as the integral of the magnitude of $D(\psi)$ squared, taken over the visible range, are also readily evaluated.

Another interesting explicit formula may be obtained by renumbering the elements so as to obtain one of the more conventional forms of the finite Fourier series. The details are slightly different according to whether the number of elements in the array is even or odd. Consider the case in which the array has an odd number of elements. Calling the center element the zeroth, the series for the pattern difference is

$$\frac{D(\psi)}{\psi} = \frac{j}{d} \sum_{n=-[(N-1)/2]}^{(N-1)/2} \Delta_n e^{jn\psi}$$

This is a standard finite Fourier series in which the quantities Δ_n are the coefficients. Thus the standard formula for the determination of such coefficients gives an explicit formula relating the departures from equal spacing Δ_n and the array-factor difference

$$\frac{j\Delta_n}{d} = \frac{1}{2\pi} \int_{-\pi}^{\pi} \frac{D(\psi)}{\psi} e^{-jn\psi} d\psi \qquad (40)$$

This latter form has an interesting alternative interpretation. It says, essentially, that a specified small pattern difference $D(\psi)$ defines a set of spacing differences. That is, it defines an unequally spaced array. Harrington recognized that this fact could be employed to reduce side lobes in an array, without appreciably changing the gain. To see this, simply imagine $AF(\psi)$ to be a *desired* pattern which, for example, is the same as the array factor with equal element spacing, $AF_e(\psi)$, except in the region of the side lobes. It is clear that under this assumption, $D(\psi)$ is zero except in the side-lobe region, and there it is known (selected so as to reduce the side-lobe levels). Then with known $D(\psi)$, Eq. (40) specifies the set of Δ_n's which are required to bring about the side-lobe reductions. For example, suppose the array factor $AF(\psi)$ is one which, in the region of the main lobe, is the same as $AF_e(\psi)$, but is zero elsewhere (no side lobes). Then $D(\psi)$ is simply the portion of the function, $\sin N\frac{1}{2}\psi / \sin \frac{1}{2}\psi$, for equal element currents, that lies outside of the main lobe, that is the portion in which ψ lies in the range $2\pi/N < |\psi| < \pi$. Recall that this development is based on the assumption that $\psi \Delta_n/d \ll 1$. Also note that since the integral in Eq. (40) extends only from $-\pi$ to π in ψ, in this analysis the basic element spacing is limited to values such that βd is approximately π, at most.

Harrington points out a way to approximate the integral in Eq. (40), in the event that the actual function is difficult to integrate. Each side-lobe residual is replaced by an impulse function of appropriate strength located at the position ψ_k of the side-lobe maximum:

$$\frac{D\psi}{(\psi)} \approx \frac{1}{\psi} \sum_k a_k \, \delta(\psi - \psi_k)$$

so that by Eq. (40)

$$\frac{j \Delta_n}{d} \approx \frac{1}{2\pi} \sum_k e^{-jn\psi_k} \frac{a_k}{\psi_k}$$

The coefficients a_k are found by trial, but since the side lobes fall off approximately as $1/\psi$,

$$\frac{j \Delta_n}{d} \approx \frac{K}{2\pi} \sum_k \frac{e^{-jn\psi_k}}{\psi_k^2}$$

only one level, K, need be determined. The fact that the series is finite smooths out the impulse function representation.

Problem 3.8 A five-element uniformly excited array with a phase shift of 90°, element to element, has unequal element spacing. Relative to an equally spaced array with quarter-wavelength spacing, the spacing errors are given by $\Delta_n = \tau^{-n}\epsilon$ where $\tau = 0.95$ and $\epsilon = \lambda/32$. Find the array factor.

3.7 SIGNAL PROCESSING ANTENNAS

Particularly in receiving applications, the gain, effective beamwidth, and flexibility of antenna arrays of a given large size may be improved by a more thorough processing of the information received (i.e., more thoroughly than simply adding the rf signals from the elements in some fixed transmission-line–phase-shifter network). Several more or less distinct schemes have been described, for example, *adaptive arrays, time-modulated arrays, logical switching, pattern-multiplication techniques,* and *correlation arrays.* In certain applications, an array may be called upon to return a signal, perhaps amplified or otherwise modified, to its point of origin. An array that performs this function is called a *retrodirective array.* In this section, we shall discuss briefly the more essential features of each of these antenna types.

3.7.1 ADAPTIVE ARRAYS[1,2]

Largely because of the interaction (mutual-impedance effects) between the antennas in the array, it is frequently a very difficult task in practice to arrange the element current phases and amplitudes which are required to produce a given pattern. Moreover, these feeding problems become more severe as the size of the array and the number of elements is increased so as to obtain very high gains. In addition, for very large arrays there is sometimes another problem; namely, the phase fronts of the waves incident on a large receiving array may not be plane. Such

[1] R. C. Hansen (ed.), *IEEE Trans.*, **AP-12:** Special issue (March, 1964).
[2] R. W. Bickmore, *IEEE Spectrum*, **1:**78 (August, 1964).

phase disturbances may arise for example from irregularities in the density and composition of the atmosphere. Worse yet, these differences may be time-varying.

Difficulties such as these, together with the need for extremely high gain, steerable arrays for satellite and deep-space communication, provided the impetus for the invention and development of self-phasing antenna systems which are sometimes called adaptive arrays. The main application of these to date has been in receiving antennas, but there is a way to apply the same idea to transmitting antennas.[1] The adaptive principle is particularly useful in large antennas which must be pointed or steered. The mechanical tolerance and structural problems associated with large steerable systems are obvious; but with adaptive principles, the large single structures may be replaced with several smaller structures with an attendant decrease in structural problems and cost. The smaller structures may themselves be quite large if desired, and they need not necessarily be all the same type. The spacing between these elements in an array is of no significance; neither is the length of the feeding transmission lines, since the element outputs are automatically brought into phase, independent of the spacing, or the line or other phase shifts present. The main disadvantage is electronic complexity.

The idea of the system, but not the usual embodiment of it, is indicated in Fig. 3.29. The signal from each of the elements in the receiving array

[1] R. T. Adams, *IEEE Trans.*, **AP-12**:224 (March, 1964).

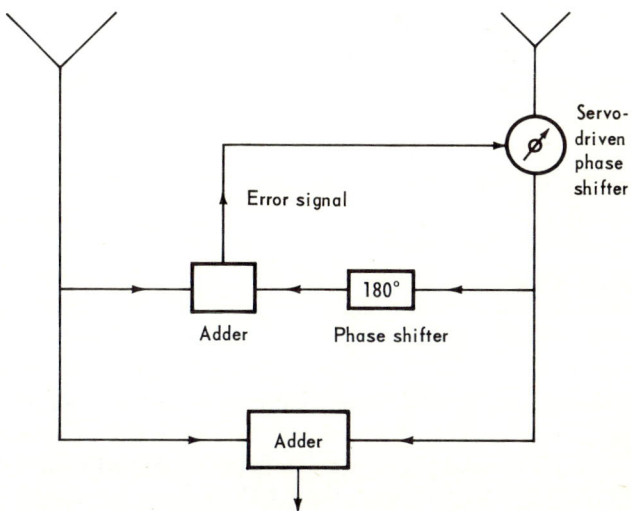

Fig. 3.29 The idea of the self-phasing array.

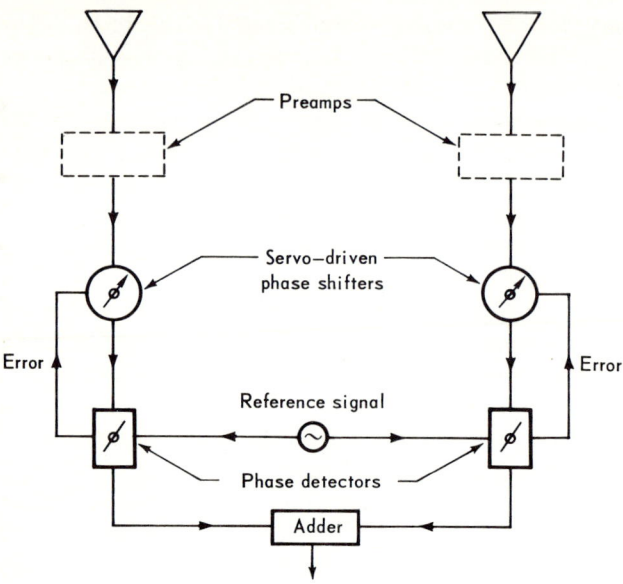

Fig. 3.30 Second version of self-phasing array. This version is too slow for most applications.

is split. Part of each of the split signals is led directly to an adder; the other parts are also added, but only after a 180° phase shift is introduced into the signal from one of the array elements. Thus if the incoming signal arrives at the two elements in phase, there will be no output signal from the phase-shifted combination. But if the signals arriving at the antennas are not in phase, the phase-shifted combination does not add to zero. In this way an error signal is created and used to drive a variable phase shifter in one of the lines. The servomotor will operate until the error signal is zero, that is, until the signals are in phase.

Another diagram indicating a method for self-phasing is indicated in Fig. 3.30. This system is closer to the actual embodiments of adaptive systems, but it is too slow for many applications. In this system, the phases in each receiving antenna element (of course there are many such elements in the array) are compared to that of a reference oscillator. If there is a phase difference, this is detected and employed as an error signal for a servosystem linked to the phase shifter in each leg.

Except for the effects due to doppler shifts in the signals (which can be handled by the addition of an AFC circuit to replace the fixed frequency reference), the diagram in Fig. 3.31 shows a typical configuration. The incoming signals are mixed with the outputs of voltage-controlled oscillators. In this mixing process, the phase is of course preserved. Thus any phase difference appears in the IF signal. This IF signal is amplified and compared with a common reference signal. Any phase difference between the signal and the reference shows up as a voltage applied to the voltage-controlled oscillator. In effect, the instantaneous frequency of the VCO is adjusted so that the composite signal in each leg has a phase which is the same as that of the reference signal. Hence, all the signals are in phase. The operation is very similar to that of a phase-locked loop in each channel.

In any case, when all the signals have been processed so that they are in phase, the summed output is maximum, and if all of the elements are of the same type, the signal level is N times that of the signal of one. Perhaps more important, this situation holds whether the propagation paths, preamplifiers, mixers, transmission lines, and other factors associated with each array element are the same or not. In fact, the individual array elements need not be the same.

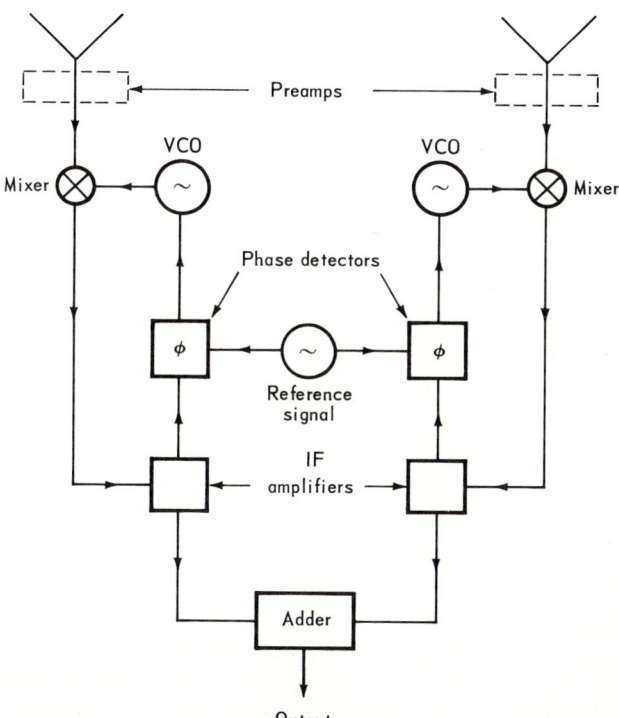

Fig. 3.31 An adaptive array; electronic self-phasing array.

3.7.2 TIME-MODULATED ARRAYS[1]

As has been demonstrated many times earlier in the chapter, the array factor for a linear array of N identical elements is [Eq. (6)]:

$$\mathrm{AF}(\phi) = \sum_{n=0}^{N-1} A_n e^{j\beta nd \cos \phi}$$

The radiation pattern itself may be written in the form

$$E = E_e \sum_{n=0}^{N-1} I_n e^{jn\beta d \cos \phi}$$

where E_e is the electric field of a single element with unit excitation, and E is the appropriate field component.

Up to now we have been concerned with single-frequency operation, with the field quantities represented as complex numbers independent of time. In this section, we shall recover the time function so that we can examine the fields as a function of time. It will be recalled that the electric field as a function of time is given by

$$E^t = \mathrm{Re}\ E e^{j\omega t} = \mathrm{Re}\ E_e \sum_{n=0}^{N-1} I_n e^{j(n\beta d \cos \phi + \omega t)} \tag{41}$$

But now suppose that we arrange things so that the current amplitudes I_n are periodic functions of time, with a period T much greater than the rf period $2\omega/\pi$. Under this restriction, Eq. (41) still gives, approximately, the pattern at any instant of time. Let us represent the modulation of the nth element in the form of a Fourier series

$$I_n = \sum_{m=-\infty}^{\infty} a_{mn} e^{jm\omega_0 t}$$

where $\omega_0 = 2\pi/T \ll \omega$. Then the pattern function is

$$E^t = \mathrm{Re}\ E_e \sum_{m=-\infty}^{\infty} e^{j(\omega+m\omega_0)t} \sum_{n=0}^{N-1} a_{mn} e^{jn\beta d \cos \phi} \tag{42}$$

where

$$a_{mn} = \frac{1}{T} \int_0^T I_n(t) e^{-jm\omega_0 t}\, dt \tag{43}$$

The form of Eqs. (42) and (43) suggests, upon close examination, that the radiation pattern at each of several distinct frequencies may be different. For example, if the received signal is processed so that all but a single fre-

[1] W. H. Kummer, A. T. Villeneuve, T. S. Fong, and F. G. Terrio, *IEEE Trans.*, **AP-11**:633 (November, 1963).

quency $\omega + p\omega_0$ is filtered out, the pattern at this frequency is

$$E(\omega + p\omega_0) = E_e \sum_{n=0}^{N-1} a_{pn} e^{jn\beta d \cos \phi}$$

where of course

$$a_{pn} = \frac{1}{T} \int_0^T I_n(t) e^{-jp\omega_0 t}\, dt$$

The particular case $p = 0$ (i.e., all but the operating frequency filtered out) has a set of excitation coefficients

$$a_{0n} = \frac{1}{T} \int_0^T I_n(t)\, dt = \text{average value of } I_n(t)$$

This latter equation implies that the pattern at the main operating frequency, when the element currents are modulated, is the same as the pattern of a steady-state array whose element currents are the time averages of the time-varying element current amplitudes.

In practice, an on-off switching is one of the easiest types of time modulations to implement. The switching may be conveniently and rapidly accomplished by diode switches. The element current functions $I_n(t)$ then simply jump in value between zero and A_n. If the ON time is represented by τ_n, $0 \leq \tau_n \leq T$, then $a_{0n} = A_n \tau_n / T$ so that the pattern is

$$E(\omega) = E_e \sum_{n=0}^{N-1} \frac{\tau_n}{T} A_n e^{jn\beta d \cos \phi}$$

That is, the pattern of the time-modulated array at the operating frequency is determined by the effective element current values, which in turn are determined by the ON time τ_n of the elements. Electronic control of the ON time constitutes electronic control of the radiation pattern. If the current amplitudes A_n are all real and positive, then any pattern that can be realized by exciting the given number of elements operating in phase can be electronically synthesized in this manner. In particular, ultra-low side-lobe Chebyshev patterns may be generated, as was demonstrated by Kummer et al.[1] Of course the quantities A_n need not be the same for all elements.

It is interesting to look back at Eq. (43) once again and realize thereby that a modulation of the current amplitudes can give rise, simultaneously, to a number of perhaps different patterns at different frequencies (the side-band frequencies). The same array may therefore be time-modulated so as to perform several functions simultaneously.

[1] *Ibid.*

Exercise A cosine-tapered current amplitude is desired in a nine-element array operating at 1 GHz. This distribution is to be synthesized by diode-switching at the elements. Determine the ON time required at each element, and suggest a practical frequency for the element modulation.

3.7.3 RETRODIRECTIVE ARRAYS[1,2]

If one or both of the terminals in a two-way communication system is moving, problems associated with pointing the antennas arise. These problems are particularly severe if the motion is rapid, random, or not well known. In such applications, an antenna that can receive a signal from any direction and immediately retransmit back toward the source of the signal is clearly appropriate. Such antennas have been called retrodirective arrays. They are suited to transponder-type marker beacons, or to satellite and space-vehicle communication systems, for example. In the latter application, the moving station may receive the pilot signal, translate it slightly in frequency, modulate it with new information, amplify it, and retransmit it in the direction from which it came.

Consider the requirements on a linear array in order that it have the retrodirective feature. If a plane wave arrives at the array from a direction ϕ_0 (Fig. 3.32), the received signals in the individual antennas have phases that are determined by this incident plane wave. That is, the phase of the signal in each array element lags behind that of its neighbor on the right by an amount $\beta d \cos \phi_0$. On the other hand, if the array acts as a transmitter, the requirement for placing the beam maximum in the direction ϕ_0 is an element-to-element phase shift of $\beta d \cos \phi_0$, with the phase of each element lagging behind its neighbor on the left. Thus, to reradiate with the beam maximum along the incident direction ϕ_0, a requirement is that the sense of the element-to-element phase shift be reversed. A method of accomplishing this was described by Van Atta. In his system, the element pairs that are symmetrically located about the center of the array are connected by equal lengths of transmission lines. The signal comes in one element and goes out the other. The equal line lengths simply add a constant phase delay to each signal so that, relatively speaking, the leftmost element reradiates a signal with the phase of the signal in the rightmost element, the second element (numbering from the left) reradiates a signal with the phase of that received by the $N - $1st, and so on. Since the $N - $1st lags in phase behind the Nth in the incoming signal, the second lags behind the first in the transmitted signal. The next step is the inclusion of amplifiers and

---

[1] L. C. Van Atta, Electromagnetic Reflector, U.S. Patent 2,909,002, Oct. 6, 1959.

[2] See also the papers by Skolnik and King, Pon, Andre and Leonard, Phillip, and Gruenberg and Johnson, in *IEEE Trans.*, **AP-12** (March, 1964). Also C. C. Cutler, R. Kompfner, and L. Tillotson, *Bell System Tech. J.*, **42**:2013 (September, 1963).

ARRAYS 129

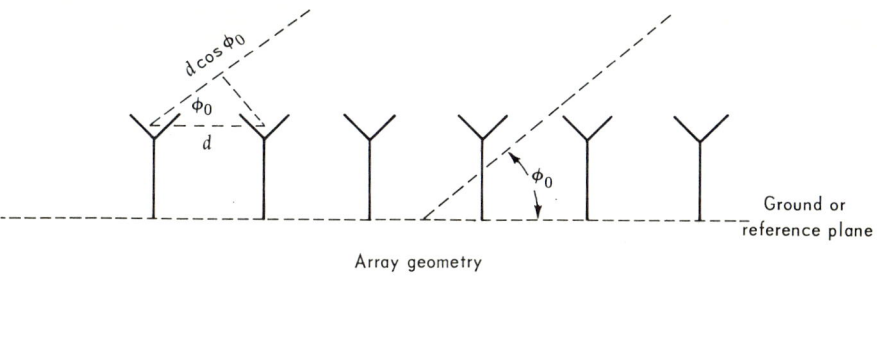

Array geometry

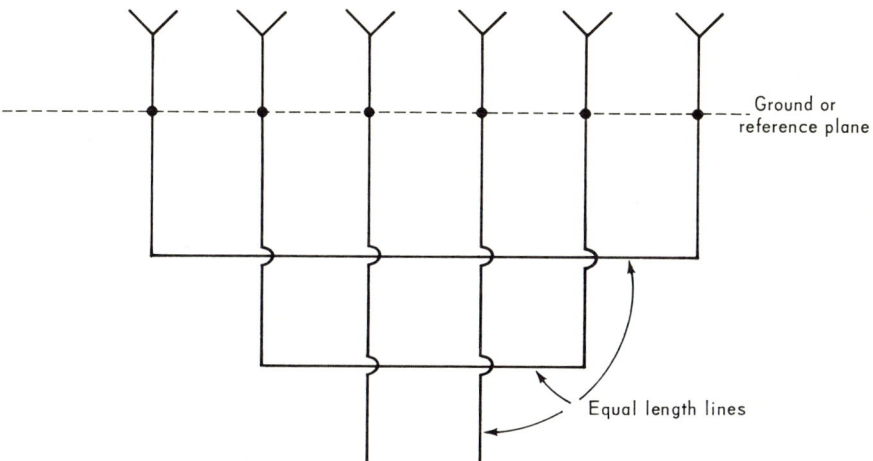

Fig. 3.32 Van Atta-type retrodirective array. Antenna elements joined with equal-length transmission lines.

modulators in the equal-length transmission lines. The idea may also be extended to two-dimensional arrays. It is assumed that the signal scattered from the individual elements is small compared with the signal received by the element.

There are other ways to accomplish the retrodirectivity. It will now be shown that if the received signals at the elements are appropriately shifted in phase before retransmitting, the retrodirective feature is accomplished. Consider a linear array with a plane-wave incident at angle ϕ_0, as indicated in Fig. 3.33. If we regard the positive direction of x as being toward the right as usual, the wave progressing toward the left will have a phase variation along x of the form $e^{j\beta_x x}$. Thus, the relative phases of the received signals at the elements vary according to $e^{jn\beta d \cos \phi_0}$. Now suppose that each of these signals is "conjugated" (the phase shift prescribed by the change of $+j$ to $-j$ in the exponential, which means a

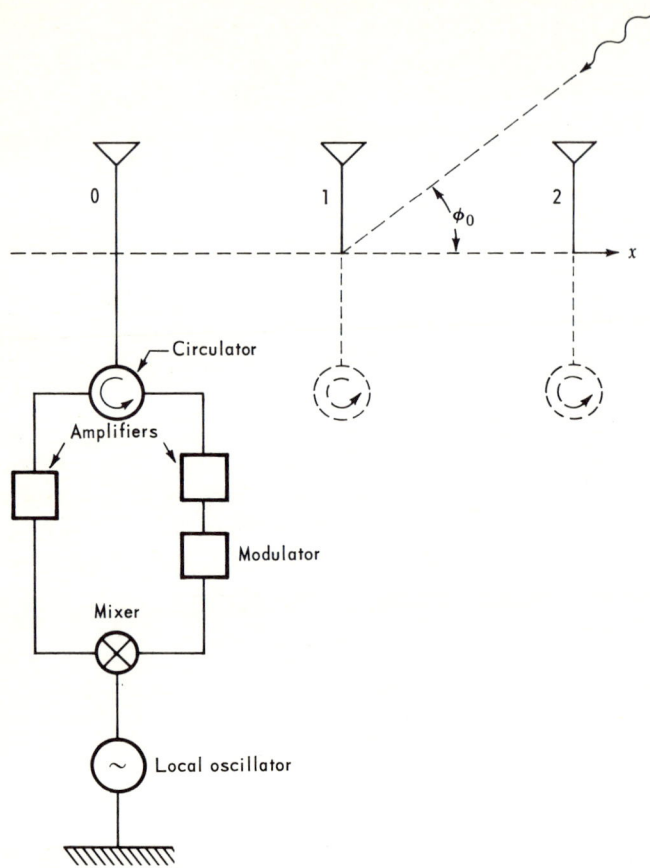

Fig. 3.33 Retrodirective array showing one of the circuits for conjugating the phase, amplifying, and modulating with new information.

change in the sign of the phase angle in the time domain) and retransmitted from the same array element. In the resulting transmitting array, the relative phase along the elements varies as $e^{-jn\beta d \cos \phi_0}$. The quantity $\psi = \beta d \cos \phi + \alpha$, in the transmitting array factor, is such that the phase shift from element to element is $\alpha = -\beta d \cos \phi_0$. Thus the variable ψ takes a form $\psi = \beta d(\cos \phi - \cos \phi_0)$ which as we have seen earlier (p. 77 and Figs. 3.8 and 3.9) means that the array maximum is in the direction ϕ_0 of the incoming signal. Thus changing the sense of the incoming phase from plus to minus at each element before retransmitting accomplishes the retrodirective function.

There are several ways in which such a phase shift may be accomplished. To illustrate one of the more useful schemes, consider a mixing operation. At least in combination with the appropriate filters, the mixer action may be approximated as a multiplication of the two signals applied to the mixer. Let the incoming signal $V_{in} \cos(\omega t + \theta)$ be mixed with a local oscillator signal $V_{LO} \cos \omega_0 t$. The output is then

$$V_{out} = kV_{in}V_{LO} \cos(\omega t + \theta) \cos \omega_0 t$$
$$= \tfrac{1}{2}kV_{in}V_{LO}[\cos(\omega_0 t - \omega t - \theta) + \cos(\omega_0 t + \omega t + \theta)]$$

Note that the lower of the two beat frequencies has a phase that is the conjugate (negative) of the phase in the original signal. Thus, if the local oscillator signal ω_0 is approximately twice the frequency of the incoming signal, the output of the mixer with a filter has a frequency of approximately that of the incoming signal, but the sign of the relative phase has been changed as required.

A schematic system is indicated in Fig. 3.33. The signal from each element passes into a circulator in order to separate the input and output channels. It is then amplified, mixed with a local oscillator, modulated, amplified, and led back through the circulator to be reradiated. If, by the mixing process, the transmitted signal is offset slightly from the incoming signal (a step which is clearly advisable in many applications), there is of course a slight error in the positioning of the beam. But the difference is readily calculated and in many applications can be kept within acceptable limits.

Problem 3.9 Design a five-element retrodirective array, which, with the necessary additional equipment, will reradiate a pattern whose array factor is derived from a Chebyshev polynomial and has side-lobe levels no greater than -20 db (coverage $\pm 60°$ from broadside).

3.7.4 LOGICAL SWITCHING, PATTERN MULTIPLICATION, AND CORRELATION ARRAYS

The individual patterns or responses of two or more antenna systems (or two or more parts of the same array) may be combined in various ways to produce greater effective directivity. A few of these methods will be described in this section. Designers should be cautious in applying these schemes since frequently the overall gain of the system is reduced significantly by the information processing.

One of the simplest schemes to describe involves what has been called logical switching. In its simplest form, this system consists of two antennas, one of which functions as a gate activator. Typically, one of the antennas (output O_1) is highly directive, while the other antenna (output O_2) has a relatively broad pattern. In the processing, the amplitudes of the signals are compared as indicated in Fig. 3.34. If O_1 is greater than O_2, the signal is passed to the detector; if O_2 is greater,

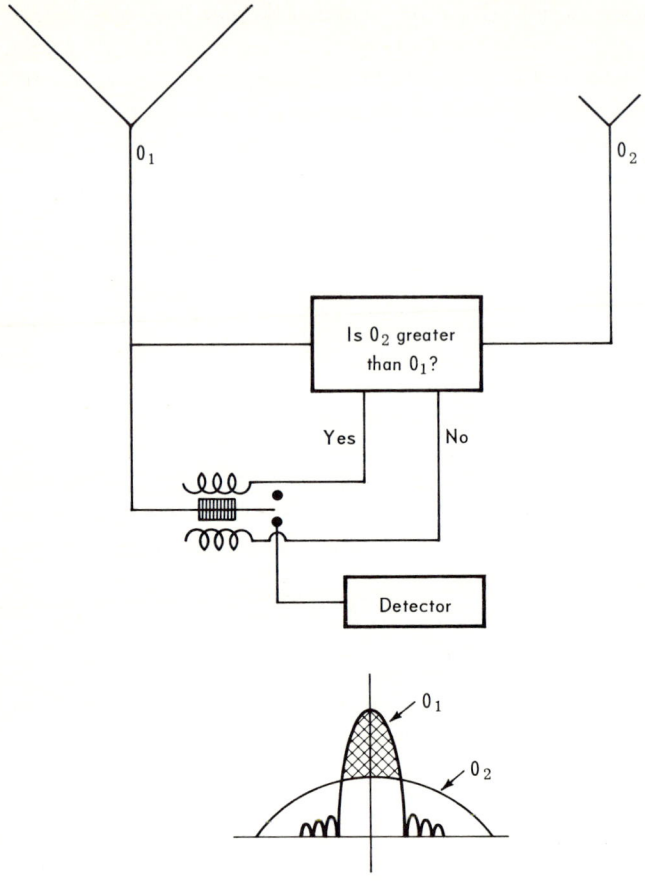

Fig. 3.34 Logical switching array for direction finding.

it is not passed. The net result is an alteration of the receiving pattern of the directive antenna. The alteration can be designed so as to result in a beam-narrowing and side-lobe depression, as indicated in the figure. When the two patterns have the form shown, only the crosshatched portion is transmitted to the detector.

Another scheme, which is a form of pattern multiplication, has been described by Mattingly.[1] The procedure is applicable to two-way patterns, such as are important in radar systems. The two-way pattern is the product of the transmitting pattern and the receiving pattern. Normally, the transmitting pattern and the receiving pattern are the same, but if the antenna system includes nonreciprocal elements, they may be

[1] R. L. Mattingly, *Proc. IRE*, **48**:795 (April, 1960).

different and the system response is thereby altered. Mattingly describes a method for obtaining a two-way Chebyshev pattern in this way. He also describes a scheme for narrowing the beam and decreasing the side lobes in a radar antenna system. The scheme makes use of two antennas as follows: Calling one of the antenna patterns (one-way) $A_1(\phi)$ and the other $A_2(\phi)$, the idea is to combine the two patterns in such a way that the two-way pattern of the system is $A_1{}^2(\phi) - A_2{}^2(\phi)$. That such a combination has lower side lobes and a narrower beam than either of the separate two-way patterns is clear by inspection of Fig. 3.35. Note that the desired response can be written mathematically as $A_1{}^2(\phi) - A_2{}^2(\phi) = [A_1(\phi) + A_2(\phi)][A_1(\phi) - A_2(\phi)]$. This equation says that the desired two-way response can be implemented by forming the sum pattern on transmission and the difference pattern on reception. The hardware for combining the patterns in this way includes directional couplers and circulators.

Another scheme, which also employs two or more antennas, has been called the correlation array.[1] The response is the product of the patterns of the antennas in a multiple antenna system. For example, suppose that two antennas in a system have patterns designated respectively by $A(\phi)$ and $B(\phi)$. If the outputs of the two antennas are combined in a multiplier, the output of the multiplier is (the phase shifts arise from the an-

[1] A. Ksienski, "Recent Advances in Signal Processing Antenna Systems," *Proc. Natl. Electron. Conf.* (October, 1964).

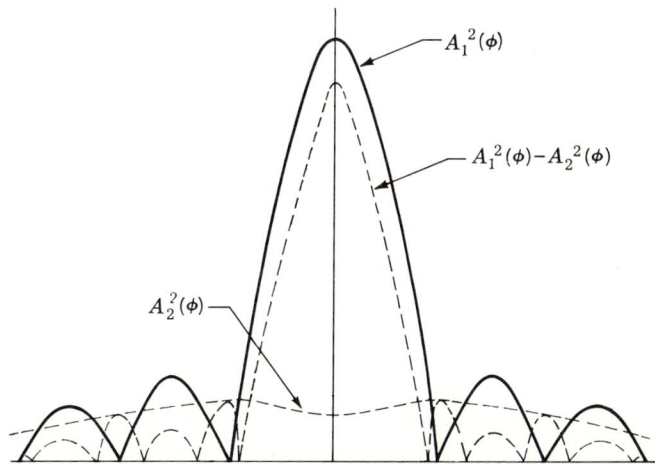

Fig. 3.35 Alteration of two-way pattern of radar antenna by forming sum pattern on transmit and difference pattern on receive. System requires at least two antennas.

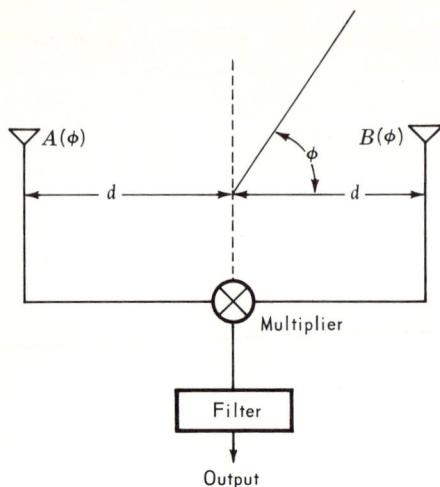

Fig. 3.36 The main parts of a correlation array.

tennas' displacement d from an origin in the center, Fig. 3.36)

$$V^t(t,\phi) = kA(\phi) \cos(\omega t - \beta d \cos \phi) \cdot B(\phi) \cos(\omega t + \beta d \cos \phi)$$
$$= \tfrac{1}{2} kA(\phi) B(\phi)[\cos 2\omega t + \cos(2\beta d \cos \phi)]$$

The double-frequency term may be filtered out, as indicated in the figure. For example, if the pattern $A(\phi)$ corresponds to a continuous linear array of length $2d$, and $B(\phi)$ corresponds to an array of two such antennas spaced $2d$ apart (often called an interferometer), the output is

$$V(\phi) = \frac{\sin(\beta d \cos \phi)}{\beta d \cos \phi} \cdot \cos(\beta d \cos \phi) \cdot \cos(2\beta d \cos \phi)$$
$$= \frac{\sin(4\beta d \cos \phi)}{\beta d \cos \phi}$$

The multiplied output thus gives a "pattern" which is the same as that of a linear array having twice the length of the actual array. The advantages of this type of antenna system are obvious. The disadvantages are more difficult to assess. The signal-to-noise ratio is degraded by the processing. The nonlinearity introduced means that the simple superposition of multiple signals (as in ordinary arrays) is lost. The net result of these peculiarities seems to be that this type of array is most applicable to signals which are uncorrelated (such as the signals from different radio stars, or the radar reflections from an object which is so extended, so moving, and/or so generally complex that the signals are essentially decorrelated in the complex reflection process even though they originate from a single coherent radar transmitter).

Exercise If a continuous linear array of length L is placed between the elements of a simple interferometer with base line L, show that the response of the system with multiplied outputs has the form of a continuous linear array of length $2L$.

We must now turn our attention back to the characteristics of some of the antennas that might be employed as the individual elements in arrays such as we have been discussing. One of the greatest difficulties in the engineering of arrays to produce a precise pattern is the actual establishment of the current distribution that is required by the array design; the currents in one part of the system interact with the currents in other parts of the system, and the overall result of these "mutual effects" is hard to predict.

4. WIRE ANTENNAS AND RELATED FORMS

Having studied in some detail the radiation patterns of antennas in combination (arrays), we now return to the study of individual antennas. Here we remove the restriction imposed in Chap. 2 that the antennas must be small compared to the wavelength. Without this restriction, it is much more difficult to determine the currents on the antenna structure.

There are two general procedures for obtaining theoretical solutions to wirelike antenna problems. The first method treats the antenna structure as a boundary and attempts to solve Maxwell's equations as a boundary-value problem in E and H. This method is conceptually satisfying but is limited in practice to a very few structural types having specially shaped boundaries, namely spheroids, cones, and infinite cylinders. The current distribution, if required, is determined from the solution for tangential H at the surface of the structure. (However, with this method the current distribution is not of vital engineering interest because by the time it can be found, general expressions for the fields everywhere and even the input impedance are already available in the solution.) The second method formulates and solves, approximately, an integral equation for the currents on the antenna. Unless the currents all have the same direction, this formulation is unwieldy. The integral equation is commonly obtained by equating two expressions for the magnetic vector potential; one of these expressions gives A_z as an integral of the current distribution and the other gives it as a solution to a differential equation (the latter equation comes from relating the vector potential to the known electric field at the antenna surface). Since the theory contains no surprises, in the organization of this book, it is presented in Chap. 5; this is done so that the results of engineering importance may be better collected together and identified.

After a general summary of the radiation and self-impedance properties of wirelike structures, some consideration is given to a variety of antenna feeding arrangements and balance problems. A discussion of mutual effects is included. The determination of the radiated field, given the currents on the structure, is a relatively easy problem. We shall begin with this topic.

4.1 RADIATION PATTERNS OF WIRE ANTENNAS

As was suggested in Chap. 2, the current distribution on straight-wire antennas closely resembles the current distribution on a similarly terminated transmission line. We shall prove this in the next chapter. Even if the (thin) wires are bent or curved, the current distribution is nearly sinusoidal. It is economic therefore to display some classical results for radiation patterns on the basis of an assumed current distribution which is like that on a similar transmission line.

Consider a thin straight wire of length L, excited by a point generator at the center. Conceptually, we may think of the antenna as an opened-out transmission line with voltage and current as indicated in Fig. 4.1. For convenience, we shall think of the generator as a constant-current generator I_g. The basic form for the current distribution, assuming that it is transmission-line-like, is

$$I(\zeta) = I_1 e^{\gamma \zeta} + I_2 e^{-\gamma \zeta}$$

In this form, γ is the propagation constant (approximately equal to $j\beta$ in many cases), and ζ is a position variable along the wire. For convenience, ζ is taken to be positive in both directions from the center, as it is along both sides of an ordinary transmission line. It is then like a radial coordinate r. The constants I_1 and I_2 are determined from the conditions that the current must be zero at the ends of the wire and must be equal to I_g at the center ($\zeta = 0$). Insertion of the boundary conditions at $\zeta = 0$ and $\zeta = L/2$ gives the equation

$$I(\zeta) = I_g \frac{e^{-\gamma \zeta} - e^{\gamma(\zeta-L)}}{1 - e^{-\gamma L}} \quad (1)$$

Let us suppose that the wire lies along the z axis. Each element of current radiates the field of the elemental current source [Eqs. (14) and (15), Chap. 1]. The collection (current distribution) constitutes a continuous collinear array of length L. We have seen in Chap. 3 [Eq. (4), for example] that the resultant field of such an array is the product of the unit pattern and the array factor

$$E_\theta = \frac{j\omega\mu e^{-j\beta r}}{4\pi r} \sin \theta \cdot \text{AF} \quad (2)$$

138 ANTENNA ENGINEERING

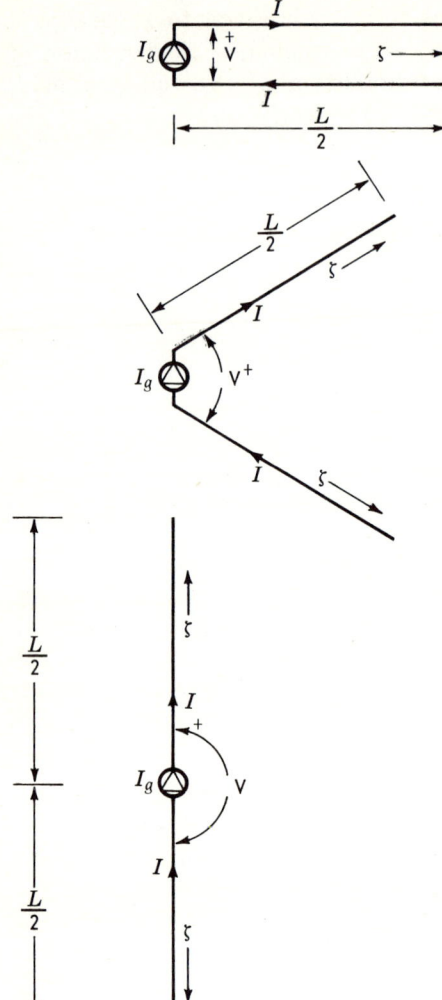

Fig. 4.1 The linear antennas as an opened-out transmission line.

where

$$\text{AF} = \int I(z) e^{j\beta \cos \theta z}\, dz$$

In these formulas θ is the usual polar angle in spherical coordinates. We have only to evaluate the array factor and we have the expression for the radiated fields. The current distribution is given by Eq. (1). The relationship between the z and ζ variables is $z = \zeta$, $z \geq 0$, and $z = -\zeta$,

$z < 0$. Thus the array factor is given by

$$\text{AF} = \frac{I_g}{1 - e^{-\gamma L}} \left[\int_0^{\frac{1}{2}L} (e^{-\gamma z} - e^{\gamma(z-L)}) e^{j\beta \cos \theta z} \, dz \right.$$
$$\left. + \int_{-\frac{1}{2}L}^0 (e^{\gamma z} - e^{-\gamma(z+L)}) e^{j\beta z \cos \theta} \, dz \right]$$

The integrations and rearrangements are straightforward but a little long and tedious so the details will not be presented here. In terms of a variable $\xi = \beta \cos \theta$ introduced earlier, the result may be put into the form

$$\text{AF} = \frac{I_g}{\sinh (\gamma L/2)} \cdot \frac{2\gamma}{\xi^2 + \gamma^2} \left(\cosh \frac{\gamma L}{2} - \cos \frac{\xi L}{2} \right) \qquad (3)$$

This expression holds for a thin center-driven antenna of any length. The only drawback is that for very long antennas, the exact value of the propagation constant γ is not given by any simple theory. In many practical situations, the propagation constant is very nearly that of plane waves in air, $\gamma = j\beta$. When this is the case, the equation for the array factor is

$$\text{AF} = \frac{I_g}{\sin \frac{1}{2}\beta L} \cdot \frac{2}{\beta \sin^2 \theta} [\cos (\tfrac{1}{2}\beta L \cos \theta) - \cos \tfrac{1}{2}\beta L]$$

and the electric field strength itself is

$$E_\theta = j \frac{60 e^{-j\beta r}}{r \sin \frac{1}{2}\beta L} I_g \left[\frac{\cos (\tfrac{1}{2}\beta L \cos \theta) - \cos \tfrac{1}{2}\beta L}{\sin \theta} \right] \qquad (4)$$

When the attenuation is small, as is frequently the case, it follows from Eq. (1) that the current distribution on the antenna is sinusoidal; the maximum value of the current on the sinusoid (whether or not that value is actually reached within the actual length of the antenna wire) is $I_m = I_g / \sin \frac{1}{2}\beta L$. Equation (4) is often written in terms of this maximum current:

$$E_\theta = j \frac{60}{r} e^{-j\beta r} I_m \left[\frac{\cos (\tfrac{1}{2}\beta L \cos \theta) - \cos \tfrac{1}{2}\beta L}{\sin \theta} \right] \qquad (4a)$$

For lengths such that $\frac{1}{2}\beta L = n\pi$, this form is certainly more esthetic than Eq. (4), mathematically speaking.

The spatial patterns represented by Eq. (4a) are somewhat difficult to visualize, and so a few typical patterns are presented for reference (Fig. 4.2). A more complete set of such patterns may be found in Jasik.[1]

Exercise Derive a formula for the radiation pattern from a pure traveling wave of current on a wire (say from half of the structure above with no reflected wave). Select some length, a substantial number of wavelengths, and sketch the pattern.

---

[1] H. Jasik, "Antenna Engineering Handbook," McGraw-Hill Book Company, New York, 1961.

140 ANTENNA ENGINEERING

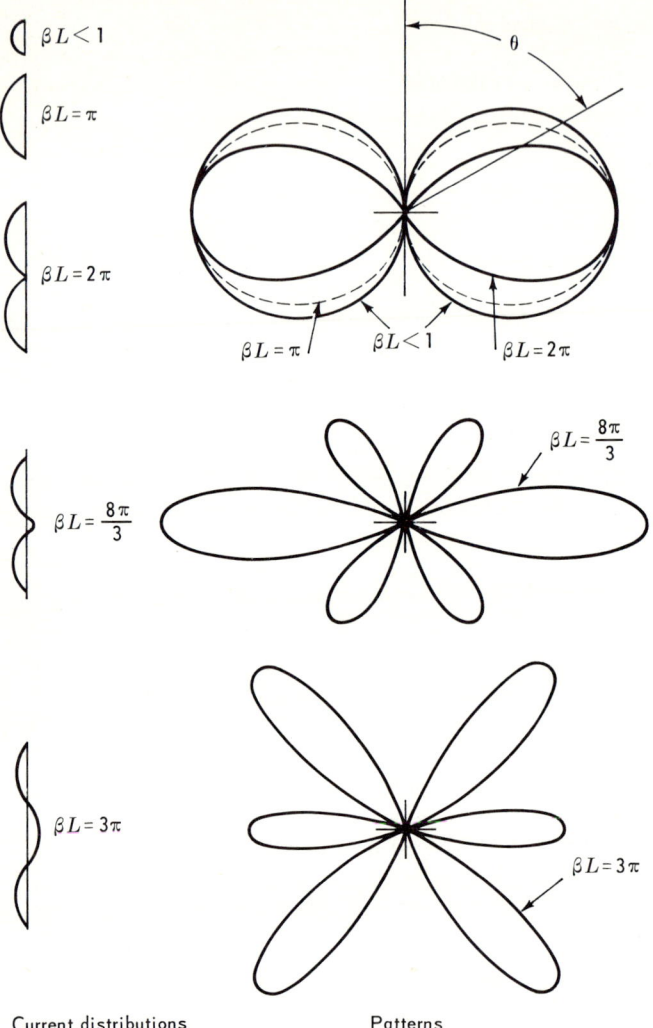

Current distributions   Patterns

Fig. 4.2 Radiation patterns of straight-wire antennas.

Problem 4.1 A five-element array is constructed from full wavelength dipoles spaced a half wavelength apart. The wires are excited so that the current is the same in each element. Sketch the radiation pattern in the three coordinate planes (clearly label the planes). Repeat for the same array phased for end fire.

4.2 BENT- AND CURVED-WIRE STRUCTURES

For various reasons it is often helpful to employ other than simple straight-wire antennas. For example, if the wires in a dipole antenna are simply inclined so as to make a V, the pattern may have greater directivity. The current distribution on the inclined wires is still nearly sinusoidal, and so the radiation patterns may be calculated in almost the usual fashion. The two halves of the antenna are treated separately and the results added. When this is done, it is found that for certain combinations of lengths and angles, an equatorial beam of increased directivity may be produced.

In order to achieve the greatest directivity for a given half-length h, the angle A of the V antenna should be tailored to the particular half-length. A useful empirical formula which gives the best V angle for a given half-length is

$$A = 60\frac{\lambda}{h} + 38°$$

When the angle and half-length h are related by the foregoing equation, the directivity may be closely estimated by the following empirical equation:

$$G_{di} = 1.2 + 2.3\frac{h}{\lambda}$$

These equations have been devised from data pertaining to antennas in the size range $0.5 < h/\lambda < 3$. The input impedance is somewhat less than that of the straight dipole of the same length. A unidirectional beam may be obtained by arraying two or more of these V antennas.

If the waves reflected from the ends of the inclined wires are low in magnitude, the beam becomes unidirectional on the inclined side as indicated in Fig. 4.3. The magnitudes of the reflected waves can be reduced by making the inclined wires relatively thick or by terminating the wires at the ends with an appropriate impedance. Thinking of the inclined wires as an opened-out transmission line, the wires are terminated in something like their characteristic impedance. As the figure indicates, this terminating impedance may be split in half and the ends joined to a nearby ground. If the reflected waves are eliminated completely, the structure is a traveling-wave array. According to the analysis presented in Chap. 3, the patterns of the individual legs of the traveling-wave array are conical beams inclined at a small angle with respect to the wire. The angle of inclination depends on the length of the wire, as pointed out earlier (also the last exercise). Thus, for certain lengths and angles of inclination, the individual patterns add to produce a unidirectional beam with side lobes.

The angles of inclination on each leg of a V or traveling-wave array may be reversed at the midpoints so that the wires form a rhombus. The resulting antenna structure is aptly referred to as a rhombic antenna

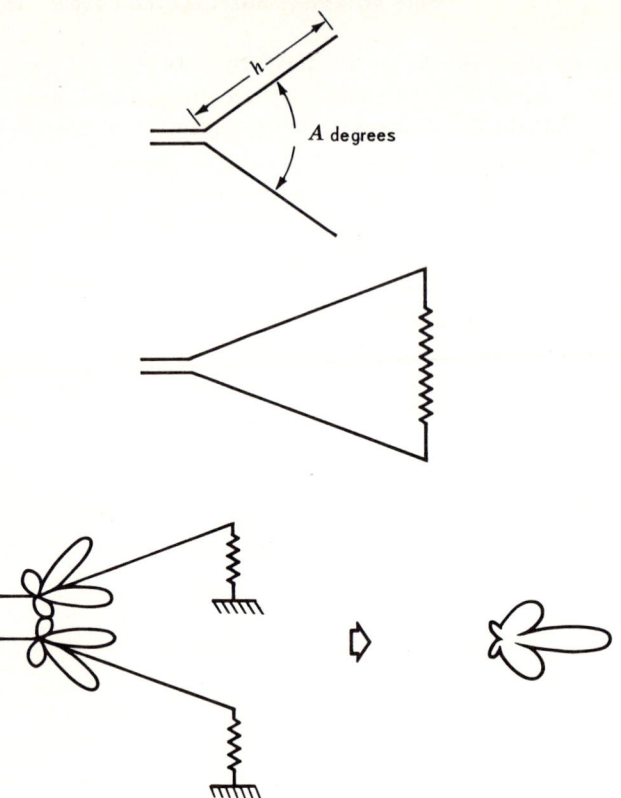

Fig. 4.3 Sloping dipole and traveling-wave V antennas.

(Fig. 4.4). This type of antenna has been widely employed for long-distance high-frequency (ionospheric) communications links. By carefully selecting the angles and lengths of the legs of the rhombus and mounting the whole structure an appropriate distance above ground, a sharp beam at the desired angle with the vertical may be formed, to take maximum advantage of the reflections from the ionosphere. Rhombic-antenna design has been worked out in considerable detail.[1,2] Typical designs call for leg lengths of 2 to 7 wavelengths, with the rhombus mounted 1 to 2 wavelengths above ground. The acute angle of the rhombus is typically between 35 and 60°. The input impedance of a wire rhombic antenna for HF is in the range of 700 to 800 ohms, although a 600-ohm value can be obtained by flaring the conductor size in the center. At higher frequencies, with construction of tubing, the impedance will be lower, perhaps 400 to 500 ohms. The side-lobe level of a

---

[1] D. Foster, *Proc. IRE*, **25**:1327 (October, 1937).

[2] A. E. Harper, "Rhombic Antenna Design," D. Van Nostrand Company, Inc., Princeton, N.J., 1941.

rhombic is rather high, typical values being -10 db although levels of -5 to -7 db are not uncommon. The half-power beamwidth for a structure with a leg length of 5 wavelengths is typically 10 to 12°.

Another useful wire antenna is the helix. Helical wire antennas are usually operated over a ground plane. The performance of the structure depends markedly on the length of each turn of wire in the helix, compared to a wavelength. Let us consider first the helical antenna operating in the "normal" mode (Fig. 4.5). In this case, the length of the individual turns is a small fraction of a wavelength. The turns of the wire add inductance, as a result of which the structure becomes self-resonant with axial heights that are substantially less than a quarter of a wavelength. Thus, such structures may be employed in situations in which the physical lengths of self-resonant monopoles must be decreased. The radiation pattern is essentially that of a straight-wire antenna whose length is the axial length of the helix. The bandwidth of these structures resembles that of a monopole in combination with a series inductance for tuning. The specific details of impedance behavior depend upon the details of pitch angle and number of turns in the helix. The inductance per unit length may be approximated by means of single-layer solenoid formulas. The structure will resonate at roughly the frequency at which the total wire length is a quarter wavelength. A better formula for the resonant height, valid when $nD^2/\lambda < \frac{1}{5}$, is

$$\frac{h}{\lambda} \doteq \frac{1}{4}\left[1 + 20(nD)^{5/2}\left(\frac{D}{\lambda}\right)^{1/2}\right]^{-1/2}$$

where n is the number of turns per unit length, and D is the diameter.

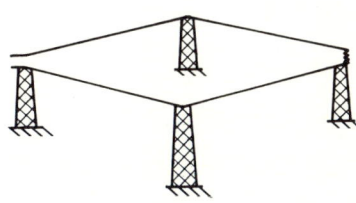

Fig. 4.4 Typical rhombic antenna construction and radiation-pattern formation.

The radiation resistance is typically that of a straight-wire antenna of the same height as the helix.

On the other hand, when the length of a single turn is in the vicinity of a wavelength, the performance of a helical antenna is markedly different. The familiar traveling wave of current along the wire is rapidly attenuated and a more complicated current distribution results from the added freedom permitted to the propagating fields. The radiation pattern in this case has its maximum along the helix axis. The helix is then said to be operating in the *axial* mode (Fig. 4.6). Typical dimensions of helices operating in the axial mode are as follows: Diameter about 0.3λ; spacing between turns, 0.25 to 0.3λ; number of turns, 6 to 12; construction from tubing is common. In this size range, the input resistance is in the vicinity of 140 ohms. An approximate formula for the input resistance is $R \doteq 140\, C/\lambda$, where C is the circumference. The directive gain depends on the number of turns, but a typical value is about 15 db with a half-power beamwidth of 35 to 40°. Perhaps most interesting of all is the fact that the radiation in the general direction of the beam maximum is

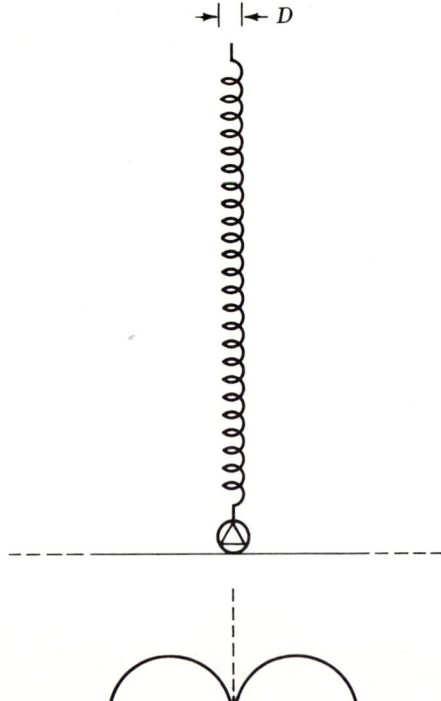

Fig. 4.5 Normal mode helix $(D \ll \lambda)$ and its radiation pattern.

WIRE ANTENNAS AND RELATED FORMS   145

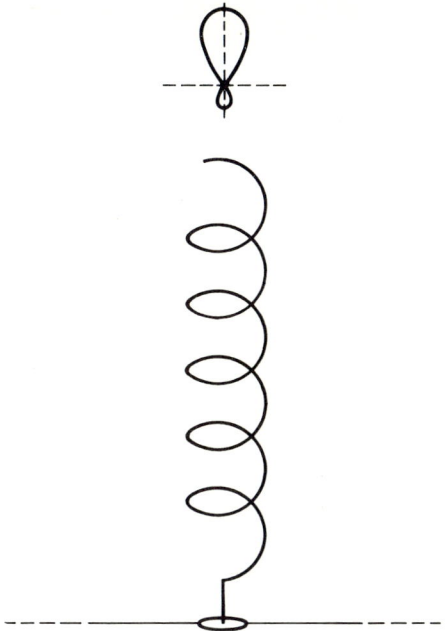

Fig. 4.6 Axial mode helix $(D\pi \sim \lambda)$ and its radiation pattern.

almost circularly polarized. Detailed data and designs for helical antennas are available in the works of Kraus.[1]

4.3 THE BICONICAL TRANSMISSION LINE

A structure consisting of two conducting cones meeting at the apex (as in Fig. 4.7) is an unusually versatile structure for conceptual purposes in the sense that special cases of it represent many interesting antenna structures. These include thin wire antennas, biconical antennas, disccone antennas, and spherical antennas to name a few. The biconical structure will support, among others, a mode that is TEM[2] to the radial direction. As will be shown, for this mode the current distribution along the radial direction is exactly like that along transmission lines.

To find the fields around a biconical transmission line, it is necessary to solve Maxwell's equations in spherical coordinates (r,θ,φ). The biconical structures of interest here have φ symmetry, so the fields are not functions of that coordinate. Placing the φ derivatives equal to zero, Maxwell's

---

[1] J. D. Kraus, *Electronics*, **20:**109 (April, 1947); *Proc. IRE*, **37:**263 (March, 1949); "Antennas," chap. 7, McGraw-Hill Book Company, New York, 1950.
[2] See footnote, page 147.

146 ANTENNA ENGINEERING

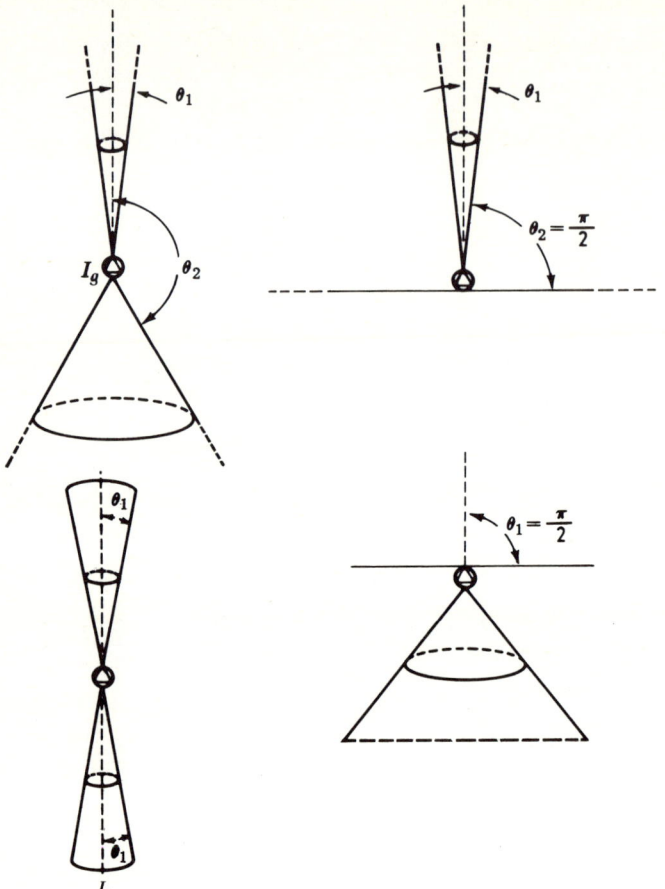

Fig. 4.7 The biconical transmission line and related structures.

equations split up into two independent sets, one of which is

$$\frac{\partial(r \sin \theta H_\varphi)}{\partial \theta} = j\omega\epsilon r^2 \sin \theta E_r$$

$$\frac{\partial(r \sin \theta H_\varphi)}{\partial r} = -j\omega\epsilon r \sin \theta E_\theta \qquad (5)$$

$$\frac{\partial(rE_\theta)}{\partial r} - \frac{\partial(E_r)}{\partial \theta} = -j\omega\mu r H_\varphi$$

The other independent set can be written down by interchanging E and H and replacing ϵ by $-\mu$ in Eqs. (5). We shall not display this latter set explicitly here since it is of no concern to us at present.

Note that the set of fields in Eqs. (5) is TM[1] to the direction r. Note further that along the surfaces of the conducting cones θ_1 and θ_2, the radial component of E must vanish. If the radial component of E were to vanish everywhere, the fields would be TEM to r; let us look for such a solution to Maxwell's equations. With $E_r = 0$, Eqs. (5) becomes

$$\frac{\partial(r \sin \theta H_\varphi)}{\partial \theta} = 0$$

$$\frac{\partial(r H_\varphi)}{\partial r} = -j\omega \epsilon r E_\theta \quad (6)$$

$$\frac{\partial(r E_\theta)}{\partial r} = -j\omega \mu r H_\varphi$$

In the variables (rH_φ) and (rE_θ), the latter pair of equations are like the differential equations for transmission lines. The solutions may be written down by inspection.

Let us define the problem more precisely by assuming that the cones are excited by a point constant current generator at the origin. As a result of this generator, currents will flow on the cones and these will set up the electromagnetic fields in the region. A voltage will exist between the two cones; this is the analog of the transverse voltage on transmission lines,

$$V = \int_{\theta_1}^{\theta_2} E_\theta r \, d\theta$$

We can introduce this voltage into Eqs. (6) by integrating the latter with respect to θ:

$$\int_{\theta_1}^{\theta_2} \frac{\partial(rH_\varphi)}{\partial r} d\theta = -j\omega \epsilon V \quad (7a)$$

$$\frac{\partial V}{\partial r} = -j\omega \mu \int_{\theta_1}^{\theta_2} rH_\varphi \, d\theta \quad (7b)$$

Next note that since the radial component of E is zero, the current on the cones at any position r is related to the magnetic field by Ampère's law

$$I = \oint \mathbf{H} \cdot \mathbf{dl}_\varphi = \int_0^{2\pi} H_\varphi r \sin \theta \, d\varphi = 2\pi r \sin \theta H_\varphi$$

---

[1] The abbreviation TM stands for transverse magnetic, which means that there is no component of the magnetic field in the stated direction, usually the direction of propagation. Similarly, TE stands for transverse electric, and TEM stands for transverse electromagnetic. In the latter case, neither E nor H has a component in the stated direction.

With this equation, we may eliminate rH_φ from Eqs. (7a and b):

$$\frac{1}{2\pi}\int_{\theta_1}^{\theta_2}\frac{1}{\sin\theta}\frac{\partial I}{\partial r}d\theta = -j\omega\epsilon V$$

$$\frac{\partial V}{\partial r} = \frac{-j\omega\mu}{2\pi}\int_{\theta_1}^{\theta_2}\frac{I}{\sin\theta}d\theta$$

Performing the indicated integrations $\left(\int_{\theta_1}^{\theta_2}d\theta/\sin\theta = \ln\tan\tfrac{1}{2}\theta\big]_{\theta_1}^{\theta_2}\right)$ gives the standard transmission-line equations for I and V:

$$\begin{aligned}\frac{\partial I}{\partial r} &= -\mathcal{Y}V & \mathcal{Y} &= \frac{j\omega\epsilon 2\pi}{\ln\tan\tfrac{1}{2}\theta\big]_{\theta_1}^{\theta_2}} \\ \frac{\partial V}{\partial r} &= -\mathcal{Z}I & \mathcal{Z} &= \frac{j\omega\mu}{2\pi}\ln\tan\tfrac{1}{2}\theta\big]_{\theta_1}^{\theta_2}\end{aligned} \quad (8)$$

The solutions are then the standard transmission-line forms

$$I = I_1 e^{\gamma r} + I_2 e^{-\gamma r}$$
$$V = Z_0(I_2 e^{-\gamma r} - I_1 e^{\gamma r})$$
$$\gamma = (\mathcal{Y}\mathcal{Z})^{1/2} = (-\omega^2\mu\epsilon)^{1/2} \quad Z_0 = \left(\frac{\mathcal{Z}}{\mathcal{Y}}\right)^{1/2} = \left(\frac{\mu}{\epsilon}\right)^{1/2}\frac{1}{2\pi}\ln\frac{\tan\tfrac{1}{2}\theta_2}{\tan\tfrac{1}{2}\theta_1} \quad (9)$$

Since the equations and solutions are those of ordinary transmission lines, the impedance (V/I) along the line transforms in the usual way for transmission lines. A terminating impedance Z_L at a position $r = L$ on the biconical line appears at the input as an impedance

$$Z_{\text{in}} = Z_0\frac{Z_L\cosh\gamma L + Z_0\sinh\gamma L}{Z_0\cosh\gamma L + Z_L\sinh\gamma L} \quad (10)$$

Of course, all the foregoing remarks are based on the assumption or supposition that only the TEM solution exists on the biconical line. Discontinuities, such as simply stopping the metal so as to make a finite-length cone, generate other modes. Nevertheless, approximate values for the input reactance of small biconical structures may be obtained from formula (10). If the biconical structure has a length L, then to a first approximation there is an open circuit at the position L; thus, a rough approximation to the input impedance is

$$Z_{\text{in}} = -jZ_0\cot\beta L$$
$$= \frac{-j}{2\pi\omega\epsilon L}\ln\frac{\tan\tfrac{1}{2}\theta_2}{\tan\tfrac{1}{2}\theta_1} \quad (11)$$

These forms neglect the fringing and end-cap capacitance. Two special cases are of interest. First consider the case that $\theta_2 = \pi/2$. This is the case of a single cone over a ground plane. An approximation for the input reactance, from (11) is then

$$X_{in} = -\frac{1}{\omega\epsilon L2\pi}\ln\frac{2L}{a} \qquad (12)$$

where a is the cone radius at the end. Next consider the case in which the cone angles are the same, $\theta_2 = \pi - \theta_1$:

$$X_{in} = -\frac{1}{\omega\epsilon L\pi}\ln\frac{2L}{a} \qquad (13)$$

if θ_1 is small.

For a better approximation, a considerably more involved field theory

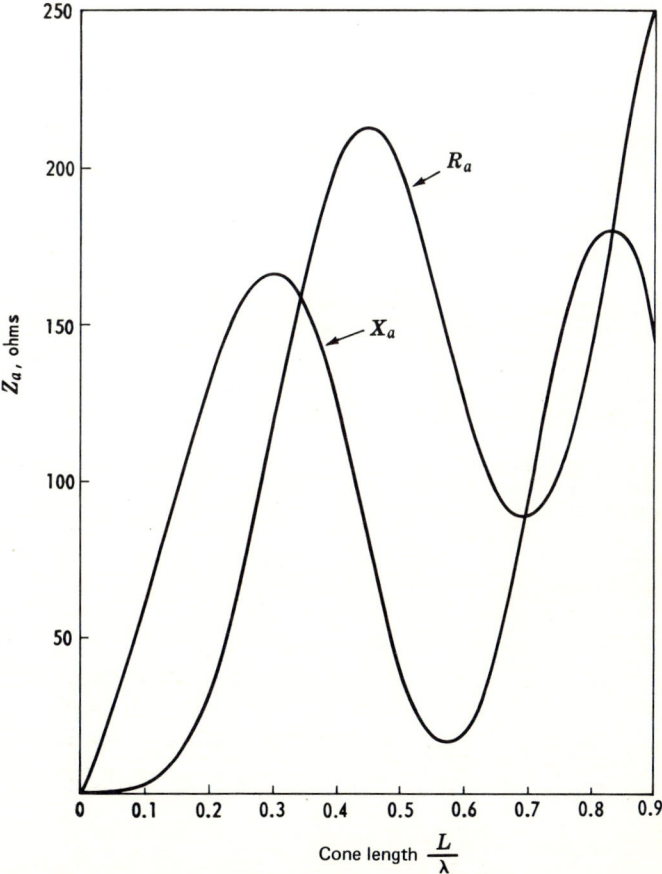

Fig. 4.8 Inverse radiation impedance, $Z_a = R_a + jX_a$, of thin biconical antennas. The terminating admittance is related to Z_a by $Y_T = Z_a/Z_0^2$.

must be carried out. This is the subject of Sec. 5.1. Basically, the field theory provides a value of the effective terminating admittance of the truncated cones. This terminating admittance may be transferred to the input or elsewhere with the aid of standard transmission-line techniques or formulas.

Schelkunoff has calculated the terminating admittance of thin biconical antennas. He presents the results in terms of a quantity he calls the inverse radiation impedance Z_a; this quantity is related to the effective terminating admittance Y_t by the relationship

$$Z_a = Z_0^2 Y_t$$

where Z_0 is given by Eq. (9). A graph of the inverse radiation impedance for thin cones is presented in Fig. 4.8.

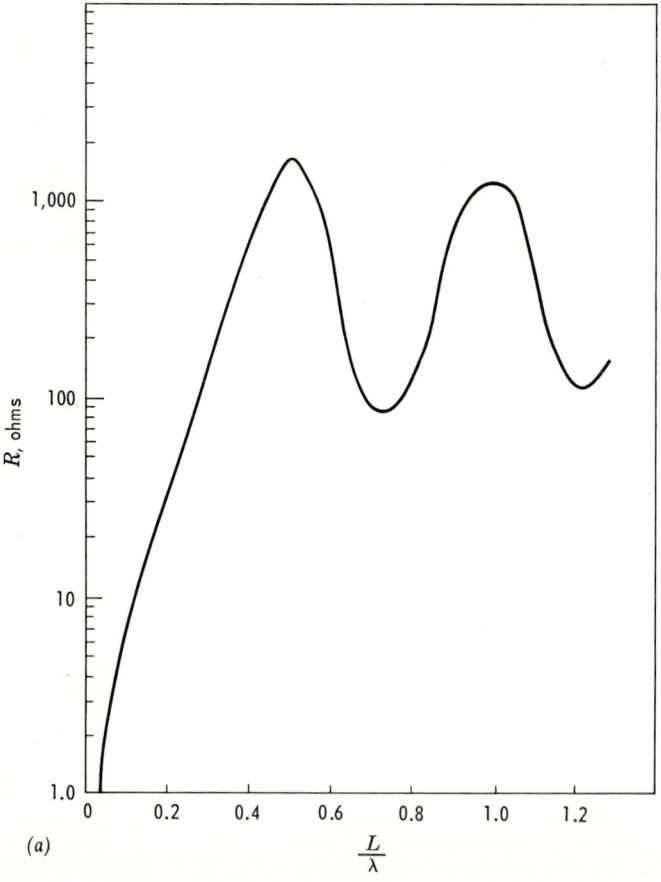

(a)

Fig. 4.9a Input resistance of a thin (one-degree) biconical dipole as a function of cone length L.

Exercise Find the resonant frequencies of a TEM biconical cavity, such as would be formed by terminating a biconical line with spherical short circuits.

4.4 SELF-IMPEDANCE PROPERTIES OF WIRELIKE STRUCTURES

Physical considerations and the foregoing results for the TEM biconical structure suggest the following observations concerning conical and wirelike antennas: the current distribution and the input impedance of these antennas should resemble those of an open-circuited transmission line of length equal to the slant length or half-length; the effective characteristic impedance should depend on the antenna thickness or half-angle.

Examination of experimental and theoretical values of input impedance for the types of structures mentioned reveals the general validity of the foregoing observation. For example, Fig. 4.9 shows the input impedance of a thin biconical antenna as the frequency is varied to give

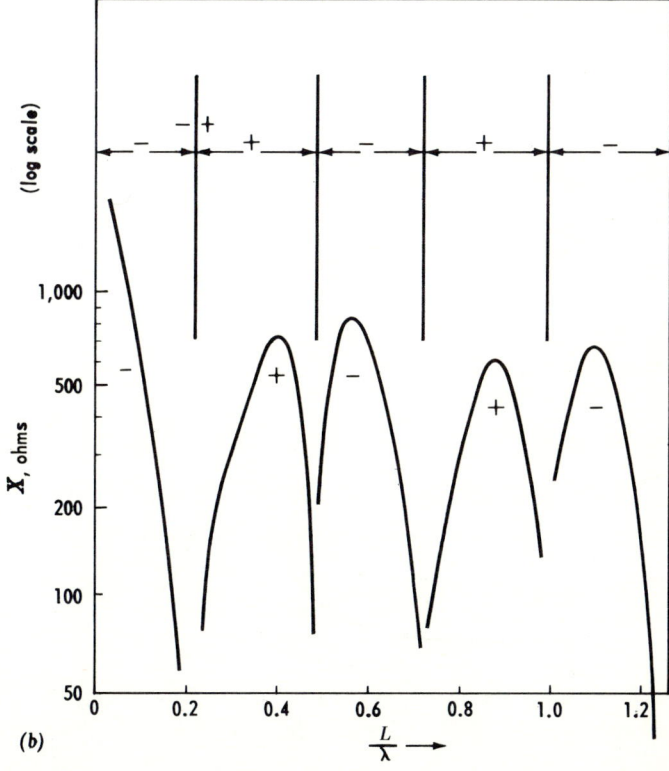

Fig. 4.9b Input reactance of a thin (one-degree) biconical dipole as a function of cone length L.

the slant length/wavelength values indicated. Note the resonance at roughly a quarter wavelength and the antiresonance at a half-wavelength slant length. Figure 4.10 shows the input impedance of a few cylindrical monopole antennas mounted over a perfect ground (double the impedance values to apply these curves to dipoles). Again note the resonance and antiresonant tendencies near frequencies at which the arms are a quarter wavelength and a half wavelength, respectively. Note also that the wider impedance swings (more like those of a low-loss line) are exhibited by the thin structures, rather than the thick ones.

To gain insight, it is worthwhile to pause and consider some of the methods that have been employed to calculate the input impedance of wirelike antennas. Accepting the approximation that the current distribution is sinusoidal, it is relatively straightforward to obtain values for the radiation resistance, as long as the length of the wire is less than a wavelength. According to Eq. (4), the power radiated is

$$P_{rad} = I_g^2 \frac{3600}{\sin^2 \frac{1}{2}\beta L} \int_0^{2\pi} \int_0^{\pi} \frac{[\cos(\frac{1}{2}\beta L \cos \theta) - \cos \frac{1}{2}\beta L]^2}{\sin^2 \theta} \sin \theta \, d\theta \, d\phi$$

and

$$R_{rad} = \frac{P_{rad}}{I_g^2}$$

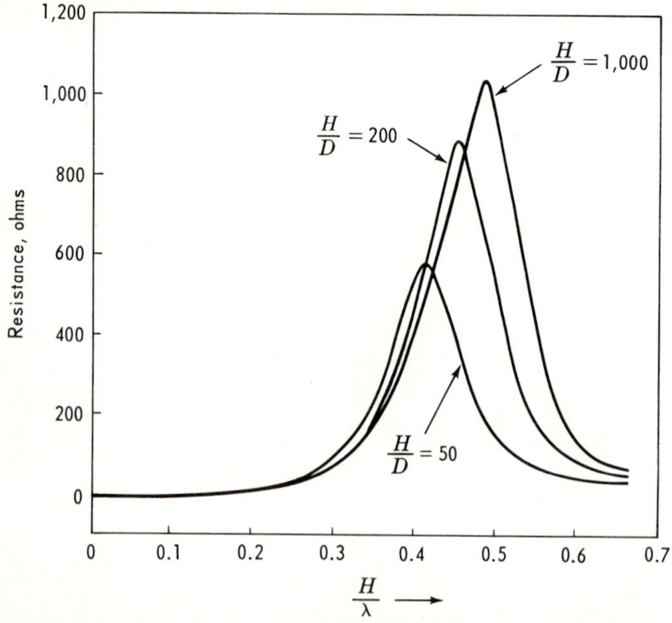

Fig. 4.10a Resistance of cylindrical monopole antennas, of height H and diameter D, over perfect ground.

The integral on θ may be evaluated numerically with the aid of a digital computer. However, historically, it was cast into the form of sine and cosine integrals, values for the latter having been tabulated (see Jahnke and Emde, for example).

4.4.1 THE INDUCED EMF METHOD

The calculation of the reactance is more involved. One of the earliest methods that gave useful results (for both input resistance and reactance) is known as the *induced emf method*. However, some of the derivations of the formula as found in the early literature have been justly criticized; thus it seems appropriate to present an alternative derivation based on the reciprocity theorem.

Consider a cylindrical antenna with a half-length L and a radius a excited by a voltage generator at its input terminals (Fig. 4.11). The voltage generator produces an induced current on the surface of the

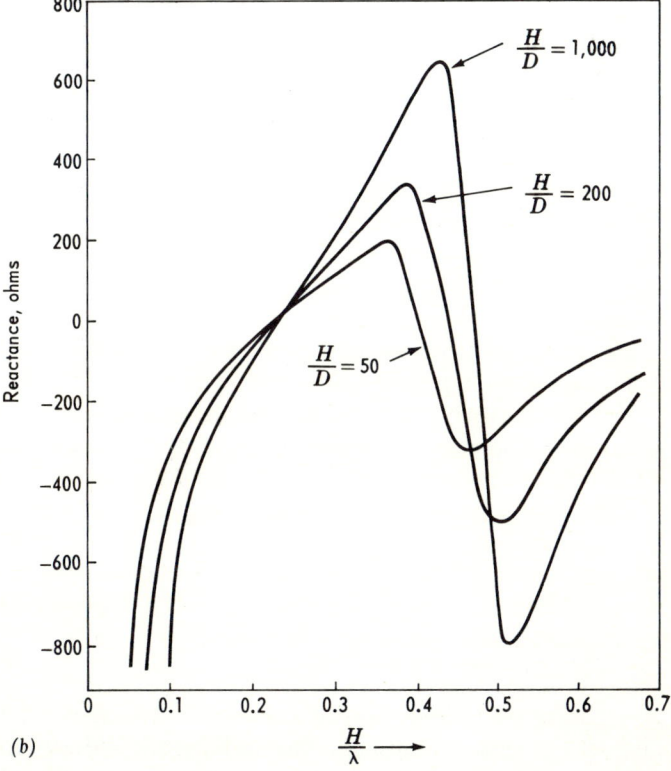

Fig. 4.10b Reactance of cylindrical monopole antennas, of height H and diameter D, over perfect ground.

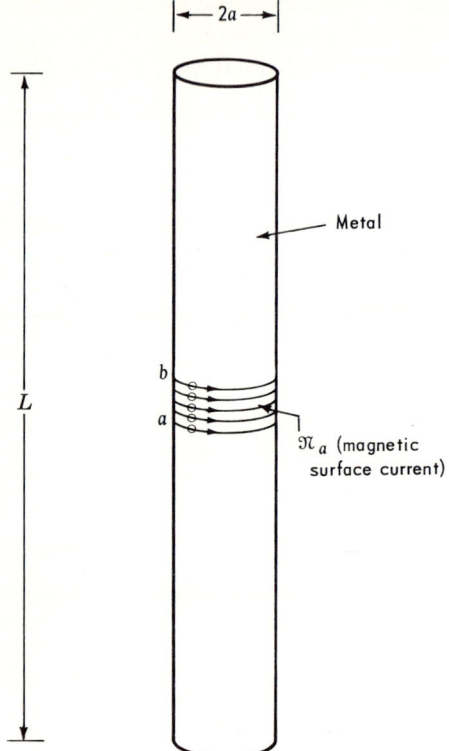

Fig. 4.11 Metal cylinder excited by magnetic surface-current equivalent of a constant-voltage generator.

antenna; call this current J_i. To find the input impedance or admittance, we must in effect find H at the constant-voltage generator, from which we may determine the input current. It is somewhat awkward to do this directly so consider the alternative approach based on the reciprocity theorem. In this approach, the voltage generator is replaced by a magnetic current source; the field-theory equivalent of an ideal voltage generator consists of a magnetic current ring $\mathfrak{M}_a$ on a perfectly conducting "plug" across the terminals of the antenna. A relationship between the magnetic surface current and the input voltage is obtained from the boundary condition on $E_{\tan}$ at the magnetic surface current

$$V_{\text{in}} = \int_a^b E_z \, dz = -\int_a^b \mathfrak{M}_\varphi \, dz$$

Next, imagine a collection of sources in free space; let these consist of magnetic currents identical to $\mathfrak{M}_a$ and a set of electric currents distributed in exactly the same way as the induced currents J_i. Call the free-space electrical currents $\mathcal{J}_i$. The sources in free space produce the same field, outside the metal region, as does the original primary source plus the

WIRE ANTENNAS AND RELATED FORMS 155

metal. Applying the reciprocity theorem between the sources $\mathfrak{M}_a$ and $\mathfrak{J}_i$ gives the result

$$-\int \mathfrak{M}_a \cdot \mathbf{H}_i \, dV = \int \mathfrak{J}_i \cdot \mathbf{E}_a \, dV$$

where of course $\mathbf{E}_a$ is generated by $\mathfrak{M}_a$ and $\mathbf{H}_i$ by $\mathfrak{J}_i$. When the terminal zone and magnetic ring are negligible in size, the integral on the left may be expressed in terms of the input current and voltage:

$$\int \mathfrak{M}_a \cdot \mathbf{H}_i \, dS = \int_0^{2\pi} H_{\varphi i} a \, d\varphi \int_a^b \mathfrak{M}_\varphi \, dz = -I(0)V(0)$$

Thus, from the fundamental definitions, we may obtain the following general formulas for the input impedance

$$Z_{\text{in}} \equiv \frac{V(0)}{I(0)}$$

$$= \frac{-1}{[I(0)^2]} \int \mathfrak{M}_a \cdot \mathbf{H}_i \, dS$$

$$= \frac{1}{[I(0)]^2} \int \mathfrak{J}_i \cdot \mathbf{E}_a \, dV \tag{14}$$

In the original problem, the tangential electric field at the metal is zero, and since $\mathfrak{M}_a$ plus $\mathfrak{J}_i$ in free space produce the same external field, it follows that at the positions of $\mathfrak{J}_i$ ($\mathfrak{J}_i$ is located at the position of the metal cylinder), the total tangential field is zero, $(E_a + E_i)_{\tan} = 0$, or $E_{a,\tan} = -E_{i,\tan}$. With this as a substitution in the last integral above, we obtain the final formula for the input impedance:

$$Z_{\text{in}} = -\frac{1}{I_{\text{in}}^2} \int \mathfrak{J}_i \cdot \mathbf{E}_i \, dV \tag{15}$$

where $\mathbf{E}_i$ is of course the field produced by $\mathfrak{J}_i$ at the position of $\mathfrak{J}_i$. This formula is exact, but it cannot usually be evaluated because $\mathfrak{J}_i$ is not known. Consequently, the impedance formulas determined by this method are approximate, because they are based on an assumed current distribution for the correct $\mathfrak{J}_i$. A sinusoidal approximation for $\mathfrak{J}_i$ is frequently employed. Fortunately, the impedance formula is stationary with respect to small errors in the assumed current distribution. Even so, Eq. (15) is not useful to us until we have explicit expressions for the *near*-field E_i, right at the surface of the antenna. We must consider the calculation of this field next.

4.4.2 NEAR FIELD OF A SINUSOIDAL CURRENT

It turns out that the field of a sinusoidally distributed current has a relatively simple form, even in the near field. This form is widely used, both in the calculation of self-impedance and of mutual impedance. As

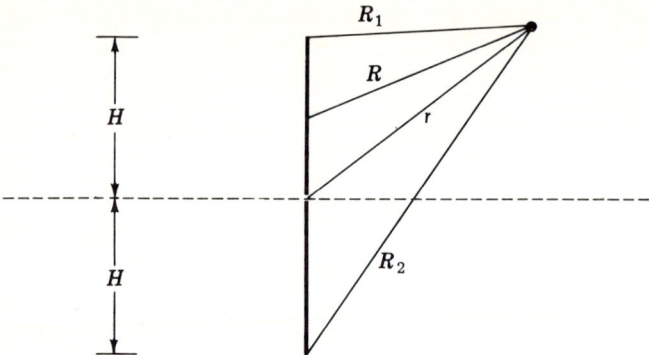

Fig. 4.12 Geometry for expression for the field generated by a sinusoidal current.

will be shown below, if the current lies along the z axis and has the form

$$I = I_m \sin \beta(H - z) \qquad z > 0$$
$$= I_m \sin \beta(H + z) \qquad z < 0$$

the interesting field component is

$$E_z = -\frac{j}{4\pi}(\mu/\epsilon)^{1/2} I_m \left(\frac{e^{-j\beta R_1}}{R_1} + \frac{e^{-j\beta R_2}}{R_2} - 2\cos \beta H \frac{e^{-j\beta r}}{r} \right) \qquad (16)$$

where the symbols are defined in Fig. 4.12. In applying this form to cylindrical antennas, the field is calculated at all near-field points outside the antenna surface on the assumption that the field is produced by a sinusoidally distributed current on the axis of the cylinder.

The derivation of the general field expression given above is a straightforward but rather complicated manipulation. The vector potential for the sinusoidally distributed current is

$$A_z = \frac{I_m}{4\pi} \left[\int_0^H \sin \beta(H - z') \frac{e^{-j\beta R}}{R} dz' + \int_{-H}^0 \sin \beta(H + z') \frac{e^{-j\beta R}}{R} dz' \right]$$

The integrals are best handled by introducing exponentials:

$$A_z = \frac{I_m}{8\pi j} \left[e^{j\beta H} \int_0^H \frac{e^{-j\beta(R+z')}}{R} dz' - e^{-j\beta H} \int_0^H \frac{e^{-j\beta(R-z')}}{R} dz' \right.$$
$$\left. + e^{j\beta H} \int_{-H}^0 \frac{e^{-j\beta(R-z')}}{R} dz' - e^{-j\beta H} \int_{-H}^0 \frac{e^{-j\beta(R+z')}}{R} dz' \right] \qquad (17)$$

The four integrals in Eq. (17) are almost the same. Looking ahead to the field calculation, it is appropriate to change variables of integration as follows: put $u = R + z' - z$. Then, since $R = [\rho^2 + (z - z')^2]^{1/2}$, where ρ is the usual cylindrical coordinate, we have a convenient relationship between the differentials, $du/dz' = u/R$; thus the integrals simplify as follows:

$$\int_0^H \frac{e^{-j\beta(R+z')}}{R}\, dz' = e^{-j\beta z} \int_{u_1}^{u_2} \frac{e^{-j\beta u}}{u}\, du$$

where $u_1 = R_1 + H - z$ and $u_2 = r - z$. The latter form may be evaluated in terms of sine and cosine integrals, but this is neither necessary nor appropriate here. We are attempting to find the E field, in particular, E_z. Recall the following:

$$j\omega\epsilon E_z = \frac{\partial^2 A_z}{\partial z^2} + \beta^2 A_z = \frac{\partial^2 A_z}{\partial z^2} - \nabla^2 A_z = -\frac{1}{\rho}\frac{\partial}{\partial \rho}\left(\rho \frac{\partial A_z}{\partial \rho}\right)$$

Accordingly, to find E_z we may differentiate the integral expression for A_z, which means differentiating the integrals; that is, we need the results of operations as follows

$$\frac{\partial}{\partial \rho} \int_{u_1}^{u_2} \frac{e^{-j\beta u}}{u}\, du$$

The mathematical rules for carrying out such differentiations are summarized in Peirce's integral tables Nos. 855 to 857; for example

$$\frac{\partial}{\partial \rho} \int_{u_1}^{u_2} \frac{e^{-j\beta u}}{u}\, du = \frac{e^{-j\beta u_1}}{u_1}\frac{\partial u_1}{\partial \rho} - \frac{e^{-j\beta u_2}}{u_2}\frac{\partial u_2}{\partial \rho}$$

Note also the results

$$\frac{\partial u_1}{\partial \rho} = \frac{\partial R_1}{\partial \rho} = \frac{\partial}{\partial \rho}[\rho^2 + (z-H)^2]^{1/2} = \frac{\rho}{R_1}$$

and

$$\frac{\partial u_2}{\partial \rho} = \frac{\rho}{r}$$

so that

$$\frac{\partial A_z}{\partial \rho} = \frac{1}{8\pi j} I_m \left[e^{j\beta(H-z)}\left(\frac{e^{-j\beta(R_1+H-z)}}{(R_1+H-z)}\frac{\rho}{R_1} - \frac{e^{-j\beta(r-z)}}{(r-z)}\frac{\rho}{r}\right) \right.$$
$$\left. + e^{-j\beta H}\frac{\partial \mathcal{I}_2}{\partial \rho} + e^{j\beta H}\frac{\partial \mathcal{I}_3}{\partial \rho} + e^{-j\beta H}\frac{\partial \mathcal{I}_4}{\partial \rho} \right]$$

where $\mathcal{I}_2$, $\mathcal{I}_3$, and $\mathcal{I}_4$ are the remaining three integrals in Eq. (17).

Consider the form of the remaining three integrals.

$$\mathcal{I}_2 = \int_0^H \frac{e^{-j\beta(R-z')}}{R}\, dz' = -e^{-j\beta z}\int_{v_1}^{v_2} \frac{e^{-j\beta v}}{v}\, dv$$

158 ANTENNA ENGINEERING

If
$$v = R - z' + z \qquad \frac{dv}{dz'} = \frac{-v}{R} \qquad v_1 = R_1 - H + z$$
$$v_2 = r + z$$

then
$$\frac{\partial g_2}{\partial \rho} = -e^{j\beta z}\left[\frac{e^{-j\beta(R_1 - H + z)}}{(R_1 - H + z)} \cdot \frac{\rho}{R_1} - \frac{e^{-j\beta(r-z)}}{(r+z)} \frac{\rho}{r}\right]$$

The other integrals are handled in a similar fashion. Note:
$$\rho^2 = R_1^2 - (H - z)^2 \qquad \text{also} \qquad \rho^2 = r^2 - z^2$$

and
$$\rho^2 = R_2^2 - (H + z)^2$$

Multiplying and dividing to introduce ρ into the expression gives
$$\frac{\partial A_z}{\partial \rho} = -H_\phi = \frac{1}{4\pi j} I_m \left(\frac{e^{-j\beta R_1}}{\rho} + \frac{e^{-j\beta R_2}}{\rho} - \frac{2\cos\beta H}{\rho} e^{-j\beta r}\right)$$

Forming the quantity
$$j\omega\epsilon E_z = -\frac{1}{\rho}\frac{\partial}{\partial \rho}\left(\rho \frac{\partial A_z}{\partial \rho}\right)$$

gives Eq. (16) for E_z.

The ρ component of E may be obtained from the expression
$$j\omega\epsilon E_\rho = -\frac{\partial^2 A_z}{\partial z\, \partial \rho}$$

4.4.3 APPROXIMATE IMPEDANCE CALCULATION

As long as the antenna wire or cylinder is thin, the fields produced by sinusoidally distributed currents on the surface of the wire are essentially the same as those produced by a sinusoidally distributed current on the axis of the cylinder. That is, the equation for E_z given in Eq. (16) may be employed as an approximation for E_i in the impedance formula, Eq. (15). Thus, according to that equation, the input impedance to a center-fed cylindrical antenna may be calculated from the integral

$$Z_{\text{in}} = -\frac{1}{I_{\text{in}}^2} \iint \mathcal{I}_{si} \frac{1}{4\pi j}\left(\frac{\mu}{\epsilon}\right)^{1/2} I_m \left(\frac{e^{-j\beta R_1}}{R_1} + \frac{e^{-j\beta R_2}}{R_2} - 2\cos\beta H \frac{e^{-j\beta r}}{r}\right) a\, d\phi\, dz$$

where
$$\mathcal{I}_{si} = I_m \frac{\sin \beta(H - z)}{2\pi a}$$

WIRE ANTENNAS AND RELATED FORMS   159

The integral expression to be evaluated is then

$$Z_{in} = j30(I_m^2/I_{in}^2) \int_{-H}^{H} \sin \beta(H - |z|) \left(\frac{e^{-j\beta R_1}}{R_1} + \frac{e^{-j\beta R_2}}{R_2} - 2 \cos \beta H \frac{e^{-j\beta r}}{r} \right) dz$$

It will be recalled from image considerations that the center-fed dipole impedance is just twice that of a monopole over a perfect ground. The field produced by the monopole is the same as that produced by the dipole in space; however, the integration over the currents extends only from 0 to H in the case of a monopole. Hence we can find the dipole impedance from the expression

$$Z_{in} = j60 \left(\frac{I_m}{I_{in}} \right)^2 \int_0^H \sin \beta(H - z) \left(\frac{e^{-j\beta R_1}}{R_1} + \frac{e^{-j\beta R_2}}{R_2} - 2 \cos \beta H \frac{e^{-j\beta r}}{r} \right) dz$$

where

$I_m/I_{in} = 1/\sin \beta H$
$R_1 = [(H - z)^2 + a^2]^{1/2}$
$R_2 = [(H + z)^2 + a^2]^{1/2}$
$r = (a^2 + z^2)^{1/2}$

Except for the cases of very small a and/or βH in the close vicinity of π, this integral may be readily evaluated by numerical integration with a digital computer. This is probably the best way to obtain values, although it is also possible to reduce the integral to a complicated expression involving sine and cosine integrals, which are tabulated. To carry out this reduction, one may proceed as follows: The sine function may be changed to exponential form

$$\sin \beta(H - z) = \frac{e^{j\beta(H-z)} - e^{-j\beta(H-z)}}{2j}$$

Inserting this into the integral, we find that we have six integrals of similar form. We shall simply indicate the type manipulation required for hand calculation. One integral has the form

$$\mathcal{I}_1 = \int_0^H e^{j\beta(H-z)} \frac{e^{-j\beta R_1}}{2jR_1} dz = \int_0^H \frac{e^{-j\beta(R_1-H+z)}}{2jR_1} dz$$

where

$R_1 = [(H - z)^2 + a^2]^{1/2}$

Put $u = \beta(R_1 - H + z)$ so that $du/dz = \beta(dR_1/dz + 1)$

$$\frac{dR_1}{dz} = \frac{\frac{1}{2} \cdot 2(H-z)(-1)}{R_1} = -\frac{H-z}{R_1}$$

$$\frac{du}{dz} = \frac{\beta(-H + z + R_1)}{R_1} = \frac{u}{R_1}$$

or

$$\mathcal{G}_1 = \int_{\beta A}^{\beta a} \frac{e^{-ju}}{2jR_1} R_1 \frac{du}{u} = \frac{1}{2j} \int_{\beta A}^{\beta a} \frac{e^{-ju}}{u} du$$

where

$$A = R_1 - H = (H^2 + a^2)^{1/2} - H = H\left\{\left[1 + \left(\frac{a}{H}\right)^2\right]^{1/2} - 1\right\} \approx \frac{1}{2}\frac{a^2}{H}$$

Alternatively

$$\mathcal{G}_1 = \frac{1}{2j}\left[\int_{\beta A}^{\beta a} \frac{\cos u}{u} du - j\int_{\beta A}^{\beta a} \frac{\sin u}{u} du\right]$$

These integrals may be expressed in terms of tabulated functions, the sine and cosine integrals

$$\mathrm{Si}(x) = \int_0^x \frac{\sin x}{x} dx \qquad \mathrm{Ci}(x) = -\int_x^\infty \frac{\cos x}{x} dx$$

as follows:

$$\int_{\beta A}^{\beta a} \frac{\sin u}{u} du = \int_0^{\beta a} \frac{\sin u}{u} du - \int_0^{\beta A} \frac{\sin u}{u} du = \mathrm{Si}(\beta a) - \mathrm{Si}(\beta A)$$

$$\int_{\beta A}^{\beta a} \frac{\cos u}{u} du = \int_{\beta A}^\infty \frac{\cos u}{u} du - \int_{\beta a}^\infty \frac{\cos u}{u} du = -\mathrm{Ci}(\beta A) + \mathrm{Ci}(\beta a)$$

or

$$\mathcal{G}_1 = \frac{1}{2j}[\mathrm{Ci}(\beta a) - \mathrm{Ci}(\beta A)] - \frac{1}{2}[\mathrm{Si}(\beta a) - \mathrm{Si}(\beta A)]$$

The other integrals are similar. The final result is an involved expression that can be found in the literature (e.g., Jordan, p. 361).

The foregoing results are quite adequate and accurate for most engineering purposes provided the input region does not depart appreciably from the biconical model and provided input impedance terminals can be accurately defined for the cylindrical structures. Frequently, however, structural requirements introduce a flat base for the antenna or other features that make it difficult to determine a unique pair of points that can be labeled input terminals. The consequence of this is usually to introduce extra capacitance near the base of the antenna; this capacitance must be accounted for in preliminary designs.

WIRE ANTENNAS AND RELATED FORMS   161

Problem 4.2 To support a cylindrical monopole for a certain application, it is necessary to hold it in a cantilever arrangement as indicated. The antenna mast is 25 ft high and 3 in. in diameter. The hole in the supporting tube is 4 in. in diameter. The relative permittivity of the insulating material is $3 - j1$. The antenna mast extends 12 in. below the ground level (there is a good ground screen at the ground level). Sketch a curve showing the input admittance and/or impedance for this antenna system in a range of frequencies such that the height varies from about 0.10λ to 0.30λ.

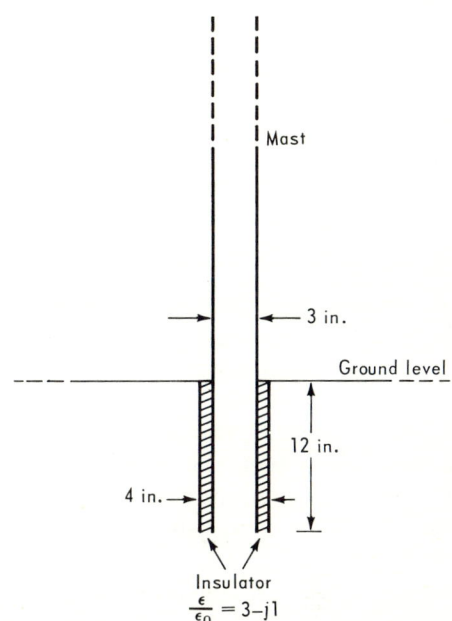

4.4.4 ASYMMETRIC FEEDING

Sometimes it is necessary or desirable to excite a wirelike antenna asymmetrically. The radiation pattern may or may not be influenced by the point of excitation. For antennas having half-lengths up to about a quarter wavelength, the pattern is essentially independent of the position of the generator; the ends of the wire force the current into almost the same distribution in any case. For longer antennas, the pattern may depend on the point of excitation. The reason is that although the current distribution is always almost sinusoidal, the relative phase at different positions on the wire may depend markedly on the position of the generator, as indicated in Fig. 4.13.

Of course the input impedance usually depends strongly on the feed position. There are as yet no exact theoretical results covering an arbi-

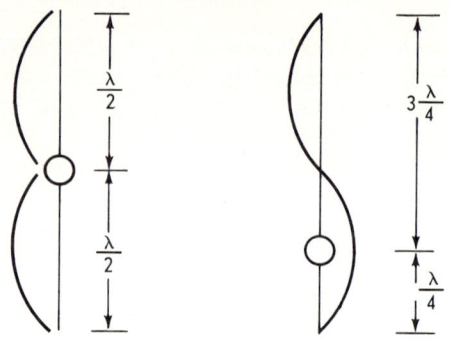

Fig. 4.13 A full-wave dipole showing the influence of point of excitation on the current distribution.

trary point of excitation. But physical considerations and experimental results have revealed an approximate formula for the input impedance of antennas fed off center, in terms of the impedances of a pair of center-fed antennas. If the point of excitation is at a position L_1 from one end and L_2 from the other end, the formula is

$$Z_{\text{in}} \doteq \tfrac{1}{2}(Z_1 + Z_2)$$

where Z_1 is the input impedance of a center-fed dipole with half-length L_1, and Z_2 is that of a center-fed dipole having half-length L_2. If better results are required, the formulas of King and Wu[1] may be employed.

This approximate formula may be made plausible by considerations as follows. Suppose the wire is excited by a current generator. The input impedance is then the ratio of the input voltage to the constant input current. The input voltage is the line integral of E from wire L_1 to wire L_2, along a field line near the input. Now suppose that L_1 and L_2 are not too different. Then it is clear that the field around each half of the asymmetric dipole is almost the same as that about the corresponding symmetric structure. From the physical point of view, to find the field of the asymmetric composite, it is necessary only to match up the half-fields of the symmetric cases. But if L_1 and L_2 are nearly equal, only minor rearrangements in the fields will be necessary in order to match up the (symmetric) half-fields. This is especially true at points near the input. Thus, in imagining the evaluation of the line integral of E from wire L_1 to wire L_2, we can see that over half of the path, the field would be almost like that about symmetric dipole L_1 (giving $\tfrac{1}{2}Z_1$), and over the other half of the path the field would be almost like that about symmetric dipole L_2 (giving $\tfrac{1}{2}Z_2$). This is suggested in Fig. 4.14. In summary, the impedance is essentially the sum of two monopole impedances to an imaginary ground plane.

Surprisingly, the approximate formula is useful even if L_1 and L_2 are quite different. In fact, it gives an approximation to the impedance of

[1] R. W. P. King and T. T. Wu, *IEEE Trans.*, **AP-13**:710 (September, 1965).

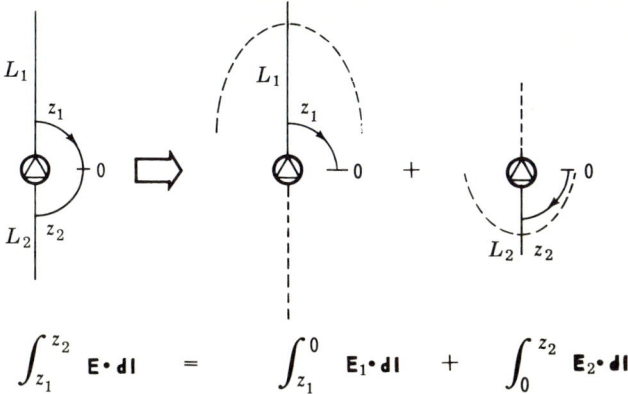

Fig. 4.14 Sketch showing physical basis for the formula for the input impedance of asymmetrically fed dipole.

antennas that are essentially end fed. For example, consider a low-frequency trailing wire antenna for aircraft. Such antennas have been designed for reeling in and out below the tail section, in flight. The long wire is essentially fed against the aircraft itself, but at low frequencies, the aircraft is dwarfed by the length of wire. A crude approximation for the input impedance may be obtained by (1) determining the input impedance of a symmetric dipole of half-length equal to that of the trailing wire, (2) determining the capacitance of the airframe (to another one like it) and thereby its input reactance, and (3) taking one-half of the sum, as before.

Closely related to the asymmetrically fed dipole is the structure known as the sleeve antenna. The sleeve construction, indicated in Fig. 4.15a, is sometimes employed to give added strength to a monopole, and also to alter impedance levels. An approximate analysis of the sleeve antenna may be carried out as follows: The sleeve monopole over a ground plane is replaced by the monopole plus its image, as indicated in Fig. 4.15b. By superposition, this doubly driven structure in Fig. 4.15b may be replaced by a pair of asymmetrically driven structures, as indicated in Fig. 4.15c. The approximate analysis ignores the change in diameter. The current at the input to the sleeve is then approximately the sum of the currents at the point $z = z_f$ on the two structures in Fig. 4.15d. But since the two structures in Fig. 4.15d are physically identical, the current at the feedpoint is $I_{in} \doteq I_{as}(z_f) + I_{as}(-z_f)$, where I_{as} is the current in the asymmetrical structure. The input admittance to the sleeve is then

$$Y_{in} \doteq \frac{I_{as}(z_f) + I_{as}(-z_f)}{V_{in}} \doteq Y_{as}\left[1 + \frac{I_{as}(z_f)}{I_{as}(-z_f)}\right]$$

164 ANTENNA ENGINEERING

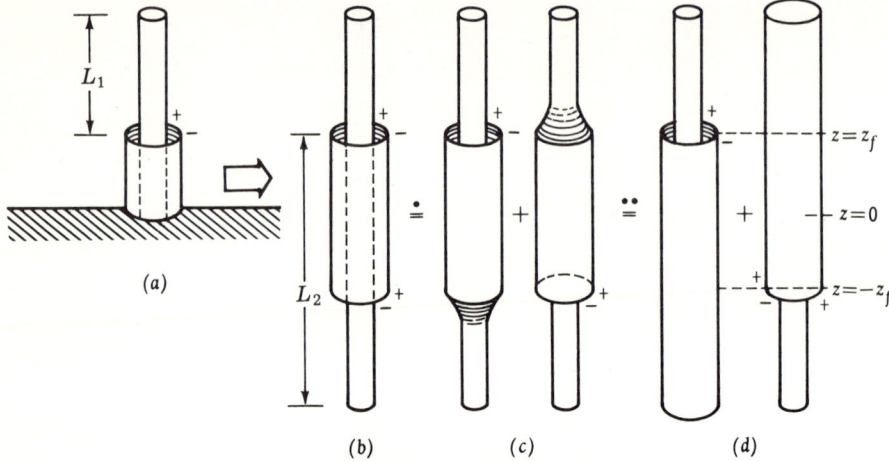

Fig. 4.15 Approximate equivalences for the sleeve-dipole analysis.

where as before $Y_{as} \doteq 2/(Z_1 + Z_2)$, where Z_1 and Z_2 are the impedances of the symmetrical antennas of half-length L_1 and L_2, respectively. Further details may be found in King.[1]

4.4.5 SHUNT FEEDING

There are at least three reasons for seeking alternative methods for exciting wirelike antenna structures. One reason is that the input impedance of an antenna fed in the manner discussed previously may be poorly matched to an appropriate transmission line. The resulting standing wave may then give rise to line losses that are intolerably high, or they may interfere with the functioning of the transmitter. A second reason may be associated with the insulators that must be employed at the base of insulated antenna towers; such a base insulator may give rise to excessive losses or undesirable impedance effects. A third reason is the hazard and vulnerability of tall insulated tower structures during electrical storms (also in areas with frequent dry, dusty winds). The system of feeding known as shunt feeding may help with all three of these problems. Typical forms of shunt feeding are indicated in Fig. 4.16. The configuration shown in Fig. 4.16a is sometimes called a gamma match. The system shown in Fig. 4.16b is commonly used for exciting broadcast antenna towers.

Shunt feeding may be understood and approximate designs obtained from arguments as follows. It is pointed out in Sec. 4.3 and shown in Sec. 5.1 that the impedance of a biconical antenna may be found by

---

[1] R. W. P. King, "Theory of Linear Antennas," pp. 407–418, Harvard University Press, Cambridge, Mass., 1956.

making use of the model indicated in Fig. 4.17a. The terminating admittance is determined from field theory; then the input impedance is determined by transforming this admittance Y_T down the biconical transmission line to the input. Now suppose that the input terminals are joined by a conducting plug and the lead-in wires are moved a short distance along the bicones toward the ends, as shown in Fig. 4.17b. Reference to the transmission-line model representing the center-fed bicone shows that the shunt excitation corresponds to the shorting of the transmission-line model at one end and exciting the line from an interior point (review Prob. 2.4). Transmission-line theory and such a model provide the following formula for the input admittance of a shunt-fed biconical antenna

$$Y_{\text{in}} = -jY_0 \cot \beta d + Y_0 \left[\frac{Y_T + jY_0 \tan \beta(L-d)}{Y_0 + jY_T \tan \beta(L-d)} \right] \tag{18}$$

The terminating impedance presented to the feeding transmission line

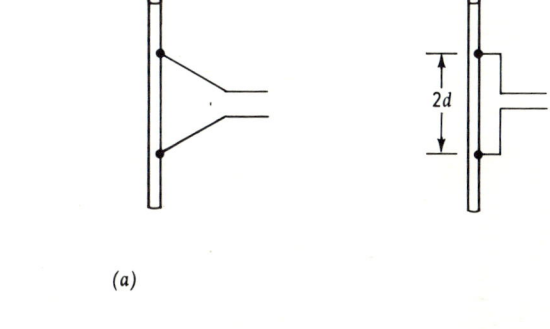

(a)

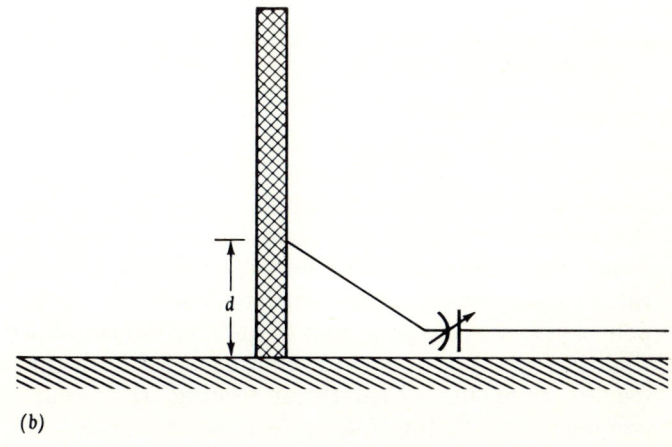

(b)

Fig. 4.16 Typical forms of shunt feeding for wires and towers.

166 ANTENNA ENGINEERING

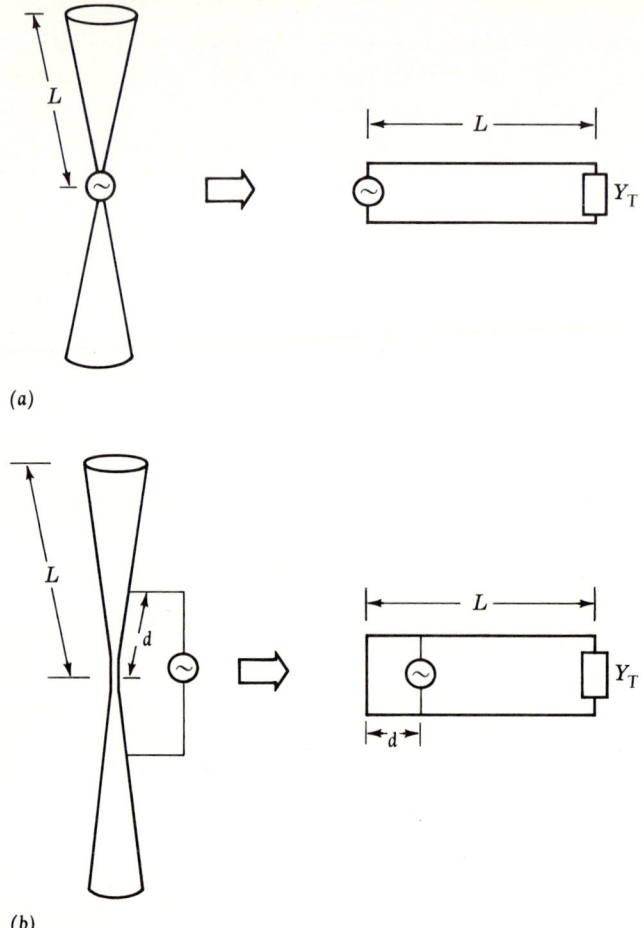

Fig. 4.17 Analogies to help in understanding the principle of shunt feeding: (a) biconical antenna center-fed, and transmission-line model; (b) shunt-fed biconical antenna and its transmission-line model.

differs from the reciprocal of Y_{in} given in Eq. (18) by the amount of inductive reactance in the loop formed by the connecting lead wires. This inductance may be tuned out with a capacitance if desired. When the antenna is near a resonant length (i.e., a quarter-wavelength monopole, a half-wave dipole, and so on), the current distribution differs only slightly from that on a similar antenna fed with the familiar series feeding arrangement. With shunt feeding, the input impedance may be increased by a factor of 4, or even as much as 8.

An alternative technique for approximating the input impedance of a shunt-fed antenna is as follows: The impedance of a cylindrical antenna of length $(L - d)$ with an appropriate length-to-diameter ratio is determined. This is combined in parallel with the impedance (inductive reactance) of the antenna structure between the tap points (the length d for monopoles and $2d$ for dipoles). This combination is added to the inductive reactance of the sloping lead wire(s) to obtain the transmission-line terminating impedance. The impedance per unit length of cylindrical structures is considered in Sec. 5.2.1.

Problem 4.3 A 400-ft tower has a length-to-diameter ratio of 100. The tower is grounded and shunt fed at a point that is 75 ft above the ground. The feeding wire angles up from a point on the ground that is 80 ft from the base of the tower. Estimate the impedance seen between this point and ground when the frequency is in the vicinity of 550 kHz.

4.5 BALANCE PROBLEMS IN TRANSMISSION LINES AND ANTENNAS

Many of the important antenna types that are used in practice are, or should be, "balanced" with respect to ground. On the other hand, there is often good reason to employ coaxial lines to feed these structures; but the conductors in coaxial lines are "unbalanced" with respect to ground. There are problems in joining balanced structures to unbalanced ones. These problems and the solutions are probably best understood after a detailed consideration of the joining of a coaxial line to a parallel-wire line. Such a discussion is presented in this section.

In circuit technology, an unbalanced system is defined as one in which the two conductors are at different potentials with respect to ground (perhaps one of the conductors is at ground potential). The capacitance with respect to ground of the individual conductors is then different, and consequently the current in the two conductors may be different. In contrast, a balanced system is one in which the two conductors are respectively above and below ground potential by the same amount. In antenna and transmission-line practice, the meanings of the terms *balanced* and *unbalanced* are often enlarged and perhaps distorted as follows: two halves of a symmetric transmission-line or antenna system are said to be balanced if they carry the same (but opposite) current, or unbalanced if their currents are unequal. This extension of the meaning is helpful in practice because the detrimental effects of unbalanced systems (i.e., extra parallel-wire transmission-line losses to ground or in radiation, and/or disturbed antenna patterns) are directly due to the current unbalance. Ordinarily, the unbalance of the potentials with respect to ground "causes" the unbalance in currents; but unfortunately, the potential with respect to ground of extended transmission and antenna systems is not always easy to define precisely. Moreover, the potentials with respect to ground in such systems may be balanced at one point but the currents at distant points may still be unbalanced.

Devices that permit connection between a balanced system and an unbalanced system, in such a way that the potentials to ground and/or the currents in the two parts of the balanced structure are equal, are known as *bal*anced-to-*un*balanced converters, or *baluns* for short.

The difficulties associated with a direct connection between a balanced and an unbalanced system can be brought out graphically and dramatically with the following exercise. Consider a coaxial cable sticking up a short distance above ground and connected to a parallel-wire line above ground as indicated in Fig. 4.18a. The parallel-wire line is terminated in an impedance Z_L. The impedance Z_s, shown dotted, is some high impedance associated with the supporting structure. As far as the parallel wire is concerned, the currents are produced by an idealized generator as indicated in Fig. 4.18b. Note from the figure that the currents in line A and line B are not in general equal since the total current flowing away from point b flows into line A, but the current flowing toward b comes both from line B and from the connection to ground. The significant structural feature is that line B is tied to the ground at one end.

The actual computation of currents is readily made in terms of a voltage generator (refer to Fig. 4.18c). For simplicity, we shall assume that the generator impedance Z_g is negligible. Also for convenience, we have altered the picture by moving the point b down to the ground plane. The new picture looks somewhat different, but it is electrically equivalent since the electrical distance from the ground plane to point b is assumed to be small. Next, the generator is split into two parts and bucking generators added in leg B, as indicated in Fig. 4.18d, with no change in the currents or voltages. Finally, we call upon the superposition theorem, according to which we can solve for the currents in Fig. 4.18d by solving, separately, for the currents in a push-pull mode (Fig. 4.18e) and a push-push mode (Fig. 4.18f), and superposing the results. Call the currents in the transmission-line system, Fig. 4.18e, $I_{t,A}$ and $I_{t,B}$, with the directions indicated; similarly label the currents in Fig. 4.18f $I_{u,A}$ and $I_{u,B}$, with the directions indicated in that figure. The currents $I_{t,A}$ and $I_{t,B}$ are equal and opposite and at the input $I_{t,A} = V/Z_{\text{in},e}$ where $Z_{\text{in},e}$ is determined by the transfer of the impedance Z_L back down the parallel-wire line to the input (Z_s is assumed or arranged to be unimportant in this process), i.e.,

$$Z_{\text{in},e} = Z_{0e} \frac{Z_L + jZ_{0e} \tan \beta L}{Z_{0e} + jZ_L \tan \beta L} \tag{19}$$

WIRE ANTENNAS AND RELATED FORMS 169

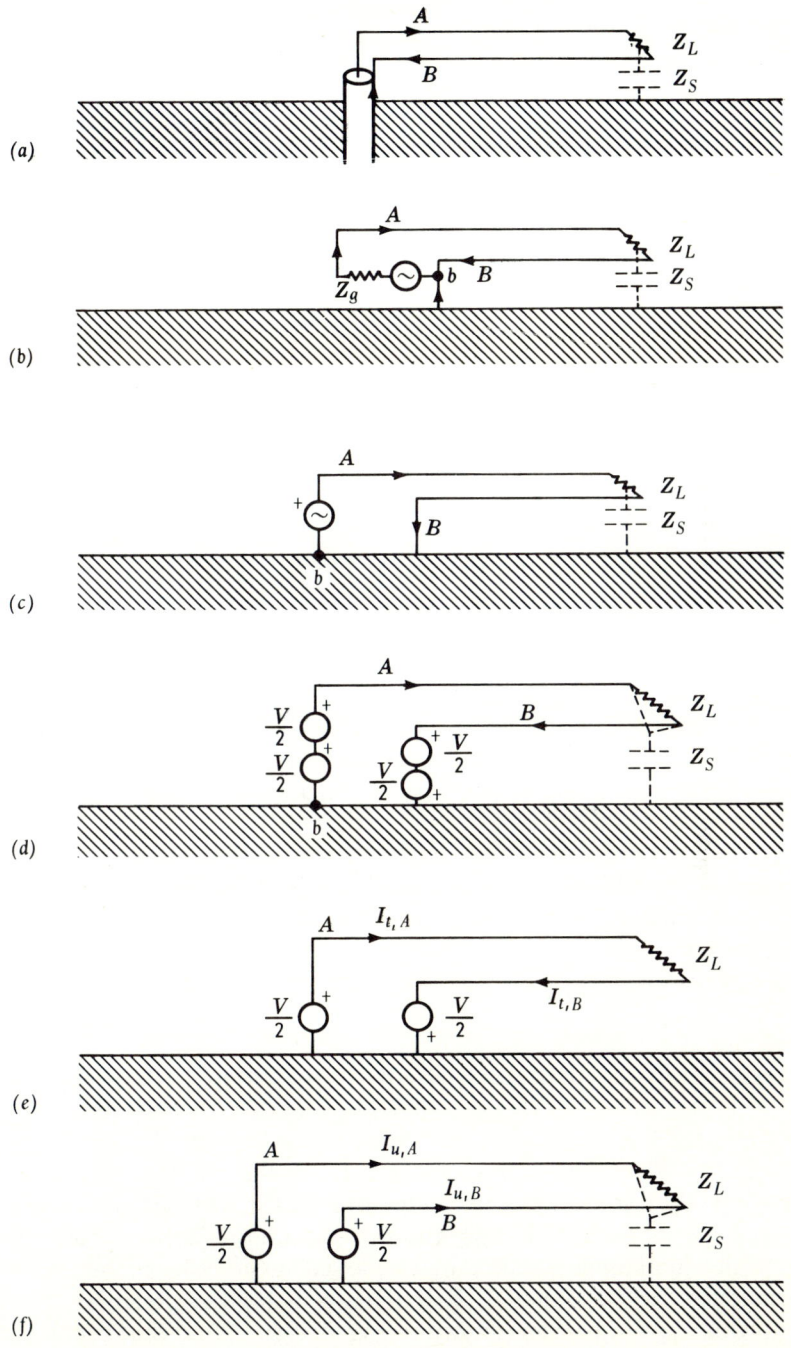

Fig. 4.18 Connection between balanced and unbalanced lines with equivalents for calculating current unbalance. Line length L.

where Z_{0e} is the characteristic impedance of the parallel-wire line. On the other hand, the currents $I_{u,A}$ and $I_{u,B}$ are equal and in the same direction and at the input $I_{u,A} = \frac{1}{2}V/2Z_{\text{in},f}$, where $Z_{\text{in},f}$ is determined by the transfer of the impedance Z_s back down the double-wire line above ground to the input (Z_L is assumed to be unimportant in this process); that is

$$Z_{\text{in},f} = Z_{0f} \frac{Z_s + jZ_{0f} \tan \beta L}{Z_{0f} + jZ_s \tan \beta L} \tag{20}$$

where Z_{0f} is the characteristic impedance of the double-wire structure over ground. The net currents in the two wires of the original system, I_A and I_B, are therefore

$$I_A = I_{t,A} + I_{u,A} \tag{21}$$

$$I_B = I_{t,B} - I_{u,B} \tag{22}$$

where the difference in sign occurs because $I_{t,A}$ and $I_{u,A}$ have the same directions but $I_{t,B}$ and $I_{u,B}$ have opposite directions. Thus, the net input currents in the two wires are given by the equations

$$I_A = \frac{V}{Z_{\text{in},e}} + \frac{V}{4Z_{\text{in},f}} = \frac{V}{Z_{\text{in},e}}\left(1 + \frac{1}{4}\frac{Z_{\text{in},e}}{Z_{\text{in},f}}\right) \tag{23}$$

$$I_B = \frac{V}{Z_{\text{in},e}}\left(1 - \frac{1}{4}\frac{Z_{\text{in},e}}{Z_{\text{in},f}}\right) \tag{24}$$

These expressions show that the currents in the two sides of a parallel-wire line excited by a coaxial line are unequal unless $Z_{\text{in},f} \gg Z_{\text{in},e}$; these currents become quite unequal as $Z_{\text{in},e} \to 4Z_{\text{in},f}$.

As this type of structure would normally appear in practice, the impedance Z_s would be quite large since it will be the capacitive reactance (and conductance) associated with insulating supports. To a good approximation

$$Z_{\text{in},f} = -jZ_{0f} \cot \beta L$$

Thus, if the line is very short, $Z_{\text{in},f}$ is so high that $I_A \doteq I_B$, as would be expected from elementary circuit considerations. As the line is lengthened toward a quarter wavelength, however, $\cot \beta L$ decreases in size; eventually it may reach the point that the term $\frac{1}{4}Z_{\text{in},e}/Z_{\text{in},f}$ is in the vicinity of one, and the currents I_A and I_B in the two wires become decidedly different. The reader will find it instructive to examine several special cases of Z_L, Z_s, and L. Equations (23) and (24) suggest one of the fundamental methods employed in practice to achieve a current balance $I_A \doteq I_B$; that is, $Z_{\text{in},f}$ is made as high as possible.

One type of balun is shown in Fig. 4.19a. Note that the device immediately introduces a symmetry with respect to ground; this *symmetry*

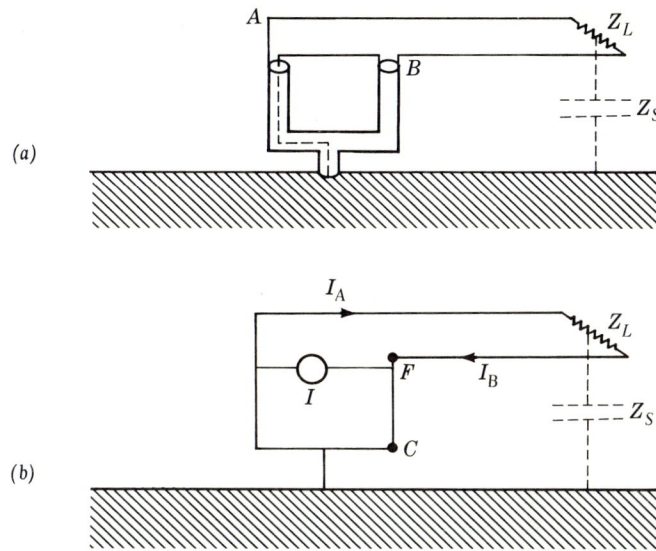

Fig. 4.19 A forked coax balun and an equivalent representation.

with respect to ground is essential in all baluns or in balanced systems. The balun has, instead of the single coaxial line leading to the balanced pair as in Fig. 4.18a, a coaxial line that forks into a pair, one of which is simply a dummy. The shields of the coaxial pair are joined to the parallel-wire pair, as indicated. The center conductor of the coaxial line is run across and joined to the shield of the dummy coaxial line. Figure 4.19b shows a representation similar to that shown in Fig. 4.19a from which the currents in the structure can be determined. It is clear that the double-wire-line-above-ground field is not excited, so that $I_A = I_B$. However, the generator is shunted by the transferred impedance of the short circuit at C, as indicated. The input impedance is the parallel combination of $Z_{in,e}$ and the impedance of the short circuit at C transferred up the parallel-wire line a distance $\overline{CF}$ to the input. Thus, the structure under consideration is an excellent balun but if the length ($\overline{FC}$) is short, the coaxial cable is almost shorted and very little of the power is delivered to the load. The impedance properties are thus unsatisfactory. Perhaps the first thought to correct this difficulty is to make the length ($\overline{FC}$) a quarter wavelength (folding if necessary). This technique is often used; however, it is inherently narrowband. In order to operate over a large bandwidth, the input impedance looking back toward C should remain high over a wide band of frequencies. This implies a high characteristic impedance for the lower (short-circuited) line, or its replacement by a high impedance device that also maintains the symmetry.

One technique for introducing a high reactance is to coil the coax as in Fig. 4.20. This scheme will maintain the high impedance over a considerable band of frequencies. The lower-frequency limit is that frequency at which the impedance of the coiled coax is too small compared to $Z_{in,e}$. It operates satisfactorily on up through the antiresonant impedance point (associated with the coils) to nearly the frequency at which the coil exhibits a series resonance (impedance too low again). Such a device may be designed to operate satisfactorily over a frequency range of 2 to 3 octaves.

The frequency range may be extended by increasing the frequency ratio between the antiresonant frequency and the series resonant frequency. One way of accomplishing this is to wind the coils of the coax on ferrite cores, as indicated in Fig. 4.21. The ferrite has the effect of maintaining a high level of impedance over a wider frequency band. It is emphasized that it is the exterior fields on the coax which see the high impedance—the interior fields are unaffected by the coiling or the ferrite. With careful construction, operation over a frequency band of 8 or even 10 to 1 may be achieved with such a structure.

To sum up, when the high impedance looking toward the fork is accomplished with the desired symmetry, the balanced feed is obtained without changing the impedance properties presented by the balanced load (the practical problem is greater for high-impedance loads than for low-impedance loads).

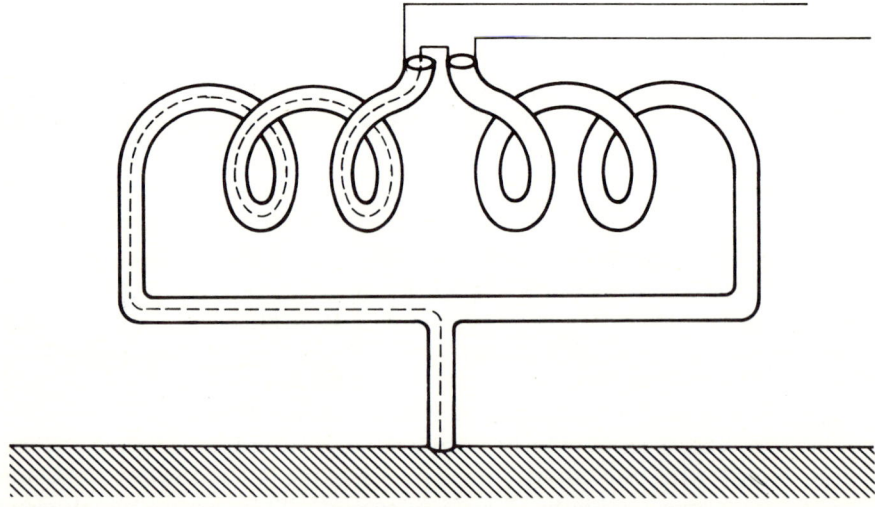

Fig. 4.20 Coiled coax balun. The coils provide the high impedance to ground over a wide-frequency band.

WIRE ANTENNAS AND RELATED FORMS 173

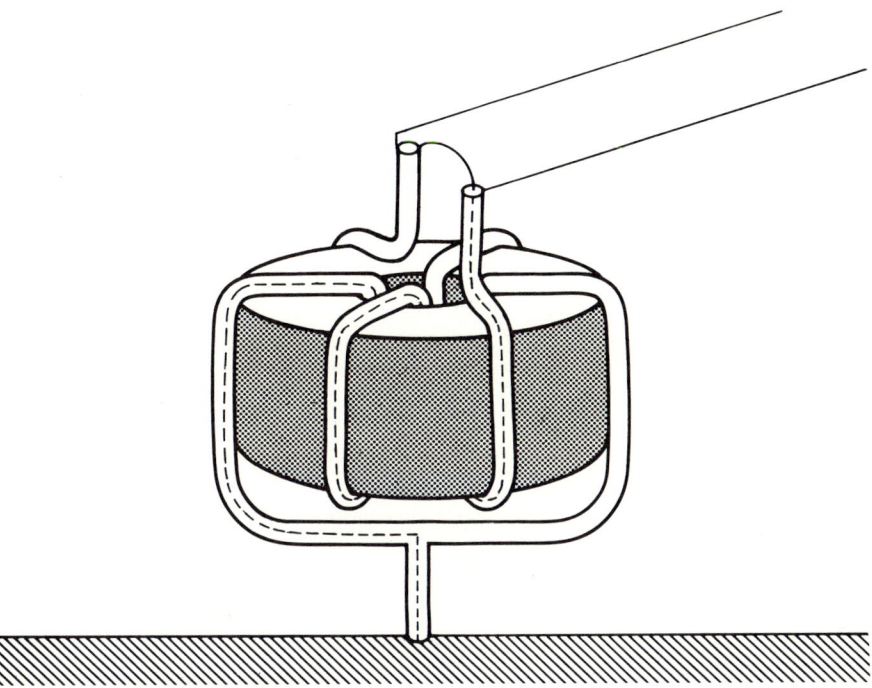

Fig. 4.21 Coiled coax on a ferrite toroid—a wideband balun. The coaxial lines may also be wound on separate toroids.

Problem 4.4 A 75-ohm coaxial line extends from a ground plane vertically upward for a distance of 1 ft, this distance being small compared to a wavelength. There it is joined to a parallel-wire line. The center conductor goes to one wire of the pair and the shield is connected to the other wire of the pair. The characteristic impedance of the parallel-wire line is 300 ohms and has wires which are 3 in. apart in a horizontal plane. The parallel-wire line is matched at a point one-eighth of a wavelength from the coaxial member. The voltage at the output of the coaxial line is 100 volts. Find the current in each side of the parallel-wire line. Design a balun to correct the unbalance (here you may assume a midband frequency of 10 MHz if desired).

4.5.1 IMPEDANCE TRANSFORMERS

An impedance transformation may be incorporated into a balun. To see how this can be done, let us start at the output terminals of the foregoing balun with balanced voltages and currents. Let these leads parallel-branch into a pair of lines as indicated in Fig. 4.22a; clearly, in order that these lines present a matched load to the first line, the characteristic impedance Z_{02} should be twice that of Z_{01}. If this is done, each load

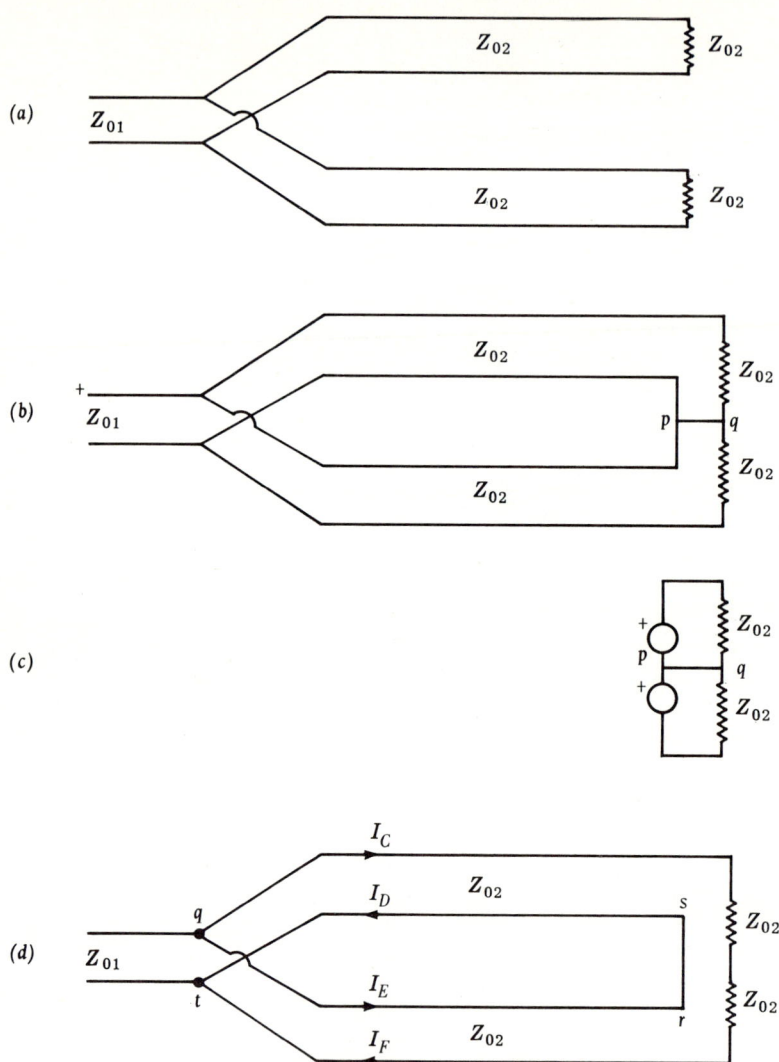

Fig. 4.22 Evolution of four-to-one impedance transformer.

shown has an impedance $2Z_{01}$ when matched. Next consider connecting the two loads together, as in Fig. 4.22b. As far as the load is concerned, it is being driven by generators connected as indicated in Fig. 4.22c. Examination of the system shows that the points p and q are at the same potential; thus the net current in the lead pq is zero; therefore this lead can be removed with no change in the currents. With the lead pq removed, the system takes on the form shown in Fig. 4.22d. Effectively,

WIRE ANTENNAS AND RELATED FORMS 175

the transmission-line output voltages have been connected in series across the two load impedances connected in series. Since each load impedance is itself $2Z_{01}$, the total load impedance is $4Z_{01}$. The device therefore brings about a four-to-one impedance transformation except for one detail. For what we have said to be true, the currents I_C and I_E must be equal and the currents I_D and I_F must be equal and opposite to I_C and I_E. The difficulty is that with the connection as indicated in Fig. 4.22d, if the line lengths are relatively short, then the path $\overline{qrst}$ is a low-impedance path to the first line. This will tend to make the currents I_D and I_E very much greater than the currents I_C and I_F; i.e., the load tends to be short-circuited. If the device is to work properly, we must somehow maintain a high impedance in the path $\overline{qrst}$.

The idealized situation can be better understood by realizing that the parallel branching of the first line at the points qt can equally well be thought of in another way, indicated in Fig. 4.23. Namely, it can be thought of as a branching into a pair of lines with unequal characteristic impedances and unequal loads. That is, there are (at least) four transmission-line modes which can be excited from the point q,t. (Note also that each of the four is initially balanced with respect to ground if q,t is balanced with respect to ground.) In order that the pair of transmission-line modes in Fig. 4.22d will predominate, we must arrange things so that the input impedance to the two modes of Fig. 4.23 is very high. This means that the characteristic impedance and/or the transferred load impedances for these modes must be very high, relative to the desired pair in Fig. 4.22d. For broadband operation, these relative impedance values must be maintained over a wide-frequency band. This can be accomplished by a construction in which the pairs $\overline{CD}$ and $\overline{EF}$ are wound, regularly and closely spaced, in a bifilar configuration into coils, or spirals, or around ferrite cores. This type of construction influences the

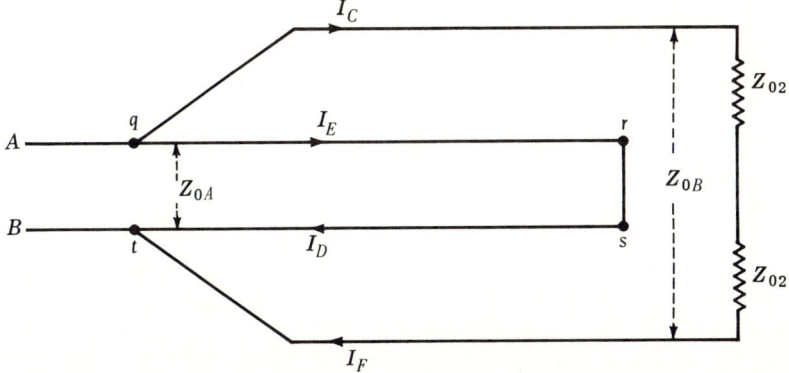

Fig. 4.23 Alternative way of viewing impedance transformer in Fig. 4.22d.

176 ANTENNA ENGINEERING

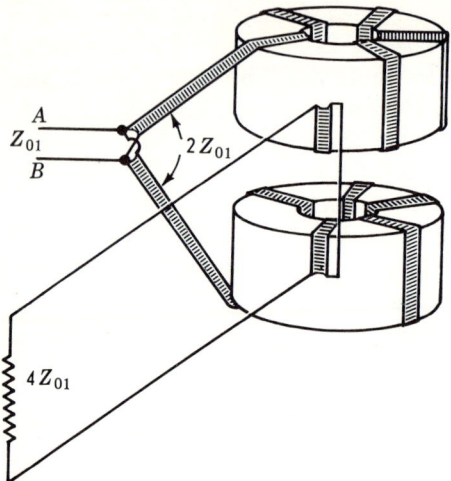

Fig. 4.24 Possible physical layout of series-parallel, parallel-wire-line, four-to-one wideband impedance transformer.

properties of the parallel-wire pairs $\overline{CD}$ and $\overline{EF}$ only slightly. However, it introduces a high series reactance into both sides of the line pairs shown in Fig. 4.23; hence the required high impedance is maintained from some low frequency (determined by the inductance of the coiled configuration) up to the frequency at which the circuit $\overline{ED}$ becomes series resonant. A possible physical layout is suggested in Fig. 4.24.

When this relatively high impedance to the undesired modes is maintained, the desired impedance transformation is accomplished. If the symmetry is preserved, the balance previously assumed on the line AB is not disturbed.

4.5.2 BALUN TRANSFORMERS

Next let us examine the effects if a series-parallel, impedance-transforming device, of the type just discussed, is employed to connect a balanced line to an unbalanced line. The currents at the inputs of the impedance transformer will be unequal if any appreciable current from ground flows up the outside of the coaxial line. To find the currents in and on this impedance transformer, we can split the coaxial excitation into the push-push and push-pull modes discussed earlier. This is indicated in Fig. 4.25. Note that if the currents at the transformer inputs are unequal, it will be, as before, because the push-push currents at the input are appreciable. However, if the structure is carefully constructed to preserve the symmetry, each point at a common distance from the push-push generators is at the same potential above ground. The push-push currents in all lines flow in the same direction; moreover, these currents have to flow through the coils, which present a high impedance to these

WIRE ANTENNAS AND RELATED FORMS   177

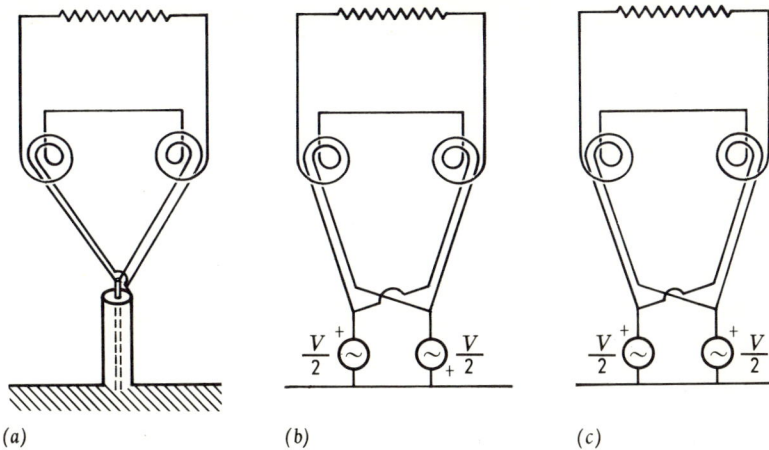

Fig. 4.25 Consideration of balancing action of impedance transformer.

push-push currents (in fact, the coils may be wrapped around a ferrite core to further increase the impedance). As mentioned earlier, this high impedance to the push-push mode is one of the requirements for a balun; the impedance-transforming device may therefore have an appreciable balancing effect and, in effect, be an impedance-transforming balun. The balancing feature will exist provided the capacitance to ground of the gross coil structure in Fig. 4.25c is not too large and provided the two line pairs are symmetric with respect to ground.

There are other ways to make an impedance-transforming balun that can have a more definite and controllable balancing feature. One device is based upon the balun of Sec. 4.5 and the impedance transformer of Sec. 4.5.1. The structure is indicated in Fig. 4.26. In this figure, the coaxial section leading in (labeled Z_{01}) has half the characteristic impedance of the coaxial sections L and M; the third line N is just a dummy

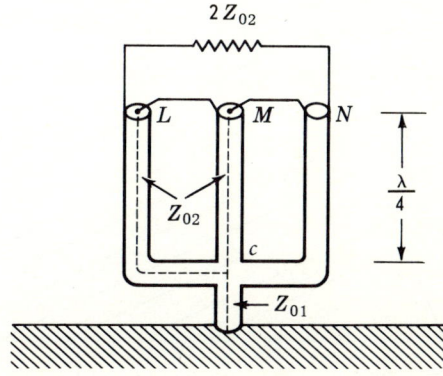

Fig. 4.26 "Candelabra" balun—a four-to-one impedance-transforming balun.

178   ANTENNA ENGINEERING

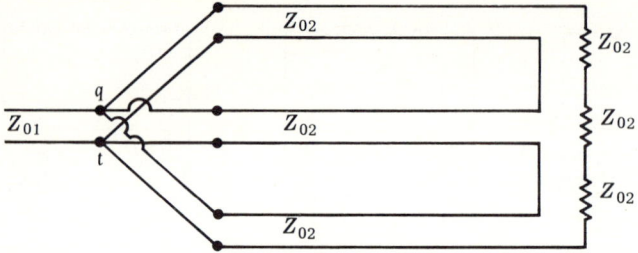

Fig. 4.27 A nine-to-one parallel-wire line, series-parallel impedance transformer.

inserted to achieve the symmetry required for balance (it has the same outer diameter as the others). The ends of the coaxial sections L and M are connected in series across the load. The system is matched when the load impedance is twice the characteristic impedance Z_{02} of lines L and M, which means that it is four times the characteristic impedance of the input line Z_{01}. The balance is perfect for any load or frequency.

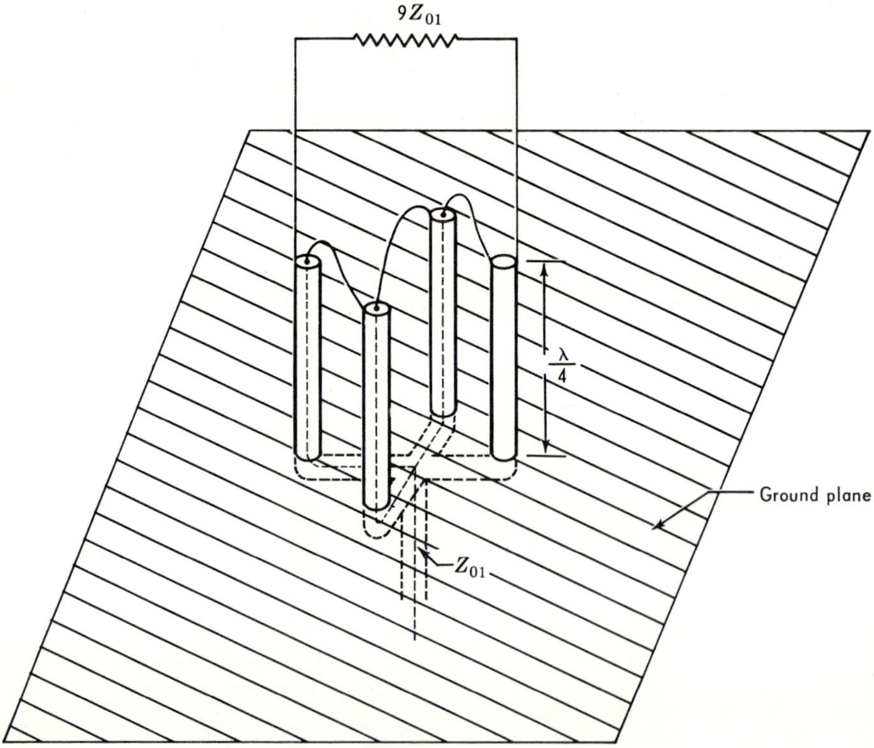

Fig. 4.28 A nine-to-one impedance-transforming candelabra balun.

However, the load impedance is shunted by the two exterior line sections that are shorted at the junction c. For narrowband operation, the length of the shorted line sections $\overline{Mc}$ may be made equal to a quarter wavelength at the operating frequency. As with the balun and transformer described earlier, wideband performance is obtained by designing the structure so that the impedance of the shorted line sections $\overline{Mc}$ remains at a high value over the entire frequency band. The techniques described earlier are applicable here. The coaxial lines L, M, and N may be coiled or wound on individual ferrite cores, provided the frequency range is such that a good ferrite material is available. Of course, this coiling or winding must be done in a way that preserves the symmetry.

Other impedance-transformation ratios may also be achieved by the series-paralleling of more lines. Parallel-wire lines may be wound bifilarly in the pairs indicated in Fig. 4.27. Here, $Z_{02} = 3Z_{01}$ so that the load impedance that is matched is nine times that of Z_{01}. The low-impedance paths must be avoided as in the four-to-one transformer; the problem is more serious here because there are two potentially troublesome low-impedance paths in shunt. A nine-to-one balun transformer, with three coaxial lines and a dummy, is sketched in Fig. 4.28. Bandwidth may be obtained in the manners described earlier.[1]

4.5.3 ANTENNA FEEDERS

The main reason for the lengthy discussion on methods of joining coaxial lines to balanced parallel-wire lines (in a book on antennas) is that the same ideas and devices are applicable to antenna excitation; however, the explanations for transmission lines are easier for most people to understand at first sight. A dipole antenna in space, or a horizontal dipole over ground, is a balanced device. Many other more complicated structures, such as Yagi arrays, and log-periodic dipole arrays, are also balanced structures. If such antennas are to be excited with coaxial lines, some sort of balun is required to maintain the balance and to prevent currents from flowing along the outside of the coaxial feeder line. All of the devices described in Secs. 4.5, 4.5.1, and 4.5.2 may be employed for exciting balanced antennas by means of unbalanced lines, with or without impedance transformation.

The basic narrowband structures and some of their variations are shown in Fig. 4.29. The split coax balun, Fig. 4.29c, is a compact construction. Here the coaxial feeder is split and a portion of the shield is snipped out on opposite sides. The exteriors of the remaining portions of the shield act as the shorted, quarter-wavelength parallel-wire line that is formed by the exteriors of the coaxials in Figs. 4.29a and b. The

---

[1] For a description of a completely different type of broadband balun transformer, see the paper by J. W. Duncan and V. P. Minerva, *Proc. IRE*, **48**:156 (February, 1960).

180 ANTENNA ENGINEERING

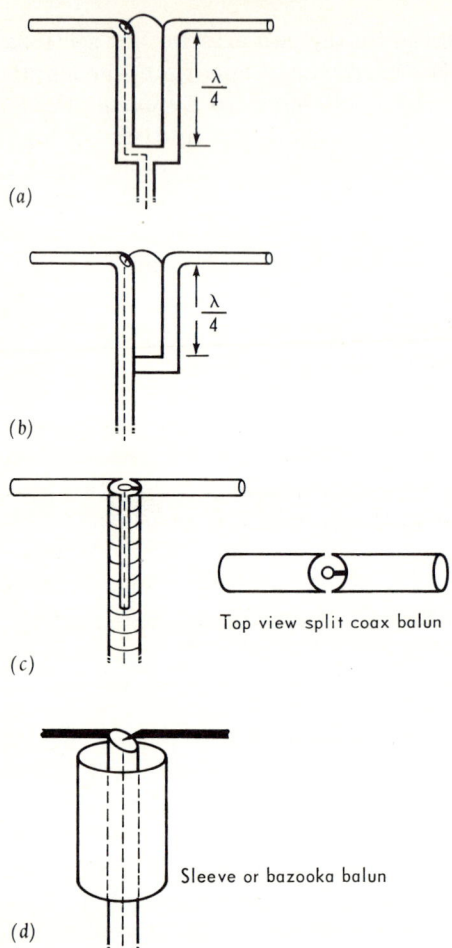

(a)

(b)

(c) Top view split coax balun

(d) Sleeve or bazooka balun

Fig. 4.29 One-to-one baluns for connecting balanced antennas to coaxial lines.

center conductor of the coaxial line is simply joined to the shield on one side of the split. The variation in Fig. 4.29d, known as the bazooka balun, provides the high exterior impedance by means of a shorted coaxial line in the form of a sleeve around the feeding coaxial. The bandwidth of the structures in Figs. 4.29a and b may be increased by incorporating the broadband methods for suppressing the undesired external currents (coiling, winding on a ferrite) described in the foregoing section on baluns.

4.6 FOLDED DIPOLES

As was pointed out in Sec. 4.5, if a parallel-wire line becomes unbalanced, it radiates. This fact is often utilized deliberately to make an antenna; the advantages of such antennas are that the input impedance levels are usually higher, and the operating bandwidth is increased slightly. One of the most common and most useful structures of this type is known as the folded dipole (see Fig. 4.30). One-half of the structure may be mounted above a ground plane so as to form a folded monopole.

The folded dipole is basically an unbalanced (and therefore radiating) transmission line. To see this, suppose that a parallel-wire line is excited in the center of one side of the line but not in the other side. This constitutes an unbalanced excitation. To find the currents and impedance presented to the generator, the unbalanced source is replaced by a pair of equivalent sources, one in the push-push configuration, and one in the push-pull, as indicated in Fig. 4.31. First, consider the push-pull or normal transmission-line excitation. For simplicity, let the terminating impedances Z_L be the same on both ends; these impedances transform along the line so that at the source position they have a new value which we shall call Z_I, that is,

$$Z_I = Z_0 \frac{Z_L + Z_0 \tanh \gamma L}{Z_0 + Z_L \tanh \gamma L}$$

It follows that the transmission-line current is $I_t = V/2Z_I$. Next consider the push-push excitation. Since the two lines and generators are

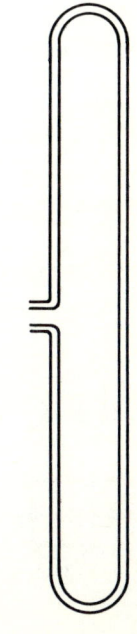

Fig. 4.30 Folded dipole made of tubing.

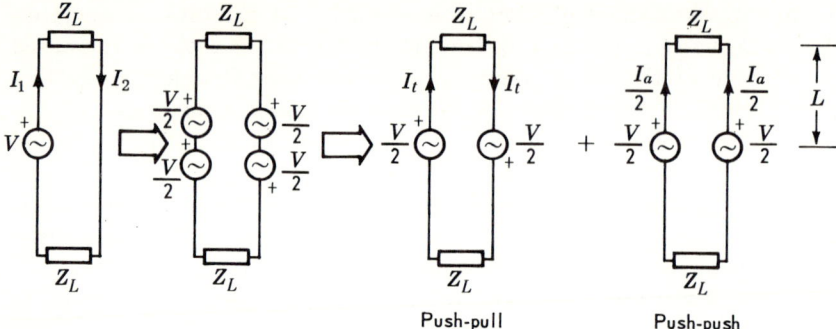

Fig. 4.31 Decomposition of unbalanced excitation of a parallel-wire line into conventional balanced excitation plus push-push excitation.

identical, at points equidistant from the generators the potential on each wire is the same. Consequently, wires could be run across from one line to the other with no change in the voltage or current. In the push-push mode the voltage generators are clearly in parallel, so we may as well think of them as a single generator, as suggested in Fig. 4.32. With this type of excitation, no current flows through the loads Z_L so these are immaterial to the push-push (or antenna) currents I_a. The push-push generators excite the antenna consisting of the double-wire, center-driven dipole. The antenna current is $I_a = V/2Z_a$, where to a first approximation, Z_a is the input impedance of a thin-wire antenna of the same length as the double-wire structure (a better approximation may be obtained by employing an equivalent L/d ratio for the double-wire structure).

With the problem broken down into balanced and unbalanced parts, it is clear that the currents in the two sides of the original parallel-wire line are unequal. The current in the left side of the line is $I_1 = I_t + \frac{1}{2}I_a$,

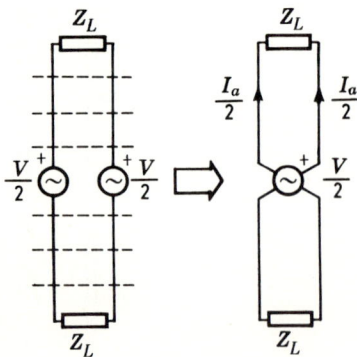

Fig. 4.32 Push-push excitation equivalent to single generator exciting an antenna comprised of two wires, each carrying one-half the current.

while that in the right side is $I_2 = I_t - \frac{1}{2}I_a$. The portion of the current I_a radiates the pattern characteristic of that particular current distribution.

Having the currents, we can now obtain a formula for the input impedance or admittance:

$$Y = \frac{I_1}{V} = \frac{I_t + \frac{1}{2}I_a}{V}$$

wherein the transmission-line current is $I_t = \frac{1}{2}VY_I$ and the antenna current is $I_a = \frac{1}{2}VY_a$. The input admittance is therefore $Y_\text{in} = \frac{1}{2}Y_I + \frac{1}{4}Y_a$, so that the input impedance is $Z_\text{in} = 4Z_I Z_a/(Z_I + 2Z_a)$. These equations show that the unbalanced excitation brings about a transformation in the antenna input impedance. The quantity Z_I can be made very large if desired, for example by making Z_L a short circuit and making each half of the line a quarter wavelength long. Note that if Z_I is very high, then the input impedance is approximately $Z_\text{in} = 4Z_a$. This fourfold increase in input impedance is often convenient in practice because for example it transforms the 70 ohms characteristic of a reso-

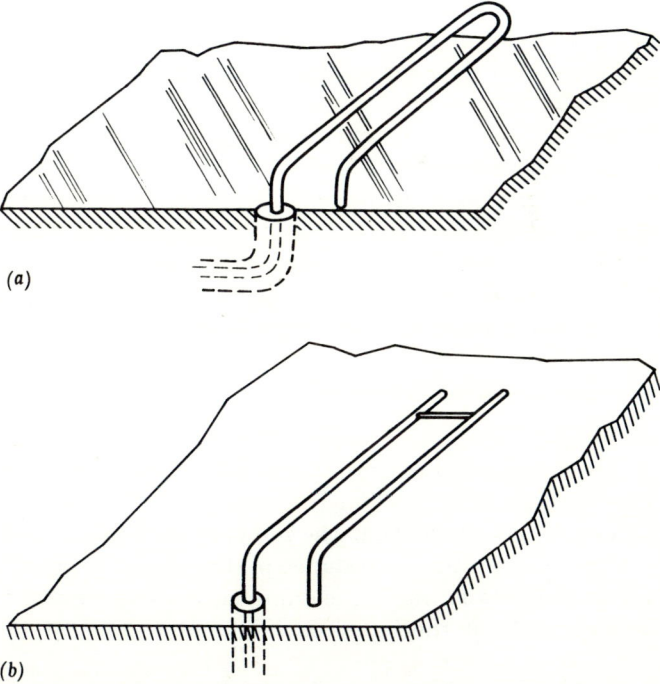

Fig. 4.33 Bent folded monopole antennas.

nant half-wave dipole into a value that is fairly well matched to a standard 300-ohm "twin-lead" transmission line.

There are other types of folded dipoles and related structures, from which a variety of impedance values are obtainable. For example, two sides of the folded dipole need not have the same diameter; or, three or more legs may be incorporated into the structure so as to obtain an impedance transformation of nine or more. Details are available in the literature.[1,2] Of course, one-half of a folded dipole structure may be operated above a ground plane so as to form a folded monopole type of structure. Monopoles of this type may be bent over if necessary to conserve height (with, consequently, a lower input impedance). See Fig. 4.33.

Problem 4.5 Employing ½-in. OD tubing at a spacing of 3 in., design a bent folded monopole antenna (Fig. 4.33b) which is as low as possible but has nonetheless an input impedance of 50 ohms (resistive). The frequency range is such that the spacing is small compared to a wavelength.

4.7 MUTUAL EFFECTS

When wire antennas are placed near one another, as they must be in order to form arrays of the types discussed in Chap. 3, their electrical characteristics change somewhat. This is so because the current in a given element depends on more than simply the source at its own input terminals; it depends on the current in neighboring antennas as well. In fact, in resonant length elements, there may be a substantial current in an element in which there is no primary source at all (these elements are called *parasitic* elements). In order to obtain a given pattern, the designer must take into account the interactions between antennas. By analogy with circuits, these interactions are called mutual effects. Unfortunately, at the time of this writing, there is no completely satisfactory way of calculating mutual effects in general.

4.7.1 MUTUAL-IMPEDANCE CALCULATIONS

As long as wire antennas are not substantially longer than a half wavelength, the current distribution produced by a source within a given wire-like element is pretty much sinusoidal. The main problem is then to determine the currents at the input of each of the elements.

Given the mutual-impedance parameters, the procedure for finding the input currents is much the same as the corresponding circuit problem. For example, if the input current and voltage at the nth element are

[1] Y. Mushiake, *IRE Trans.*, **AP-3**:163 (October, 1954).
[2] C. W. Harrison and R. W. P. King, *IRE Trans.*, **AP-9**:171 (March, 1961).

designated, respectively, I_n and V_n, the input impedance is

$$Z_{\text{in},n} = \frac{V_n}{I_n} = \frac{1}{I_n} \int_{\text{input}} \mathbf{E}_n \cdot \mathbf{dl}_n$$

The total field at the nth element $\mathbf{E}_n$ is the sum of the field produced by the current in the nth element $\mathbf{E}_{sn}$ plus the field $\mathbf{E}_{mn}$ generated by the currents in all of the other elements:

$$\mathbf{E}_n = \mathbf{E}_{sn} + \mathbf{E}_{mn}$$

$$\mathbf{E}_{mn} = \sum_{p=1}^{n-1} \mathbf{E}_{pn} + \sum_{p=n+1}^{N} \mathbf{E}_{pn}$$

or

$$\mathbf{E}_n = \sum_{p=1}^{N} \mathbf{E}_{pn}$$

where $\mathbf{E}_{pn}$ is the field at the nth element produced by the current in the pth element. The expression for the input impedance may then be written

$$Z_{\text{in}} = \frac{1}{I_n} \int \sum_{p=1}^{N} \mathbf{E}_{pn} \cdot \mathbf{dl}_n$$

From the fundamental definition of the mutual-impedance quantity

$$Z_{pn} \equiv \frac{1}{I_p} \int_{\text{input of } n} \mathbf{E}_{pn} \cdot \mathbf{dl}_n \tag{25}$$

it follows that we can write the equation for the input impedance in the form

$$Z_{\text{in},n} = \frac{1}{I_n} \sum_{p=1}^{N} I_p Z_{pn} \tag{26}$$

The current at the input of the nth element is then

$$I_n = \frac{V_n}{Z_{\text{in},n}}$$

In practice, as in circuit analysis, it is usually necessary to solve simultaneously for the set of N input currents, by means of the set of equations equivalent to Eq. (26):

$$V_n = \sum_{p=1}^{N} I_p Z_{pn} \tag{27}$$

i.e.,

$$V_1 = I_1 Z_{11} + I_2 Z_{12} + I_3 Z_{13} + \cdots$$
$$V_2 = I_1 Z_{21} + I_2 Z_{22} + I_3 Z_{23} + \cdots$$
$$\cdots\cdots\cdots\cdots\cdots\cdots\cdots\cdots$$

Of course to make use of the equations, we must have available the mutual impedances Z_{pn}, determined either from field theory or from experiments.

There is a basic difficulty hidden in these equations. It stems from the conventional definition of mutual impedance, Eq. (26), which is satisfactory for lumped circuits but inadequate in general. For with antennas and other extended systems, the integral $\int \mathbf{E}_{pn} \cdot \mathbf{dl}_n$, giving the voltage at the nth element due to the currents in the pth element, may be nonzero even when the current at the input of antenna p, I_p, is zero. In this case according to Eq. (25), Z_{pn} is infinite; in general, it is not unique. The physical reason is that the currents in the nth element may induce appreciable currents in the pth structure even though the terminals of the pth structure are open; these induced currents then in turn react back on the nth structure, inducing a voltage. These second-order effects are difficult to calculate, but of course they are present in experimental determinations of the impedance quantities in Eq. (27). For example, if Z_{11} is determined by open-circuiting all terminals except the first and measuring the ratio V_1/I_1, then Z_{11} includes the effects of the induced currents in any and all of the other elements that can flow with the terminals open. Thus, the quantity Z_{11} *is not exactly the isolated self-impedance, and in general it has a different value for each member of each different array, even though the individual elements may be identical.* Similarly, Z_{ij} is slightly different as elements i and j take up different positions in the same array, or in different arrays. There seems to be no easy way out of these difficulties. The only potentially accurate method known is the simultaneous solution of a set of integral equations for the currents; this is a new problem for each array with different numbers and/or positions of elements. To actually obtain a solution is quite difficult.

Frequently, in practice, the effects discussed above are ignored. Z_{ii} is taken to be the isolated self-impedance, and Z_{ij} the mutual impedance as if no other elements were present. Unfortunately, ignoring a difficulty does not eliminate it.

Consider the derivation of a useful formula for the calculation of the mutual impedance between a pair of wire antennas. It is based on the conventional definition of mutual impedance, namely, the ratio of the voltage induced in one antenna to the current at the input of the other. Let us label the individual antennas in the pair a and b; also let us select the form of excitation for maximum convenience and minimization of second-order effects. At the input to a we can place a point ideal-current generator $\mathfrak{J}_a$, so as to be able to conveniently determine the open-circuit voltage there. At the input to antenna b we can place an elemental magnetic current ring $\mathfrak{M}_b$, as the field-theory equivalent of a constant-voltage generator; this type of excitation places a short circuit at the terminals of antenna b so that any induced second-order currents will tend to have the same current distribution as that produced by the primary excitation, and hence will influence the input current. These sources interact according to the reciprocity theorem as follows:

$$\int \mathbf{E}_b \cdot \mathfrak{J}_a \, dV = -\int \mathbf{H}_a \cdot \mathfrak{M}_b \, dV$$

or since $\mathfrak{J}_a$ is a point source

$$\int \mathbf{E}_b \cdot \mathfrak{J}_a \, dV = V_{oc,b}(a) I_a$$

where $V_{oc,b}(a)$ is the open-circuit voltage at a produced by antenna b. Now imagine replacing the metal of antenna b by a distribution of currents $\mathfrak{J}_{ib}$ in space, without the metal. The distribution $\mathfrak{J}_{ib}$ is identical to that of the induced currents $\mathbf{J}_{ib}$ on the metal of antenna b. The field produced by $\mathfrak{M}_b$ plus $\mathfrak{J}_{ib}$ is identical to that produced by $\mathfrak{M}_b$ plus the metal of antenna b; that is,

$$\int \mathbf{E}_{bi} \cdot \mathfrak{J}_a \, dV = V_{oc,b}(a) I_a$$

According to the reciprocity theorem

$$\int \mathbf{E}_{bi} \cdot \mathfrak{J}_a \, dV = \int \mathbf{E}_{a0} \cdot \mathfrak{J}_{ib} \, dV - \int \mathbf{H}_{a0} \cdot \mathfrak{M}_b \, dV$$

The last term in the foregoing equation is usually negligible since the field H_{a0} is not large except for very small spacings; the magnetic ring is small and in an almost uniform field. Combining equations we obtain a formula for the mutual impedance:

$$\pm Z_{12} = \frac{V_{oc}(a)}{I_b} = \frac{1}{I_a I_b} \int \mathbf{E}_{a0} \cdot \mathfrak{J}_{ib} \, dV \tag{28}$$

where $\mathbf{E}_{a0}$ is the field produced by antenna a in free space (i.e., with the metal of antenna b removed), and I_a and I_b are the currents at the inputs of the antennas. The sign depends upon a polarity-mark convention, the negative sign being the more common.

The foregoing mutual-impedance formula will give an exact value for the mutual impedance if the integration is performed with the correct current distribution $\mathfrak{J}_{ib}$ and the correct field quantity $\mathbf{E}_{a0}$. But in practice, neither the current distribution nor the field is known exactly; consequently, the integral cannot be evaluated exactly. In practice, the integral is usually evaluated by (1) assuming a sinusoidal form for $\mathfrak{J}_{ib}$, (2) computing an approximation for $\mathbf{E}_{a0}$ based on an assumption of a sinusoidal distribution of current on antenna a. With these approximations, the integral for Z_{12} can be evaluated. Fortunately, this formula is stationary with respect to errors in the assumed current distributions, so that useful approximations can be obtained with the sinusoidal current assumptions.

Even with this useful approximate formula, the amount of calculation that is involved to describe mutual-impedance effects is discouragingly large, since the results depend upon antenna lengths (which may be different in a continuum of ratios) and orientations (in which also the number of possibilities is infinite). The results for a pair of thin antennas, a half wavelength long, in the two most important orientations, are shown in Fig. 4.34. A few other situations are covered by the data displayed in Fig. 4.35. Note that the mutual-impedance forms for identical side-by-side antennas at close spacings approach the values of the self-impedances. In fact, the mutual-impedance forms may be employed to obtain useful values for the self-impedance of wire antennas.

As an example, let us consider the mutual impedance between a pair of short antennas that have uniform currents along their lengths. With unit currents at both inputs we have

$$Z_{12} = -\int \mathbf{E}_{a0} \cdot \mathcal{J}_{ib} \, dV$$
$$\doteq -\int E_{a0} \, dz$$

To proceed, we must have an expression for the near field E_{a0} produced by a short uniform current element. Let the currents be directed along the z axis and be separated by a distance s. Then, as long as the lengths of the elements are small in comparison to s and to the wavelength, their fields may be calculated by employing the field expressions for elemental currents. Also, if the centers of the elements lie in the $z = 0$ plane, then E_z may be approximated by $-E_0\left(\dfrac{\pi}{2}\right)$:

$$E_{a0,z} = -E_\theta\Big]_{\theta=\frac{1}{2}\pi}$$
$$= -\frac{e^{-j\beta r}}{4\pi} l_1 \left[\frac{j\omega\mu}{r} + \left(\frac{\mu}{\epsilon}\right)^{\frac{1}{2}} \frac{1}{r^2} + \frac{1}{j\omega\epsilon r^3} \right]$$
$$= -j\left(\frac{\mu}{\epsilon}\right)^{\frac{1}{2}} \frac{\beta^2 e^{-j\beta r}}{4\pi} l_1 \left[\frac{1}{\beta r} - \frac{j}{(\beta r)^2} - \frac{1}{(\beta r)^3} \right]$$

If we assume that this field is constant over the length of the small antennas, and that r is approximately equal to s, then the result for the mutual impedance between the short antennas of lengths l_1 and l_2 is

$$Z_{12} = j30 \frac{l_1 l_2}{\lambda^2} 4\pi^2 e^{-j\beta s} \left[\frac{1}{\beta s} - \frac{j}{(\beta s)^2} - \frac{1}{(\beta s)^3} \right]$$

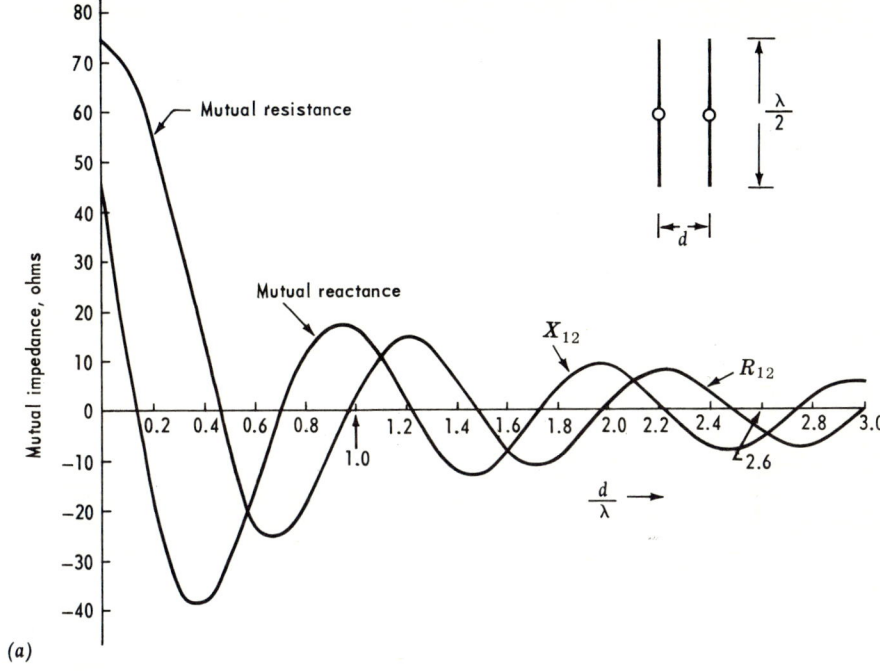

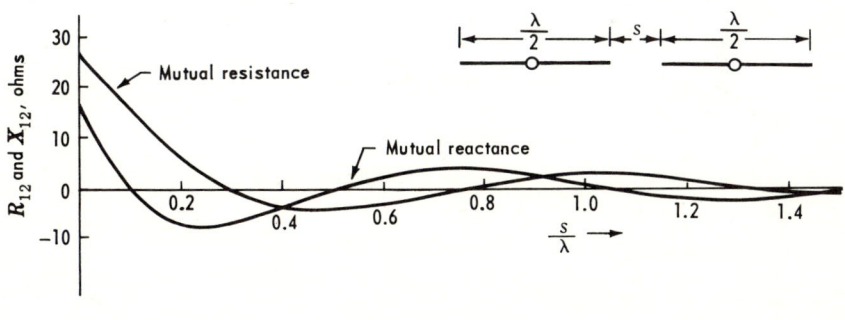

Fig. 4.34 (a) Mutual impedance between half-wave antennas placed side by side, as function of separation in wavelengths; (b) mutual impedance between half-wave antennas arranged collinearly, as function of separation in wavelength.

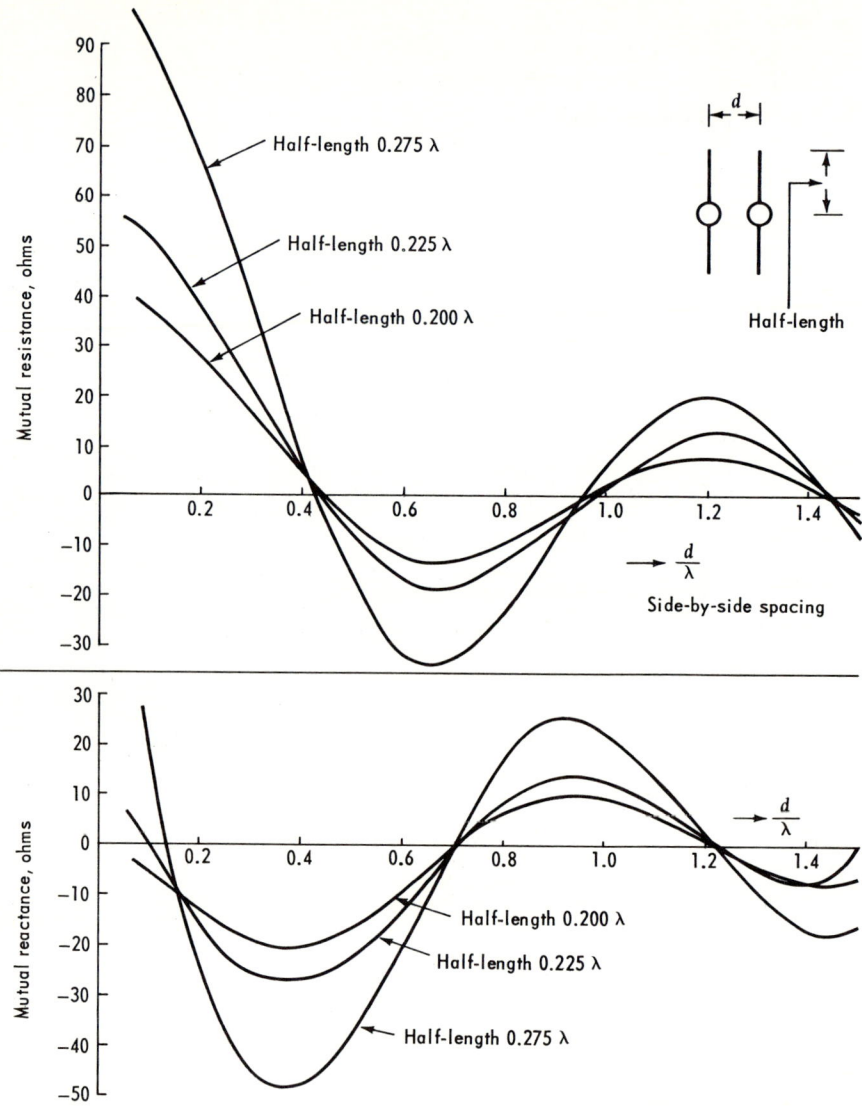

Fig. 4.35 Mutual impedance of thin dipoles, side by side, as a function of spacing in wavelengths.

As an example of this result consider a transmission-line-loaded small antenna, such as discussed in Chap. 2. Let the line separation be d, the line length a half wavelength, $L = \lambda/2$, $d \ll L$, and let the system be fed at one end, with a shorting wire at the other end of the line (Fig. 2.3a). Suppose that the radiation is coming primarily from the vertical wires.

In this particular case, since $s = L$, $\beta s = \pi$, so the mutual impedance between the two vertical elements is

$$Z_{12} = -120 \left(\frac{d}{\lambda}\right)^2 \left[1 + j\left(\pi - \frac{1}{\pi}\right)\right]$$

$$= -120 \left(\frac{d}{\lambda}\right)^2 (1 + j2.8) \quad \text{ohms}$$

The input impedance of each of the elements is identical if the currents are identical and for example

$$Z_{in,1} = \frac{V_1}{I_1} = Z_{11} + Z_{12}\frac{I_2}{I_1}$$

At the input to the line (since the solution for the current has the form $I(z) = C_1 e^{j\beta z} + C_2 e^{-j\beta z}$), the current is $I(0) = C_1 + C_2$, while at the end of the line $L = \lambda/2$, it is $I(\lambda/2) = -C_1 - C_2$. Thus in view of the change in reference direction for the current in the load wire, we see the result $I_{in} = I_{load}$. It follows that at either element the input impedance to the element is $Z_{in} = Z_{11} + Z_{12}$. The total input impedance seen by a generator at the input end is (assuming that the transformation of impedance is that along a half-wavelength lossless transmission line)

$$Z_{in}(0) = 2(Z_{11} + Z_{12})$$

or the real part of the input impedance is

$$R_{in}(0) = 2(R_{11} + R_{12})$$

$$= 2\left[80\pi^2 \left(\frac{d}{\lambda}\right)^2 - 120\left(\frac{d}{\lambda}\right)^2\right]$$

Next let us investigate the directive gain of this two-element array. It will be simpler if we first get the directivity of the array relative to a single element of the same type. From the definitions

$$\frac{G_{d,array}}{G_{d,single}} = \frac{PD_{array}}{PD_{single}} \frac{(I_{in}^2 R)_{single}}{(I_{in}^2 R)_{array}}$$

With the same currents, the fields produced by the array in the direction of the maximum are twice as great as those of a single element, so the power density PD is four times as great. From the results above, the ratio of the resistances is

$$\frac{R_{single}}{R_{array}} = \frac{80\pi^2 (d/\lambda)^2}{2[80\pi^2 (d/\lambda)^2 - 120(d/\lambda)^2]} = \frac{1}{2(1 - 120/80\pi^2)}$$

which gives for the directive gains

$$\frac{G_{d,array}}{G_{d,single}} = \frac{4}{2(1 - 1.5/\pi^2)} = 2.4$$

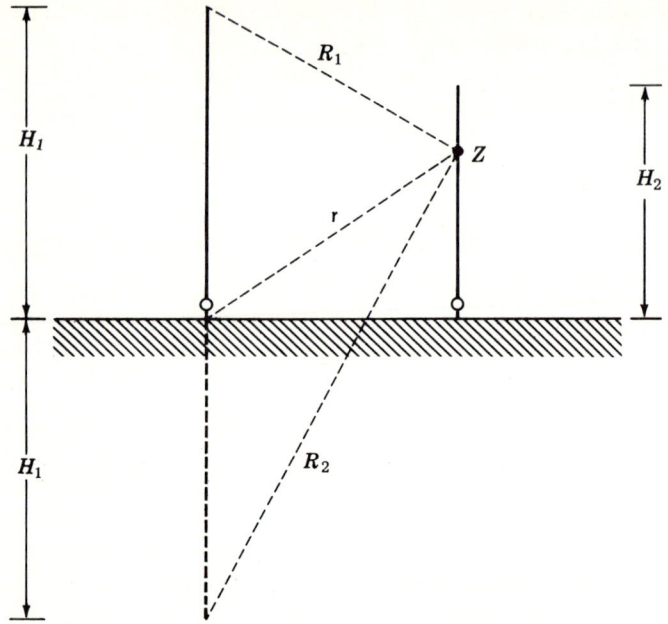

Fig. 4.36 Geometry for mutual-impedance calculation.

It will be recalled that the directive gain of a single element is 1.5, so the directivity of this two-element array is 3.6.

Problem 4.6 Find the mutual impedance and directive gain of a collinear couplet of elemental currents at half-wavelength spacing.

Problem 4.7 Find the mutual impedance between a quarter-wavelength monopole over a perfect ground and a nearby very short vertical monopole over ground.

As a second example, consider the calculation of the mutual impedance between a pair of monopoles over ground, having sinusoidally distributed currents $I_a = I_{m1} \sin \beta(H_1 - z)$, $I_b = I_{m2} \sin \beta(H_2 - z)$. The field at an arbitrary position z at antenna 2 due to the current in antenna 1 is given in Eq. (16). The formula for the mutual impedance is therefore (see Fig. 4.36)

$$Z_{12} = \frac{I_{m1} I_{m2}}{I_1 I_2} j30 \int_0^{H_2} \left[\left(\frac{e^{-j\beta R_1}}{R_1} + \frac{e^{-j\beta R_2}}{R_2} - 2 \cos \beta H_1 \frac{e^{-j\beta r}}{r} \right) \cdot \sin \beta(H_2 - z) \right] dz$$

Recall that I_1 and I_2 are input currents while I_{m1} and I_{m2} are currents at the "loops," which means $I_1/I_{m1} = \sin \beta H_1$, $I_2/I_{m2} = \sin \beta H_2$.

This mutual-impedance integral may be evaluated numerically with the aid of a digital computer. Alternatively, it can be reduced to a formula in terms of tabulated functions. It will be noticed that the mutual-impedance integral has essentially the same form as the integral for the self-impedance, page 159. The integrations may be carried out in the same general way as those described on pages 159–160. The final result, in the special case that the monopoles have the same height and are a distance d apart, can be put into the form

$$R_{12} = \frac{30}{\sin^2 \beta H} \bigg\{ \sin \beta H \cos \beta H [\text{Si}(u_2) - \text{Si}(v_2) - 2\text{Si}(v_1) + 2\text{Si}(u_1)]$$
$$- \frac{\cos 2\beta H}{2} [2\text{Ci}(u_1) - 2\text{Ci}(u_0) + 2\text{Ci}(v_1) - \text{Ci}(u_2) - \text{Ci}(v_2)]$$
$$- [\text{Ci}(u_1) - 2\text{Ci}(u_0) + \text{Ci}(v_1)] \bigg\}$$

$$X_{12} = \frac{-30}{\sin^2 \beta H} \{ \sin \beta H \cos \beta H [2\text{Ci}(v_1) - 2\text{Ci}(u_1) + \text{Ci}(v_2) - \text{Ci}(u_2)]$$
$$- \tfrac{1}{2} \cos 2\beta H [2\text{Si}(u_1) - 2\text{Si}(u_0) + 2\text{Si}(v_1) - \text{Si}(u_2) - \text{Si}(v_2)]$$
$$- [\text{Si}(u_1) - 2\text{Si}(u_0) + \text{Si}(v_1)] \}$$

where

$u_0 = \beta d$
$u_1 = \beta[(d^2 + H^2)^{\frac{1}{2}} - H]$
$v_1 = \beta[(d^2 + H^2)^{\frac{1}{2}} + H]$
$u_2 = \beta[(d^2 + 4H^2)^{\frac{1}{2}} + 2H]$
$v_2 = \beta[(d^2 + 4H^2)^{\frac{1}{2}} - 2H]$

$\text{Si}(x) = \int_0^x \frac{\sin u}{u} \, du$

$\text{Ci}(x) = \int_\infty^x \frac{\cos u}{u} \, du$

Some specific values are plotted in Figs. 4.34a and 4.35.

Problem 4.8 An antenna system consists of two quarter-wave monopoles above a perfect ground, with a spacing of a quarter wavelength. The effective length-to-diameter ratio is 100:1. The object is to radiate 10 kW in a cardioidlike pattern, so the antennas are to be fed with currents which are equal in amplitude but 90° out of phase. Find the magnitude and phase of the required driving-point voltages for the two antennas.

Problem 4.9 A linear array is to be made up of three quarter-wavelength vertical monopoles above ground, $l/d = 100$, spaced at quarter wavelength. These radiators are to have equal-amplitude currents that have a linear progressive

194 ANTENNA ENGINEERING

phase shift from one to the next of 120°. (1) Design a feeding system to give the desired currents. (2) Sketch the radiation pattern in the horizontal plane. (3) Estimate the efficiency, and calculate the field strength at 1 mile in the direction in which it is a maximum for 1 kW of power output from the transmitter (1 mile is in the distant field).

4.7.2 IMPEDANCE OF ANTENNAS HELD ABOVE GROUND

Frequently it is necessary to mount horizontal antennas in the vicinity of the earth, or other ground plane. Both the pattern and the input impedance are influenced by the proximity to ground. The effects of the ground in both cases can be accounted for by the introduction of an image. If the antenna is at height h, the field everywhere above the ground level is the same as would be produced by the original antenna plus an image antenna located at a distance h below ground level. The current in the image antenna is equal and opposite to the current in the actual antenna; this means that the input impedance to the elevated antenna is $Z_{\text{in}} = Z_{11} - Z_{12}$, where Z_{12} is the mutual impedance between

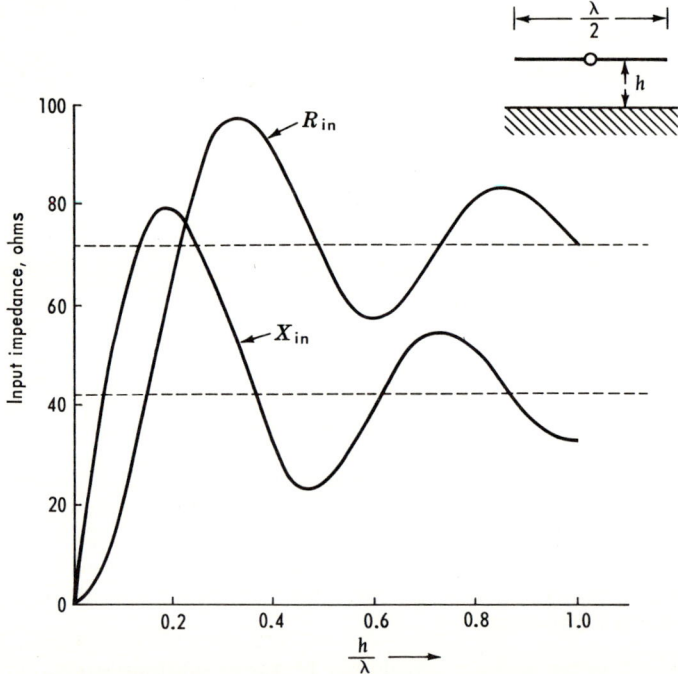

Fig. 4.37 Input impedance of horizontal half-wave antenna over perfect ground, as function of height in wavelengths.

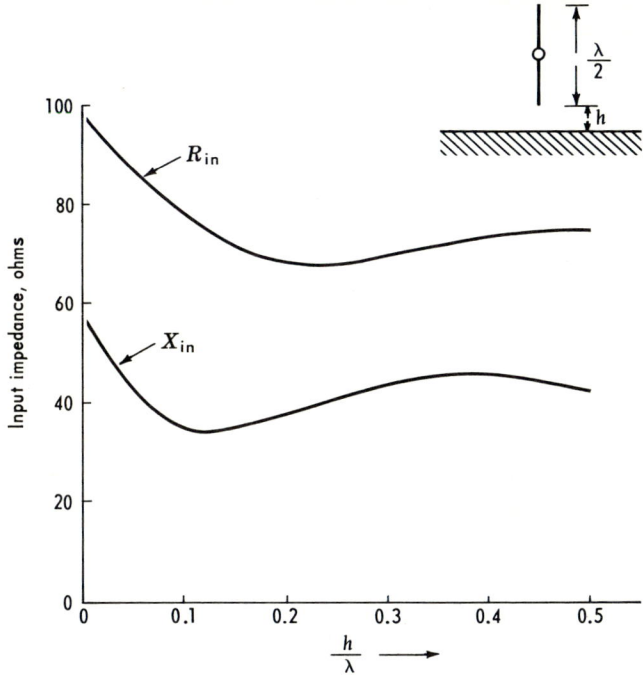

Fig. 4.38 Input impedance of vertical half-wave antenna over perfect ground, as function of height above ground in wavelengths.

like antennas at distance $2h$. The impedance data for a horizontal half-wavelength dipole above ground is displayed in Fig. 4.37.

Similar effects occur when vertical antennas are mounted above ground, but the variation in impedance for vertical dipoles supported above ground is very much less than for horizontal ones except for very close spacing. The vertical image current is in phase with the current in the actual antenna so the input impedance is $Z_{in} = Z_{11} + Z_{12}$, where if the center of the original antenna is a height h above ground, then Z_{12} is the mutual impedance between like collinear antennas with a center-to-center spacing of $2h$. Impedance data for vertical half-wavelength dipoles over ground are plotted in Fig. 4.38.

Problem 4.10 Study the directivity of a half-wavelength dipole lying parallel to the surface of a large plane reflector. Consider the direction in the plane of the antenna and its image, with antenna-plane separations of a quarter wavelength and less.

4.7.3 PARASITIC ELEMENTS AND YAGI ARRAYS

The feeding systems to produce the required currents in large arrays can become very complex. This fact led antenna designers early into the study of arrays with parasitic elements, in which only one or a few of the array elements are connected to the primary excitation. The magnitude and the phase of the currents in the parasites depends on their sizes and spacings with respect to the driven element. The actual relative currents may be determined from the circuitlike equation, (27), together with mutual impedance data. For example, if an array consists of a driven element and a single parasite, the equations that determine the currents are

$$V = I_1 Z_{11} + I_2 Z_{12}$$
$$0 = I_1 Z_{21} + I_2 Z_{22}$$

From the second equation, the currents are related by

$$I_2 = -I_1 \frac{Z_{21}}{Z_{22}}$$

Thus, given the spacing and self-impedance data, the relative currents can be obtained and the radiation pattern calculated in a straightforward fashion. Many possibilities of relative currents are represented by the ratio Z_{21}/Z_{22}. One fact is clear from physical considerations; namely, if I_2 (parasite current) is to be comparable in magnitude to the driven current I_1, the parasite must be close to the driven element. A detailed study shows that a single parasitic element may act either as a *reflector* or a *director* depending on spacing and relative size. For example, if the driven element is a half-wave dipole, a parasite acts as a reflector when it is spaced in the range from 0.05 to 0.15 wavelength from the driven element and has a length in the range from 0.51 to 0.52 wavelength. On the other hand, a parasite acts as a director if it is spaced in the range mentioned above and has a length in the range from 0.48 to 0.38 wavelength. For a reflector, the spacing for maximum gain is in the vicinity of 0.15 wavelength while for a director, the spacing for maximum gain is about 0.11 wavelength. The input impedance to the driven element is

$$Z_{in} = Z_{11} - \frac{Z_{12}^2}{Z_{22}}$$

The real part of the input impedance tends to be low when a parasite is close to the driven element.

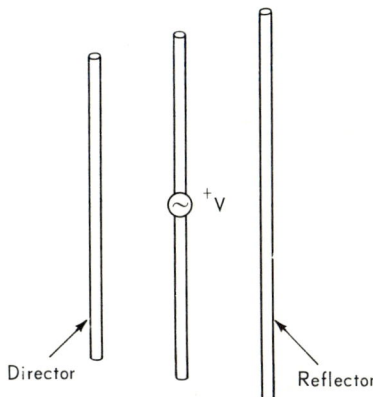

Fig. 4.39 A three-element Yagi array.

Two parasites, one reflector and one director, may be employed simultaneously for increased directive gain. Such an array is the simplest of a class of antennas known as Yagi arrays, or Yagi-Uda arrays. The equations that determine the currents are (numbering the reflector 1, the driven element 2, and the director 3)

$$0 = I_1 Z_{11} + I_2 Z_{12} + I_3 Z_{13}$$
$$V = I_1 Z_{21} + I_2 Z_{22} + I_3 Z_{23}$$
$$0 = I_1 Z_{31} + I_2 Z_{32} + I_3 Z_{33}$$

From these equations, the currents are in the ratios

$$I_1 : I_2 : I_3 = -(Z_{12} Z_{33} - Z_{23} Z_{13}) : (Z_{11} Z_{33} - Z_{13}^2) : -(Z_{11} Z_{23} - Z_{13} Z_{12})$$

The input impedance to the driven element is

$$Z_{\text{in}} = Z_{22} + \frac{I_1}{I_2} Z_{12} + \frac{I_3}{I_2} Z_{23}$$

where the current ratios are given above. Sizes and spacings are often in the ranges mentioned for single parasites. The input resistance of such arrays tends to be rather low if the parasites are close to the driven element. The input-impedance level can be increased by making the driven element in the form of a folded dipole. Typical gain for such structures is 6 to 8 db. Bandwidths of 1 to 2 percent are common.

198 ANTENNA ENGINEERING

More directors may be incorporated into the Yagi-Uda array for increased directive gain, within limits. Gains in excess of 15 to 20 db are difficult to achieve in this way.

Problem 4.11 Design a three-element Yagi array, giving all the electrical parameters of interest. Take the frequency of operation to be 100 MHz. In the absence of complete enough mutual-impedance data, lumped elements may be included in the parasites if desired.

5. INTRODUCTION TO THE MATHEMATICAL THEORY OF ANTENNAS

In this chapter, a review of the theory of conical and cylindrical antennas is presented. The biconical antenna provides an illustration of an antenna theory based on the solution of differential equations with stated boundary conditions. Some results for idealized cylindrical structures are obtained in a similar way. Then, cylindrical antennas are treated in terms of an integral equation for the current.

5.1 THEORY OF THE BICONICAL ANTENNA

There are many antenna shapes whose boundaries fit nicely into a spherical coordinate system. Some of these are illustrated in Fig. 5.1. All these structures are φ symmetric. It was pointed out in Chap. 4, in the discussions centered around Eq. (5), that if the condition of φ symmetry is imposed on Maxwell's equations written in spherical coordinates, the equations split into two independent sets, one of which is

$$\frac{\partial}{\partial \theta}(r \sin \theta\, H_\varphi) = j\omega\epsilon r^2 \sin \theta\, E_r$$

$$\frac{\partial}{\partial r}(r \sin \theta\, H_\varphi) = -j\omega\epsilon r \sin \theta\, E_\theta \qquad (1)$$

$$\frac{\partial}{\partial r}(rE_\theta) - \frac{\partial}{\partial \theta}(E_r) = -j\omega\mu r H_\varphi$$

It will be observed that these fields are TM to the direction r. The theory of symmetrically excited (in φ) radiators, such as those shown in Fig. 5.1, can be based on these equations.

Let the radius of the antenna be L. As we proceed, it will be necessary to determine the fields in the exterior region $r > L$ and to reexamine the solutions for the transmission-line region $r < L$. In both regions,

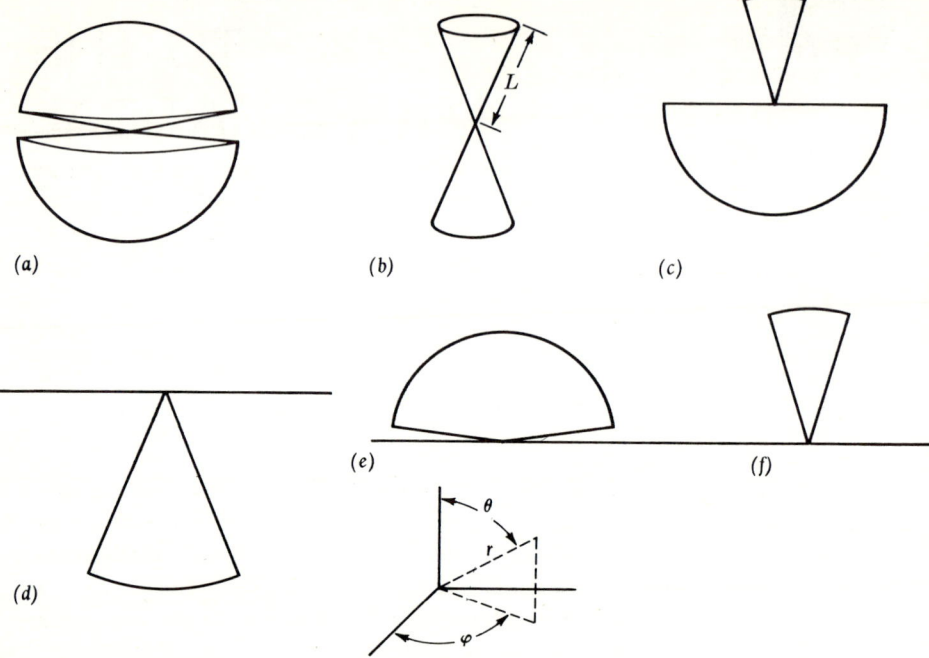

Fig. 5.1 Antenna shapes whose boundaries fit a spherical coordinate system: (a) spherical; (b) biconical; (c) cone-hemisphere; (d) discone; (e) hemispherical; (f) monoconical.

the fields must satisfy Eqs. (1) with the appropriate boundary conditions. To proceed with the solution, we may eliminate the E components in Eqs. (1) by differentiating the first two equations and substituting into the last. Then, if a new variable, $A = r \sin \theta \, H_\varphi$, is introduced, the latter equation takes the form

$$r^2 \frac{\partial^2 A}{\partial r^2} + \sin \theta \frac{\partial}{\partial \theta} \left(\frac{1}{\sin \theta} \frac{\partial A}{\partial \theta} \right) + r^2 \omega^2 \mu \epsilon A = 0$$

It is then assumed that A has a separable form $A = R(r)\Theta(\theta)$, and the equation may be put into the form

$$\frac{r^2}{R} \frac{\partial^2 R}{\partial r^2} + \frac{\sin \theta}{\Theta} \frac{\partial}{\partial \theta} \left(\frac{1}{\sin \theta} \frac{\partial \Theta}{\partial \theta} \right) + r^2 \omega^2 \mu \epsilon = 0$$

This partial differential equation, being the sum of terms depending only on r and a term depending only on θ, may be split into the ordinary differential equations

$$r^2 \frac{d^2R}{dr^2} + (r^2\omega^2\mu\epsilon - k)R = 0 \qquad (2)$$

$$\sin\theta \frac{\partial}{\partial\theta}\left(\frac{1}{\sin\theta}\frac{\partial\Theta}{\partial\theta}\right) + k\Theta = 0 \qquad (3)$$

where k is the separation constant.

For those with sufficient mathematical experience, all that need be said is that (1) Eq. (2) is a differential equation whose solution is well known to involve Bessel functions; (2) a substitution of variables $v = \cos\theta$, and $\Theta = (1 - v^2)^{1/2} K_n^1$ changes Eq. (3) into Legendre's associated differential equation of degree 1. The differential equations and their solutions may be found for example in Jahnke and Emde, *Tables of Functions*. A detailed discussion of the solution of the equations is presented in the next section for the benefit of those with less mathematical experience.

5.1.1 THE SOLUTIONS TO THE DIFFERENTIAL EQUATIONS

Consider the solutions that hold for the exterior region $r > L$. This region includes the z axis. Then note that on the z axis, the components E_θ and H_φ must vanish, independent of r, in order to avoid a discontinuity of the fields on the axis. These components depend upon the Θ-function portion of the solution, i.e., the solution to Eq. (3); it follows that the Θ functions must vanish at $\theta = 0$ and $\theta = \pi$. This fact constitutes a boundary condition on the functions and essentially determines the permissible values of the separation constants k, sometimes called the eigenvalues.

To get into the details of the solution, consider the solution of Eq. (3). A change of variables $v = \cos\theta$ puts the equation into a simpler form

$$(1 - v^2)\frac{d^2\Theta}{dv^2} + k\Theta = 0$$

To solve this equation, we shall try a power series in v

$$\Theta = \sum_{m=0}^{\infty} a_m v^m$$

where the coefficients a_m are to be determined. Performing the indicated differentiations and substituting back into the differential equation gives

$$\sum_{m=0}^{\infty} m(m-1)a_m v^{m-2} + \sum_{m=0}^{\infty} a_m[k - m(m-1)]v^m = 0$$

For this equation to be satisfied, the coefficients of like powers of m must be equal, that is

$$(m+2)(m+1)a_{m+2} + [k - m(m-1)]a_m = 0$$

or

$$a_{m+2} = -a_m \frac{[k - m(m - 1)]}{(m + 2)(m + 1)} \tag{4}$$

Inspection of this relation shows that given a_0, all even-order coefficients are specified; also, given a_1, all odd-order coefficients are determined. The even-order a's determine the even functions and the odd-order a's determine the odd functions; these even and odd functions may be calculated separately.

Note also that if the separation constant k has the value $k = m(m - 1)$, then the coefficient a_{m+2} is zero and therefore *all higher-order* coefficients of that family are also zero. This means that the assumed series form for Θ reduces to a *polynomial*.

A general formula for the even coefficients in terms of a_0 is

$$a_m = \frac{(-1)^{\frac{1}{2}m}}{m!} [k - (m - 2)(m - 3)] \times [k - (m - 4)(m - 5)] \cdots [k - 2 \cdot 1] k a_0$$

that is,

$$a_2 = \frac{-k a_0}{2} \qquad a_4 = \frac{(k - 2) k a_0}{4 \cdot 3 \cdot 2}$$

for the odd coefficients the formula is

$$a_m = \frac{(-1)^{\frac{1}{2}(m-1)}}{m!} k a_1 [k - (m - 2)(m - 3)] \times [k - (m - 4)(m - 5)] \cdots [k - 3 \cdot 2]$$

that is,

$$a_3 = \frac{-a_1 k}{3 \cdot 2} \qquad a_5 = k a_1 \frac{(k - 3 \cdot 2)}{5!}$$

As pointed out above, we are particularly interested in those Θ functions that vanish at $\theta = 0$ and π, that is, at $v = \cos\theta = \pm 1$. Thus, for the even functions, we would like to find those k values which satisfy the following equations:

"even functions"

$$\Theta_e(\pm 1) = 0 = a_0 \left[1 - \frac{k}{2!} + \frac{k(k - 2)}{4!} - \frac{k(k - 2)(k - 4 \cdot 3)}{6!} + \cdots \right]$$

"odd functions"

$$\Theta_o(\pm 1) = 0 = a_1 \left[1 - \frac{k}{3!} + \frac{k(k - 3 \cdot 2)}{5!} - \frac{k(k - 3 \cdot 2)(k - 5 \cdot 4)}{7!} + \cdots \right]$$

Looking first at the even functions, we note that $k = 2$ not only terminates the series but also makes the function vanish at $v = \pm 1$ as required; also, $k = 4 \cdot 3$ also does the job, as does $k = 6 \cdot 5$. Each of these values of k is then an eigenvalue which defines a characteristic function, in this case a polynomial. Next looking at the odd functions, we see that $k = 3 \cdot 2$ terminates the series and makes the function vanish at $v = \pm 1$ as required; the values $k = 5 \cdot 4$, $7 \cdot 6$, and so on also do the job. In summary, what we have found is that the set of values $k = n(n + 1)$ generates a set of polynomials that satisfy the conditions that the functions vanish at $\theta = 0$ and π.

The first several functions are the following polynomials. Each has an arbitrary constant that may be selected for convenience.

$\Theta_1 = a_{01}(1 - v^2)$

$\Theta_2 = a_{12}v(1 - v^2)$

$\Theta_3 = a_{03}(1 - 6v^2 + 5v^4)$

$\Theta_4 = a_{14}v\left[1 - \dfrac{4 \cdot 5}{3 \cdot 2}v^2 + \dfrac{4 \cdot 5}{5!}(4 \cdot 5 - 3 \cdot 2)v^4\right]$

$\Theta_5 = a_{05}\left[1 - \dfrac{5 \cdot 6}{2!}v^2 + \dfrac{5 \cdot 6}{4!}(5 \cdot 6 - 2)v^4 \right.$
$\left. \qquad\qquad - \dfrac{5 \cdot 6}{6!}(5 \cdot 6 - 2)(5 \cdot 6 - 4 \cdot 3)v^6\right]$

$\Theta_6 = a_{16}v\left[1 - \dfrac{6 \cdot 7}{3!}v^2 + \dfrac{6 \cdot 7}{5!}(6 \cdot 7 - 3 \cdot 2)v^4 \right.$
$\left. \qquad\qquad - \dfrac{6 \cdot 7}{7!}(6 \cdot 7 - 3 \cdot 2)(6 \cdot 7 - 5 \cdot 4)v^6\right]$

Each function satisfies the differential equation and the boundary conditions for any value of the constant. The constant may be selected, for example, so that the functions are *normalized*, namely, $\int_{-1}^{1} \Theta_n^2 \, dv = 1$. However, at this time, we shall not select the constants in this way. Rather, we shall select the constants in such a way that the functions are closely related to a set of mathematical functions about which there is a great deal of information in the literature. In particular, we shall select the constants so that the functions are related to the associated Legendre polynomials of degree 1:

$\Theta_n = (1 - v^2)^{1/2} P_n^1$

This identification is particularly convenient for computational purposes in view of the recurrence relationship

$P_{n+1}^1 = \dfrac{1}{n}[(2n + 1)vP_n^1(v) - (n + 1)P_{n-1}^1(v)]$

Of course, the associated Legendre polynomials form a complete set of orthogonal functions (as do the Θ functions) as may be proved as follows:

$$(1 - v^2) \frac{d^2}{dv^2} \Theta_n + k_n \Theta_n = 0$$

$$(1 - v^2) \frac{d^2}{dv^2} \Theta_m + k_m \Theta_n = 0$$

Multiply the first of these equations by Θ_m, the second by Θ_n, subtract one from the other and integrate over the range of orthogonality (in this case from $v = -1$ to $v = +1$), then use the condition $\Theta(\pm 1) = 0$ to show that $\int_{-1}^{1} \Theta_n \Theta_m \, dv = 0$, if $n \neq m$. The orthogonality conditions for the Legendre polynomials may also be found on page 116 of Jahnke and Emde; the result is

$$\int_{-1}^{1} P_n^1(x) P_m^1(x) \, dx = \begin{cases} 0 & \text{if } m \neq n \\ \dfrac{2}{2n+1} \dfrac{(n+1)!}{(n-1)!} & \text{if } m = n \end{cases} \tag{5}$$

This completes the discussion of the solution of the differential equation giving the variation with θ in the external region of the biconical antenna. The same general technique may be employed to find the functions in the region $r < L$. However, the boundary condition in the internal region is $(1/\sin \theta) \, \partial A / \partial \theta = 0$ at $\theta = \theta_1$ and $\theta = \theta_2$ (i.e., the radial component of E must vanish at the conducting surface). Thus the condition $\partial \Theta / \partial v = 0$ at $v = \cos \theta_1$ and $v = \cos \theta_2$ may be employed to determine the eigenvalues k_m and the functions characteristic of a particular cone angle. The actual determination of the functions in the general case is rather involved.

Let us now return to the differential equation, (2), for the radial variation. In the exterior region, the separation constants k_n have already been determined above: $k = n(n + 1)$. Thus the differential equation may be written in the form

$$\frac{d^2 R}{dr^2} + \left[\beta^2 - \frac{n(n+1)}{r^2} \right] R = 0 \qquad \beta^2 = \omega^2 \mu \epsilon$$

This differential equation may be found on page 146 in Jahnke and Emde (the third one in the list), from which it appears that the solution can be expressed in terms of Bessel functions. However, the particular (half-integral order) Bessel functions involved here have a particularly simple form; thus it is instructive to carry out a series solution of the differential equation. Note, first of all, that for $n = 0$, the functions are trigonometric or exponential. Also note that if n is not too large, the solutions in regions of large r (asymptotic solutions) are approximately trigonometric or exponential. These facts, combined with the form of the approximate solutions to the differential equation in the range such that ω and/or r tends to zero, lead us to try, as a solution, a power series multiplied by an exponential

$$R = e^{\pm j\beta r} \sum_{m=0}^{\infty} b_m (r)^{-m} \qquad \beta = \omega(\mu\epsilon)^{1/2} \tag{6}$$

Placing this form of solution and its second derivative with respect to r into the differential equation (2) and equating the coefficients of like powers of r gives the result

$$m(m+1)b_m + 2j\beta(m+1)b_{m+1} - n(n+1)b_m = 0$$

or

$$b_{m+1} = \frac{[n(n+1) - m(m+1)]}{2j\beta(m+1)} b_m \tag{7}$$

Thus, given the constant b_0, the remainder of the constants are all determined. Note that since n and m are integers, the series will terminate when $m = n$, which means that the R_n functions will be given by a finite series. Moreover, each R_n function will be associated with a Θ_n or P_n^1 function of the same order. The first few coefficients are

$$b_1 = \frac{n(n+1)}{2j\beta} b_0 \qquad b_2 = \frac{n(n+1)}{2(2j\beta)^2} [n(n+1) - 2]b_0$$

$$b_3 = \frac{n(n+1)}{3 \cdot 2 \cdot (2j\beta)^3} [n(n+1) - 2][n(n+1) - 2 \cdot 3]b_0$$

or

$$b_m = \frac{n(n+1)}{m!(2j\beta)^m} [n(n+1) - 2]$$
$$\times [n(n+1) - 2 \cdot 3] \cdots [n(n+1) - (m-1)m]b_0$$

or since $n(n+1) - (m-1)m = (n+m)[n - (m-1)]$

$$b_m = \frac{(n+m)!}{m!(n-m)!(2j\beta)^m} b_0 \tag{8}$$

or finally,

$$R_n^{(1)}{}^{(2)} = b_{0n} e^{\pm j\beta r} \sum_{m=0}^{n} \frac{(n+m)!}{m!(n-m)!(2j\beta r)^m} \tag{9}$$

As far as the differential equation is concerned, the constants b_{0n} are arbitrary. We can make the functions equal to Schelkunoff's spherical Bessel functions by taking $b_{0n} = (j)^{n+1}$. A more or less standard notation for these functions is then

$$R_n{}^{(2)}(x) = \hat{H}_n{}^{(2)}(x) = (\tfrac{1}{2}\pi x)^{\frac{1}{2}} H^{(2)}_{n+\frac{1}{2}}(x)$$
$$R_n{}^{(1)} = \hat{H}_n{}^{(1)}(x) = (\tfrac{1}{2}\pi x)^{\frac{1}{2}} H^{(1)}_{n+\frac{1}{2}}(x)$$

where the functions $H^{(2)}_{n+\frac{1}{2}}(x)$ are the Bessel functions of the third kind of half-integral order:

$$H^{(2)}_{n+\frac{1}{2}}(x) = J_{n+\frac{1}{2}}(x) \pm j N_{n+\frac{1}{2}}(x)$$

where the J's and N's are Bessel functions of the first and second kind, respectively. Similarly, the functions $\hat{H}_n$ for real argument have real and imaginary parts as follows:

$$\hat{H}_n{}^{(2)}(x) = \hat{J}_n(x) \pm j \hat{N}_n(x)$$

Note:

$$\hat{J}_0(x) = \sin x \qquad\qquad \hat{N}_0(x) = -\cos x$$
$$\hat{J}_1(x) = \frac{\sin x}{x} - \cos x \qquad \hat{N}_1(x) = -\sin x - \frac{\cos x}{x}$$
$$\hat{H}_0{}^{(2)}(x) = je^{-jx} \qquad\qquad \hat{H}_1{}^{(2)}(x) = -e^{-jx}\left(1 + \frac{1}{jx}\right)$$

and these functions satisfy recurrence relations as follows:

$$\hat{Z}_{n+1}(x) = \frac{2n+1}{x}\hat{Z}_n(x) - \hat{Z}_{n-1}(x)$$
$$\hat{Z}'_n(x) = \frac{n+1}{x}\hat{Z}_n(x) - \hat{Z}_{n+1}(x)$$
$$= \frac{1}{2n+1}[(n+1)\hat{Z}_{n-1}(x) - n\hat{Z}_{n+1}(x)]$$

This completes the brief study of the mathematical functions that enter into the calculations for biconical antennas. About the only way to become familiar with these functions, from an engineering point of view, is to study graphs of the functions, such as are available in Jahnke and Emde.

5.1.2 THE FIELDS

In the preceding section, we have found the solutions to the differential equations for the separable parts of the function $A = R(r)\Theta(\theta) = r \sin\theta\, H_\varphi$. Each order of the functions is a solution, and so a sum of them is also a solution; or, for the external region, the general solution is

$$A = r \sin\theta\, H_\varphi = \sum_{n=1}^{\infty} A_n \hat{H}_n{}^{(2)}(\beta r)(1-v^2)^{\frac{1}{2}} P_n{}^1(v) \qquad (10)$$

or

$$H_\varphi = \frac{1}{r}\sum_{n=1} A_n \hat{H}_n^{(2)}(\beta r) P_n^1(v) \qquad (11)$$

The $\hat{H}_n^{(2)}$ functions rather than the $\hat{H}_n^{(1)}$ functions are selected so that H_φ will have the form of *outgoing* waves as $r \to \infty$. Also, from Eqs. (1), we find for the electric field

$$E_\theta = \frac{-1}{j\omega\epsilon r}\frac{\partial(rH_\varphi)}{\partial r} = \frac{-1}{j\omega\epsilon r}\sum_{n=1} A_n \frac{d\hat{H}_n^{(2)}(\beta r)}{dr} P_n^1(v) \qquad (12)$$

where, recall, $v = \cos\theta$.

To see the manner in which an explicit solution is obtained, consider a biconical antenna made up of such thick cones that only a small slot is left in the spherical caps which terminate the cones (Fig. 5.1a). Suppose a primary source maintains a constant voltage V between the two sectors. Although the electric field in the slot is not known exactly, since the slot is narrow, a good first approximation to the field in the slot is surely $E_\theta = V/s$, where s is the width of the slot $[s = L(\theta_2 - \theta_1)]$. On the surface of the conducting sphere, the tangential component of the electric field is then known everywhere. It is zero everywhere except at the slot opening; since we know the value at the slot, we can represent the field over the whole surface as a pulse function

$$E_{\theta,r=L} = \begin{cases} \dfrac{V}{s} & v_1 < v < v_2 \\ 0 & \text{elsewhere} \end{cases}$$

This known function may be expanded into a series of associated Legendre polynomials of degree 1, since the latter constitute a complete set of orthogonal functions in the interval $-1 \leq v \leq 1$:

$$E_{\theta,r=L} = \sum_{n=1} a_n P_n^1(v)$$

The coefficients may be determined in the usual way:

$$a_n = \frac{\int_{v_1}^{v_2} E_\theta P_n^1(v)\, dv}{\int_{-1}^{1} [P_n^1(v)]^2\, dv}$$

$$\int_{-1}^{1} [P_n^1(v)]^2\, dv = \frac{2n(n+1)}{2n+1}$$

and with a small slot:

$$a_n = \frac{V}{s}\frac{2n+1}{2n(n+1)}\int_{v_1}^{v_2} P_n^1(v)\, dv \qquad (13)$$

Writing out this known expansion makes it easy to see the way we can evaluate the coefficients A_n in the general solutions given by Eqs. (11) and (12). The general solution for the electric field, when it is evaluated at the surface of the sphere $(r = L)$, must be equal to the expansion of the known pulse function

$$-\frac{\beta}{j\omega\epsilon L}\sum_n A_n \hat{H}'_n(\beta L) P_n^1(v) = \sum_n a_n P_n^1(v)$$

where

$$\hat{H}'_n(\beta L) = \frac{d\hat{H}_n^{(2)}(\beta r)}{d(\beta r)}\bigg]_{r=L}$$

The two series will be equal if the corresponding orders of the Legendre polynomials are equal term by term. Equating the coefficients of corresponding orders, we obtain an expression for the unknown coefficients A_n in terms of the known coefficients a_n [given by Eq. (13)]

$$A_n = \frac{-j\omega\epsilon L a_n}{\beta \hat{H}'_n(\beta L)}$$

Thus the fields outside the sphere of radius L are given by the expressions

$$H_\varphi = -\frac{j\omega\epsilon L}{\beta r}\sum_{n=1}^\infty \frac{a_n}{\hat{H}'_n(\beta L)} \hat{H}_n(\beta r) P_n^1(v) \qquad (14)$$

$$E_\theta = \frac{L}{r}\sum_{n=1}^\infty a_n \frac{\hat{H}'_n(\beta r)}{\hat{H}'_n(\beta L)} P_n^1(v) \qquad (15)$$

In the distant field region (βr large), an asymptotic approximation for the Hankel functions is

$$\hat{H}_n^{(2)}(\beta r) \underset{\beta r \to \infty}{\longrightarrow} j^{n+1} e^{-j\beta r}$$

This approximation is good for arguments βr greater than about $4n(n+1)$. The coefficients $a_n/H'_n(\beta L)$ approach zero quite rapidly for large values of n; thus the series converges rapidly. The radiation patterns for spherical antennas of radii of the order of a wavelength or less is then the superposition of only a few orders of the associated Legendre polynomials. It will be recalled that the first few orders of these polynomials are $P_1^1(v) = (1-v^2)^{1/2} = \sin\theta$, $P_2^1 = \frac{3}{2}\sin 2\theta$, and $P_3^1 = \frac{3}{8}(\sin\theta + 5\sin 3\theta)$.

The foregoing mathematics has further implications. In particular, if an antenna is so small that it can be enclosed by a spherical surface of small radius, it will be very difficult to cause it to radiate a field having anything other than a sinusoidal (with θ) field-strength pattern. This follows because the fields of the antenna can be computed from the fields on the surface of the small sphere and these can be represented as a series of spherical functions as described above; but since with small arguments, the higher orders of the functions $\hat{H}_n(\beta L)$ are very much larger than the lower orders, the contributions of the higher-order terms are very small compared to the lower-order terms. Thus *there is an inherent limitation on the gain of small antennas.*

The expression for the magnetic field, evaluated at the surface of the sphere, may be employed to find the surface current density **K** on the spherical antenna caps:

$$\mathbf{K} = \hat{\mathbf{n}} \times \mathbf{H} = \hat{\mathbf{r}} \times \hat{\boldsymbol{\varphi}} H_\varphi = -\hat{\boldsymbol{\theta}} H_\varphi$$

From this, the total current starting up along the spherical cap can be found from $I = \int \mathbf{K} \cdot \mathbf{dS}$. For example, if the current flowing in the direction of $-\hat{\boldsymbol{\theta}}$ (up along the sphere) is desired

$$I = \int K(-\hat{\boldsymbol{\theta}}) L \sin \theta \, d\varphi = 2\pi r \sin \theta \, H_\varphi \tag{16}$$

The ratio of the current to the voltage across the slot may be called the terminating impedance. It represents the effect of the exterior region on the biconical transmission line.

5.1.3 IMPEDANCE OF BICONICAL ANTENNAS

In most cases of practical interest, the azimuthal slot formed by the termination of the biconical transmission line into spherical end caps is not negligible in width, as was assumed in the foregoing example. In such a case, although the analysis is formally the same, the electric field across the slot is not known, which means that the coefficients a_n and A_n are not known. In principle, this more general biconical antenna problem may be solved by writing down the solutions for the fields in the interior region along with those for the exterior region and forcing the tangential components of E and H to be equal in the slot at $r = L$. In practice, this analysis is very difficult to carry out because the functions for the interior region are (1) relatively poorly tabulated, and (2) characterized by different eigenvalues and eigenfunctions from the functions for the exterior region. This latter fact prohibits term-by-term matching of the solutions.

Symbolically at least, the interior magnetic field may be represented as the sum of the TEM field plus all the other solutions (modes) for the interior region

$$H_{\varphi,\text{inside}} = \frac{I_0(r)}{2\pi r \sin \theta} + \sum_\nu \frac{C_\nu R_\nu(\beta r)}{r \sin \theta} \Theta_\nu(v) \tag{17}$$

where the Θ_ν functions are those that fit the interior boundary conditions. This field component, evaluated at $r = L$, must be equal to the exterior field component given by Eq. (14), with the constants determined by the equation

$$a_n = \frac{2n+1}{2n(n+1)} \int_{v_1}^{v_2} E_\theta P_n^1(v)\, dv$$

where

$$E_{\theta,\text{inside}} = \frac{-1}{j\omega\epsilon r}\left\{\frac{1}{2\pi \sin\theta}\frac{\partial}{\partial r}I_0(r) + \sum_\nu C_\nu \frac{\partial}{\partial r}[R(\beta r)]\frac{\Theta_\nu(v)}{\sin\theta}\right\} \quad (18)$$

As an approximation, if the bicone angles are not too small, we may assume that the electric field is given to a good approximation by the TEM field alone:

$$E_{\theta,\text{inside}} = -\frac{1}{j\omega\epsilon_i r 2\pi \sin\theta}\frac{d}{dr}I_0(r)$$

With this approximation

$$a_n = -\frac{2n+1}{2n(n+1)}\frac{d}{dr}I_0(r)\frac{1}{j\omega\epsilon_i L 2\pi}\int_{v_1}^{v_2}\frac{P_n^1(v)}{\sin\theta}\,dv \quad (19)$$

In view of the relationships between the Legendre polynomials P_n and the associated Legendre polynomials of degree 1

$$P_n^1 = (1-v^2)^{1/2}\frac{d}{dv}P_n$$

and the relation $(1-v^2)^{1/2} = (1-\cos^2\theta)^{1/2} = \sin\theta$, it follows that the coefficients are given by

$$a_n = \frac{-(2n+1)}{2n(n+1)j\omega\epsilon_i L 2\pi}[P_n(v_2) - P_n(v_1)]\frac{d}{dr}I_0(r) \quad (20)$$

The argument can be carried out for the asymmetrical type of antenna covered by Eq. (20); but let us simplify things and restrict the detailed treatment to biconical antennas having equal cone angles: $\theta_2 = \pi - \theta_1$. For symmetric bicones, $v_2 = -v_1$; this fact coupled with the symmetry properties of the Legendre polynomials $P_n(-v) = (-1)^n P_n(v)$ gives the following results for the coefficients

$$a_n = \begin{cases} \dfrac{2n+1}{n(n+1)j\omega\epsilon_i L 2\pi}P_n(v_1)\dfrac{d}{dr}I_0(L) & \text{for } n \text{ odd} \\ 0 & \text{for } n \text{ even} \end{cases} \quad (21)$$

To calculate the impedance or admittance at the slot, we must form and evaluate the ratio $I_0(L)/V(L)$ for the TEM mode. A useful expression for $I_0(L)$ may be found with the aid of Eq. (17); in particular, divide the equation by $\sin\theta$ and integrate

$$\int \frac{H_{\varphi,\text{inside}}}{\sin\theta}\,dv = \frac{I_0}{2\pi r}\int \frac{dv}{\sin^2\theta} + \sum_{\nu} C_\nu \frac{R_\nu(\beta r)}{r}\int \frac{\Theta_\nu}{\sin^2\theta}\,dv$$

The last term in this expression (the integral over the higher-order modes) vanishes. That this is so may be shown as follows: From the differential equation for the Θ functions in the interior region, we have

$$\int_{v_1}^{v_2} \frac{d}{dv}\Theta'_\nu\,dv + k\int_{v_1}^{v_2}\frac{\Theta_\nu}{(1-v^2)}\,dv = 0$$

or

$$\Theta'_\nu\Big]_{v_1}^{v_2} + k\int_{v_1}^{v_2}\frac{\Theta_\nu}{(1-v^2)}\,dv = 0$$

But from the boundary conditions in the internal region, $\Theta'_\nu(v_2) = \Theta'_\nu(v_1) = 0$, so

$$\int_{v_1}^{v_2}\frac{\Theta_\nu}{(1-v^2)}\,dv = \int_{v_1}^{v_2}\frac{\Theta_\nu}{\sin^2\theta}\,dv = 0 \tag{22}$$

so the integral over the higher-order modes vanishes, as stated. Since the magnetic field in the exterior region must match that in the interior region, we can employ this simplified equation and obtain the relation

$$\frac{-j\omega\epsilon_0 L}{\beta_0 L}\sum_{n=1} a_n \frac{\hat{H}_n(\beta_0 L)}{\hat{H}'_n(\beta_0 L)}\int_{v_1}^{v_2}\frac{P_n^1(v)}{\sin\theta}\,dv = \frac{I_0}{2\pi L}\int_{v_1}^{v_2}\frac{dv}{\sin^2\theta} \tag{23}$$

We can make certain simplifications in this equation as follows: as before

$$\int_{v_1}^{v_2} P_n^1 \frac{dv}{\sin\theta} = \begin{cases} 2P_n(v_1) & n \text{ odd} \\ 0 & n \text{ even} \end{cases}$$

Also, the integral on the right of Eq. (23) may be evaluated

$$\int_{v_1}^{v_2}\frac{dv}{\sin^2\theta} = -\int_{v_1}^{v_2}\frac{d\theta}{\sin\theta} = -\ln\tan\tfrac{1}{2}\theta\Big]_{\theta_1}^{\theta_2} = -Z_0 2\pi\left(\frac{\epsilon_i}{\mu_i}\right)^{1/2}$$

Next consider the voltage between the bicones:

$$V(L) = \int_{\theta_1}^{\theta_2} E_\theta L\,d\theta = -L\int_{v_1}^{v_2}\frac{E_\theta}{\sin\theta}\,dv$$

E_θ is given by Eq. (18), but from Eq. (22), we see that the integral over all modes except the TEM mode vanishes; thus

$$V(L) = -L\int_{\theta_1}^{\theta_2}\frac{1}{j\omega\epsilon_i L}\frac{d}{dr}I_0(L)\frac{d\theta}{2\pi\sin\theta}$$

$$= \frac{-1}{j\omega\epsilon_i}\frac{d}{dr}I_0(L)\frac{1}{2\pi}\ln\tan\tfrac{1}{2}\theta\Big]_{\theta_1}^{\theta_2} = -\frac{Z_0}{j\beta_i}\frac{d}{dr}I_0(L) \tag{24}$$

where Z_0 is the characteristic impedance of the biconical transmission-line portion of the antenna. Using Eq. (24) in Eq. (21) for a_n, we find

$$a_n = \begin{cases} -\left(\frac{\mu_i}{\epsilon_i}\right)^{1/2} \frac{2n+1}{n(n+1)} \frac{V(L)}{Z_0 L 2\pi} P_n(v_1) & n \text{ odd} \\ 0 & n \text{ even} \end{cases} \quad (25)$$

thus from Eq. (23) and the intermediate results:

$$j\frac{V(L)}{Z_0} \frac{\epsilon_0}{\epsilon_i} \frac{\beta_i}{\beta_0} \sum_{n=1} \frac{(2n+1)}{n(n+1)} [P_n(v_1)]^2 \, 2 \frac{\hat{H}_n(\beta_0 L)}{\hat{H}'_n(\beta_0 L)} = -2\pi Z_0 \left(\frac{\epsilon_i}{\mu_i}\right)^{1/2} I_0(L)$$

The terminal admittance Y_t is the ratio $I_0(L)/V(L)$ given by this equation

$$Y_t \equiv \frac{I_0(L)}{V(L)} = -\frac{j\left(\frac{\epsilon_0}{\mu_0}\right)^{1/2} \frac{\mu_i}{\epsilon_i}}{Z_0^2 \pi} \sum_{n=1} \frac{(2n+1)}{n(n+1)} \frac{\hat{H}_n(\beta_0 L)}{\hat{H}'_n(\beta_0 L)} [P_n(v_1)]^2 \quad n \text{ odd} \quad (26)$$

where $Z_0 = \left(\frac{\mu_i}{\epsilon_i}\right)^{1/2} \frac{1}{\pi} \ln (\cot \tfrac{1}{2}\theta_1)$.

The functions P_n are the Legendre polynomials

$P_0 = 1$

$P_1 = v$

$P_2 = \tfrac{1}{2}(3v^2 - 1)$

The higher-order Legendre polynomials may be computed with the aid of the recurrence formula

$$P_n(v) = \frac{2n-1}{n} v P_{n-1} - \frac{n-1}{n} P_{n-2}$$

Note that Eq. (26) gives a formula for the terminating admittance in the form $Y = \Sigma Y_n$; that is, it is the summation of the admittances of many modes in parallel. The terminating admittance may be transferred back to the input with the aid of the usual formulas for transferring admittances along a transmission line. The formulas for the input admittance so obtained are useful for bicones having radii up to something less than a half wavelength, with cone angles of about 30° or more. Equation (26) is readily programmed for computation by a digital computer.

Note, however, that the current which actually flows up around the lip of the slot onto the spherical end cap depends upon the cone angle. For cones having very small cone angles, there is essentially no cap and consequently no current flow onto the end cap. In such a case it is better to base the development more specifically in terms of the interior mode sets. The current on the cones is then given in terms of a summation over these modes. One of these interior modes is of course the TEM mode. Close to the end of the biconical line (that is, r near L) the current in the TEM mode is the part that flows onto the end caps. The total current is the sum of the currents in all modes $I(r) = I_0(r) + \tilde{I}(r)$, where I_0 is the TEM mode current and $\tilde{I}$ is the sum over all the other modes. As shown above, the line integral of E from one cone to the other vanishes except for the E field associated with the TEM mode. Hence, in the biconical region, $V(r) = V_0(r)$. The admittance at any point may be thought of as the sum of two parts

$$Y(r) = \frac{I_0(r)}{V_0(r)} + \frac{\tilde{I}(r)}{V_0(r)}$$

Close to the origin there is current only in the TEM mode. The reason for this is that the higher-order modes are generated by the discontinuity at the end of the biconical line, not by the source at the input. The higher-order modes are localized to points near the end of the line since their variation with r is as r^n, where n is greater than one. Consequently, they die out in amplitude at points near the origin, as if they were cut off. This means that at the input $Y_{in} \doteq I_0(0)/V(0)$, the admittance depends only on the current in the TEM mode. Near the end of the line, some or all of the current effectively leaves the TEM mode; thus it appears to this mode as if there were a shunt admittance in the vicinity of the end of the line. The terminating admittance may be thought of as the parallel combination of two parts

$$Y(L) = \frac{I_0(L)}{V_0(L)} + \frac{\tilde{I}(L)}{V_0(L)}$$

One part represents the current which flows up into the end cap and the other represents that going into the higher-order modes.

Schelkunoff and others have calculated the terminating admittance of biconical antennas and tabulated them to some extent.[1] The results of the computations are presented in terms of a quantity called the inverse-radiation impedance Z_a

$$Z_a = Z_0^2 Y_t$$

The terminating impedance, found from this formula, may be employed to find the input admittance at the apex of the cones or elsewhere. The transformations may be done with either the Smith chart or the formula

$$\frac{Y_{in}}{Y_0} = \frac{Y_t \cos \beta L + jY_0 \sin \beta L}{Y_0 \cos \beta L + jY_t \sin \beta L}$$

[1] S. A. Schelkunoff, "Advanced Antenna Theory," John Wiley & Sons, Inc., New York, 1952.

Alternatively, the input impedance may be calculated directly with the relationship

$$\frac{Z_{\text{in}}}{Z_0} = \frac{Z_0 \cos \beta L + j Z_a \sin \beta L}{Z_a \cos \beta L + j Z_0 \sin \beta L}$$

Some results for a thin cone are displayed in Figs. 4.8 and 4.9.

5.2 IDEALIZED CYLINDRICAL STRUCTURES

Most of the practical radiators of cylindrical cross section cannot be treated conveniently by boundary-value-problem techniques of the type just described; these techniques are well suited only to spheroidal and conical structures. A method based on an integral equation for the current will be described in Sec. 5.3. In this section, some useful facts will be obtained by carrying out solutions for some rather artificial problems in cylindrical coordinates.

As a first example, consider a structure like the capacitor-plate antenna of Fig. 2.1, but let the radius of the plates be infinite. The metallic post, extending from one of the conducting planes to the other, has radius a. Currents in the post will be excited in a manner to be described in a moment. As a result, currents will be induced in the conducting planes; but the fields of these latter currents might be calculated from image currents in line with the post (the z direction). Consequently, we expect that the fields can be calculated from a magnetic vector potential having a single component A_z. That is, the mathematical problem is to solve the equation

$$\nabla^2 A_z = \gamma^2 A_z$$

presumably working in a cylindrical coordinate system. Then the detailed equation to be solved is

$$\frac{1}{\rho} \frac{\partial}{\partial \rho}\left(\frac{\rho \partial A_z}{\partial \rho}\right) + \frac{1}{\rho^2} \frac{\partial^2 A_z}{\partial \varphi^2} + \frac{\partial^2 A_z}{\partial z^2} - \gamma^2 A_z = 0$$

In the problem at hand, $\dfrac{\partial A_z}{\partial \varphi} = 0$. Thus, the equation may be solved by assuming that A_z has a separable form $A_z = R(\rho)Z(z)$. With this assumption, it is easily determined that the equation satisfied by $Z(z)$ is

$$\frac{d^2 Z}{dz^2} = \gamma_z{}^2 Z$$

where γ_z is an arbitrary separation constant. The solutions to this differential equation may be written in the form of sines, cosines, or exponentials. The equation satisfied by $R(\rho)$ is found to be

$$\frac{d}{d\rho}\left(\rho \frac{dR}{d\rho}\right) + (\gamma_z{}^2 - \gamma^2)\rho R = 0$$

The solution to this latter equation may be found by assuming a power series, in a procedure as described earlier in Sec. 5.1.1. If this is done, it is found that in general the solution may be expressed in terms of any of the orders or types of the functions known as *Bessel functions;* in the case at hand, because of the φ symmetry, the solutions consist of the Bessel functions of order zero. In particular, the solution is a linear combination of the following functions (see Jahnke and Emde for detailed information about the functions):

$J_0(\beta\rho)$	Bessel function of the first kind
$N_0(\beta\rho)$	Bessel function of the second kind
$H_0{}^{(1)}(\beta\rho) = J_0 + jN_0$	Bessel function of the third kind
$H_0{}^{(2)}(\beta\rho) = J_0 - jN_0$	Bessel function of the third kind

where $\beta = (\gamma_z{}^2 - \gamma^2)^{1/2}$. The function $H_0{}^{(2)}(\beta\rho)$ represents outgoing cylindrical waves.

Thus the solution for the vector potential in the region exterior to the metallic post may be written (setting $\gamma_z = j\beta_z$)

$$A_z = \sum_{n=0} H_0{}^{(2)}(\beta_n\rho)(a_n \cos \beta_{zn}z + b_n \sin \beta_{zn}z)$$

and the fields found from this vector potential by the relationships

$$E_z = \frac{1}{j\omega\epsilon}(\gamma_z{}^2 - \gamma^2)A_z \qquad H_\varphi = -\frac{\partial A_z}{\partial \rho}$$

If the post is not a perfect conductor, then solutions interior to the post may be written in the form

$$A_{zi} = \sum_{n=0} J_0(\beta_{in}\rho)(a_{in} \cos \beta_{zn}z + b_{in} \sin \beta_{zn}z)$$

where

$$\beta_{in} = (\gamma_z{}^2 - \gamma_i{}^2)^{1/2}$$

5.2.1 UNIFORM EXCITATION

Suppose first that the separation h between the planes is small compared to a wavelength. Suppose also that a voltage V_g is applied between the planes. The total field in the region between the planes must consist of the superposition of the applied field and the induced currents in the post and planes $E_T = E_{\text{ap}} + E_{\text{ind}}$.

216 ANTENNA ENGINEERING

Suppose the post is perfectly conducting. Then, on the surface of the post, the tangential electric field must vanish:

$$E_T = E_{ap} + E_{ind}\Big]_{\rho=a} = 0$$

or

$$E_{ap} = -E_{ind} = -\frac{(\gamma_z^2 - \gamma^2)}{j\omega\epsilon} \sum_{n=0} H_0^{(2)}(\beta_n a)(a_n \cos \beta_{zn} z + b_n \sin \beta_{zn} z)$$

But E_{ap} is uniform along z; therefore, E_{ind} must be independent of z. This implies the separation constant $\beta_{zn} = 0$; thus

$$E_{ap} = \frac{V_g}{h} = -\frac{\beta^2}{j\omega\epsilon} a_0 H_0^{(2)}(\beta a)$$

or

$$a_0 = \frac{V_g}{j\omega\mu h H_0^{(2)}(\beta a)}$$

The fields are then

$$E_{z,ind} = -\frac{V_g H_0^{(2)}(\beta\rho)}{h H_0^{(2)}(\beta a)}$$

$$H_{\varphi,ind} = -\frac{V_g \beta H_0^{(2)'}(\beta\rho)}{h j\omega\mu H_0^{(2)}(\beta a)}$$

To find the current in the perfectly conducting post, we may employ Ampère's law (since $E_z = 0$ in the post):

$$\oint \mathbf{H} \cdot \mathbf{dl} = \int_0^{2\pi} H_\varphi a \, d\varphi = I$$

or

$$I = H_\varphi(a) 2\pi a$$
$$= j\left(\frac{\epsilon}{\mu}\right)^{1/2} \frac{2\pi a V_g H_0^{(2)'}(\beta a)}{h H_0^{(2)}(\beta a)}$$

The impedance seen by the source of the exciting voltage is then

$$Z = \frac{V_g}{I} = \frac{-j\left(\frac{\mu}{\epsilon}\right)^{1/2} h H_0^{(2)}(\beta a)}{2\pi a H_0^{(2)'}(\beta a)}$$

Separating real and imaginary parts by replacing $H_0(\beta a)$ by $J_0 - jN_0$,

$$Z = -j\left(\frac{\mu}{\epsilon}\right)^{1/2} \frac{h}{2\pi a} \left[\frac{j(J_0 N_0' - N_0 J_0') + J_0 J_0' + N_0 N_0'}{J_0'^2 + N_0'^2}\right]$$

We may now recall the following properties of the Bessel functions: the quantity $J_0(\beta a) N_0'(\beta a) - N_0(\beta a) J_0'(\beta a)$ is the wronskian determinant which is equal to a constant, $2/\pi\beta a$. Also, the derivatives of the zero-order functions are related to the first-order functions by $J_0' = -J_1$, $N_0' = -N_1$. With these equalities, the equation for the impedance is

$$Z = \left(\frac{\mu}{\epsilon}\right)^{\frac{1}{2}} \frac{h}{2\pi a} \left\{ \frac{(2/\pi\beta a) + j[J_0(\beta a)J_1(\beta a) + N_0(\beta a)N_1(\beta a)]}{J_1^2(\beta a) + N_1^2(\beta a)} \right\}$$

The imaginary part of this expression may be called the external inductance of the cylindrical post. The real part may be called the radiation resistance.

In most cases to which the theory above could be applied, the argument βa is very small, and the Bessel functions may be approximated as follows:

$$J_0(\beta a) \doteq 1 \qquad J_1(\beta a) \doteq \tfrac{1}{2}\beta a$$
$$N_0(\beta a) \doteq -\frac{2}{\pi} \ln \frac{2}{1.781\beta a} \qquad N_1(\beta a) \doteq -\frac{1}{\pi}\frac{2}{\beta a}$$

thus

$$J_1^2 + N_1^2 \doteq \left(\frac{2}{\pi\beta a}\right)^2 \qquad J_0 J_1 + N_0 N_1 \doteq \left(\frac{2}{\pi}\right)^2 \frac{1}{\beta a} \ln \frac{2}{1.781\beta a}$$

and

$$Z \doteq \left(\frac{\mu}{\epsilon}\right)^{\frac{1}{2}} \left(\frac{1}{4} + \frac{j}{2\pi} \ln \frac{2}{1.781\beta a}\right) \beta h$$

This quantity is an approximation for the external impedance of a short length h of a cylindrical post of small radius. The radiation resistance differs from that of a small dipole in space because in this case the fields generated are cylindrical waves in a restricted region having a distant field dependence of $r^{-\frac{1}{2}}$ rather than r^{-1}.

The impedance quantities derived above are referred to as external impedances because the fields within the post were zero since the post was said to be perfectly conducting. Next let us remove that assumption, presuming instead that the post is made of a typical metal, such that $\gamma_i^2 = -\omega^2\mu\epsilon_i \doteq j\omega\mu\sigma$. Within the post, the field must also be uniform in z, since the applied field is uniform in z. Thus an appropriate form for the field within the post is (the N_0 and H_0 functions are infinite at $\rho = 0$)

$$E_{z,\text{int}} = \frac{-\gamma_i^2 a_{0i} J_0(\beta_i \rho)}{j\omega\epsilon_i}$$
$$= -j\omega\mu a_{0i} J_0(\beta_i \rho)$$
$$H_{\varphi,\text{ind}} = -a_{0i}\beta_i J_0'(\beta_i \rho)$$

where

$$\beta_i = (-\gamma_i^2)^{\frac{1}{2}} \doteq (-j\omega\mu\sigma)^{\frac{1}{2}}$$

218 ANTENNA ENGINEERING

We assume that the planes are held at potential V_g, as before, and that the field is uniform along z. The external electric field is then

$$E_z = \frac{V_g}{h} + a_0(-j\omega\mu)H_0^{(2)}(\beta_0\rho)$$

wherein the latter term is the field generated by the induced currents and the term V_g/h is the applied field. Both tangential E and tangential H must be continuous across the surface of the cylinder; this implies the equations

$$a_{0i}(-j\omega\mu)J_0(\beta_i a) = \frac{V_g}{h} + a_0(-j\omega\mu)H_0^{(2)}(\beta_0 a)$$

$$a_{0i}\beta_i J_0'(\beta_i a) = a_0\beta_0 H_0^{(2)\prime}(\beta_0 a)$$

These equations may be solved for a_{0i}:

$$a_{0i}\left[(-j\omega\mu)J_0(\beta_i a) + j\omega\mu\frac{\beta_i}{\beta_0}\frac{J_0'(\beta_i a)}{H_0'(\beta_0 a)}H_0^{(2)}(\beta_0 a)\right] = \frac{V_g}{h} \qquad (27)$$

or

$$a_{0i} = \frac{V_g}{h[D]}$$

where $[D]$ is the bracket in Eq. (27).

To find the impedance in this case, we first find the total current in the cylinder I_T

$$I_T = \oint H_\varphi a\, d\varphi = \frac{-2\pi a V_g \beta_i J_0'(\beta_i a)}{h[D]}$$

The ratio of the voltage V_g to this current is the impedance of the post:

$$Z_{\text{in}} = \frac{V_g}{I_T} = \frac{-h[D]}{2\pi a \beta_i J_0'(\beta_i a)}$$
$$= \frac{h}{2\pi a}\left[j\left(\frac{\mu}{\epsilon_i}\right)^{1/2}\frac{J_0(\beta_i a)}{J_0'(\beta_i a)} - j\left(\frac{\mu_0}{\epsilon_0}\right)^{1/2}\frac{H_0^{(2)}(\beta a)}{H_0^{(2)\prime}(\beta a)}\right]$$

The impedance is seen to consist of a sum of two terms. The second term is the input impedance that was found earlier in the case that the post was perfectly conducting; note that the constant a_{0i} and the first term in the foregoing equation vanish in the case that the conductivity is infinite. The first term in the foregoing impedance expression thus clearly arises from the finite conductivity of the cylinder and the fields within the post and hence is commonly called the internal impedance

$$Z_{\text{int}} = \frac{h}{2\pi a}j\left(\frac{\mu}{\epsilon_i}\right)^{1/2}\frac{J_0(\beta_i a)}{J_0'(\beta_i a)}$$

When the wire is very thin or the frequency is low enough that the quantity $\beta_i a$ is small, we may employ the following approximations for the Bessel functions

$$J_0(\beta_i a) \doteq 1 - (\tfrac{1}{2}\beta_i a)^2$$
$$J_0'(\beta_i a) \doteq -J_1(\beta_i a) = \tfrac{1}{2}\beta_i a - \tfrac{1}{2}(\tfrac{1}{2}\beta_i a)^3$$

With these approximations, the equation for the internal impedance assumes a familiar form

$$Z_{\text{int}} \doteq \frac{h}{\pi a^2 \sigma} + \frac{j\omega\mu h}{8\pi} \qquad (\beta_i a \ll 1)$$

There is another interesting approximation that can be made; suppose that the conductivity and the frequency are both high enough so that $\beta_i a \gg 1$. In this case, large argument approximations for the Bessel functions may be employed. Consulting Jahnke and Emde, for example, for such asymptotic forms for the functions, we find, as the argument z tends to infinity

$$\frac{J_0(z)}{J_0'(z)} = -\frac{J_0(z)}{J_1(z)} \doteq -j$$

With this approximation, the internal impedance is

$$Z_{\text{int}} = \frac{\left(\frac{\mu}{\epsilon_i}\right)^{1/2} h}{2\pi a} = \frac{\left(\frac{\omega\mu}{2\sigma}\right)^{1/2}(1+j)h}{2\pi a}$$

In the event that $\beta_i a$ is neither large nor small compared to unity (say in the range from 0.25 to 10), the tables in Jahnke and Emde, pages 246 to 251, may be employed. In those pages, the following functions are tabulated: $\operatorname{Re} J_0(x\sqrt{i})$, $\operatorname{Im} J_0(x\sqrt{i})$, $\operatorname{Re} J_1(x\sqrt{i})$, $\operatorname{Im} J_1(x\sqrt{i})$. To make use of these tables, note that

$$J_0(x\sqrt{-j}) = \operatorname{Re} J_0(x\sqrt{i}) - j \operatorname{Im} J_0(x\sqrt{i})$$

and

$$J_1(x\sqrt{-j}) = \operatorname{Re} J_1(x\sqrt{i}) - j \operatorname{Im} J_1(x\sqrt{i})$$

The foregoing impedance formulas are useful in finding the circuit parameters of sections of round wires, or cylindrical posts or masts. They can be used in situations in which the fields over the length of the structure are almost independent of the length variable.

5.2.2 RING EXCITATION

Although the impedance formulas obtained in Sec. 5.2.1 are useful approximations for circuit-type problems, the uniform excitation condition is not a good approximation in general. A more realistic approximation for some situations is to say that the post is cut in two and excited symmetrically in φ by a voltage applied across the narrow gap. (The post is still presumed to be terminated in a pair of infinite parallel planes, but the plane separation h is no longer required to be small in comparison to the wavelength.) With the cut of width w placed around an origin at the center, the electric field on the surface of the perfectly conducting post can be written

$$E_z = \begin{cases} \dfrac{V_g}{w} & -\tfrac{1}{2}w < z < \tfrac{1}{2}w \\ 0 & \text{otherwise} \end{cases} \qquad (27a)$$

This function may be expanded in a Fourier series

$$E_z = c_0 + \sum_{n=1}^{\infty} c_n \cos 2n\pi \frac{z}{h} \qquad (27b)$$

with

$$c_0 = \frac{V_g}{h} \qquad c_n = \frac{V_g}{w} \int_{-\frac{1}{2}w}^{\frac{1}{2}w} \cos 2n\pi \frac{z}{h}\, dz$$

$$= -V_g \frac{2}{n\pi w} \sin n\pi \frac{w}{h}$$

With no dependence on φ, the general form for the electric field in the cylindrical region exterior to the post is

$$E_z = \sum_{n=0}^{\infty} \left(\frac{\beta_0^2 - \beta_{zn}^2}{j\omega\epsilon}\right) H_0^{(2)}(\beta_n \rho)(a_n \cos \beta_{zn} z + b_n \sin \beta_{zn} z)$$

where $\beta_n = (\beta_0^2 - \beta_{zn}^2)^{1/2}$. On the surface of the post this expression for the electric field must match the electric field expressed above in Eqs. (27). The two expressions are equal if

$$\beta_{zn} = \frac{2n\pi}{h} \quad \text{and} \quad a_n = \frac{j\omega\epsilon c_n}{(\beta_0^2 - \beta_{zn}^2) H_0^{(2)}(\beta_n a)}$$

The magnetic field in the exterior region is then

$$H_\varphi = -\beta_0 \left[\frac{j}{\omega\mu} c_0 \frac{H_0^{(2)\prime}(\beta_0 \rho)}{H_0^{(2)}(\beta_0 \rho)} + \sum_{n=1}^{\infty} \frac{j\omega\epsilon}{(\beta_0^2 - \beta_{zn}^2)} c_n \frac{H_0^{(2)\prime}(\beta_n \rho)}{H_0^{(2)}(\beta_n a)} \cos 2n\pi \frac{z}{h}\right]$$

The current in the post at $z = \tfrac{1}{2}w$ is $2\pi a H_\varphi \big|_{\rho=a}$. The input admittance may be found from the ratio I_{in}/V_g, which will clearly be an infinite series. The first term in the series is the same as the external imped-

5.2.3 THE FIELD OF A CIRCULAR ARRAY OF SHORT DIPOLES IN SPACE

Finally, let us determine the vector potential for a continuous array of z-directed point dipoles arranged in a circle of radius a. The result will be useful in the next section, as well as being of practical interest in itself.

The equation for the vector potential in cylindrical coordinates is

$$\frac{1}{\rho}\frac{\partial}{\partial \rho}\left(\rho \frac{\partial A_z}{\partial \rho}\right) + \frac{\partial^2 A_z}{\partial z^2} = \gamma^2 A_z - \mathcal{J}_z$$

where

$$\mathcal{J}_z = \frac{I}{2\pi a}\delta(z - z_0)\,\delta(\rho - a)$$

In this case, it is helpful to remove the z variation by means of a Fourier transform technique:

$$a_z(\rho,k) = \int_{-\infty}^{\infty} A_z(\rho,z) e^{-jkz}\,dz$$

The transformed differential equation is then

$$\frac{1}{\rho}\frac{\partial}{\partial \rho}\left(\rho \frac{\partial a_z}{\partial \rho}\right) - \gamma^2 a_z - k^2 a_z = -\frac{I}{2\pi a}\delta(\rho - a)e^{-jkz_0}$$

The source on the right-hand side of this equation is now essentially an infinite cylindrical sheet of current. In the region $\rho > a$, there are no sources; consequently, we expect, as in Sec. 5.2.1, that the field will vary as $a_{0,2}H_0^{(2)}(\beta_k\rho)$, while in the interior region (since the other Bessel functions become infinite at $\rho = 0$) we expect the solution to have the form $a_{0,1}J_0(\beta_k\rho)$. At the current sheet itself, $\rho = a$, the tangential electric field should be continuous while the tangential magnetic field should be discontinuous by the amount of the surface current density in the sheet. The application of these boundary conditions gives the following equations for the constants $a_{0,1}$ and $a_{0,2}$:

$$a_{0,1}J_0(\beta_k a) + a_{0,2}H_0^{(2)}(\beta_k a) = 0$$

$$-a_{0,1}J_0'(\beta_k a) + a_{0,2}H_0^{(2)\prime}(\beta_k a) = -\frac{I}{2\pi a \beta_k}e^{-jkz_0}$$

where

$$\beta_k^2 = \beta^2 - k^2$$

Solving this pair of equations for a_{02} and employing the wronskian relationship

$$J_0(x)H_0^{(2)'}(x) - H_0^{(2)}(x)J_0'(x) = -j\frac{2}{\pi x}$$

we find

$$a_{0,2} = \frac{IJ_0(\beta_k a)}{4j}$$

or

$$a_{z,\text{ext}} = \frac{I}{4j}e^{-jkz_0}J_0(\beta_k a)H_0^{(2)}(\beta_k \rho)$$

We can now apply the inverse transform to find $A_{z,\text{ext}}$

$$\begin{aligned}A_{z,\text{ext}} &= \frac{1}{2\pi}\int_{-\infty}^{\infty} a_z(\rho,k)e^{jkz}\,dk \\ &= \frac{I}{8\pi j}\int_{-\infty}^{\infty} J_0(\beta_k a)H_0^{(2)}(\beta_k \rho)e^{-jk(z_0-z)}\,dk\end{aligned} \quad (28)$$

where $\beta_k{}^2 = \beta^2 - k^2$. The evaluation of this inverse transformation is not trivial but it can be done numerically or with approximations for the distant-field region.

5.3 CYLINDRICAL ANTENNA THEORY

As mentioned in the introduction of this chapter, the theoretical alternative to the spherical-mode theory of antennas is based on an approximate solution of an integral equation. We shall consider this method now. The credit for the development of the theory belongs mainly to Hallén, and R. W. P. King, among others. An interesting survey paper, with a Fourier series solution of the Hallén equation, has been given by Duncan and Hinchey.[1] The method presented here is similar to the latter, but differs from it slightly in the system of numerical calculation.

In the case of a hollow cylindrical antenna which is symmetrically excited (Fig. 5.2), the surface current has only a z component, so the magnetic vector potential **A** has likewise only a z component. Then since **H** = $\nabla \times$ **A** and **E** = $(1/j\omega\epsilon)\nabla \times$ **H** = $(1/j\omega\epsilon)\nabla \times \nabla \times$ **A**, it follows that

$$j\omega\epsilon E_z = \frac{\partial^2}{\partial z^2}A_z + \beta^2 A_z \quad (29)$$

We may employ this equation in the development of an integral equation for the current distribution. For on the surface of the cylindrical antenna

[1] R. H. Duncan and F. Hinchey, *J. Res. Natl. Bur. Std.* **64D**:569 (September–October, 1960).

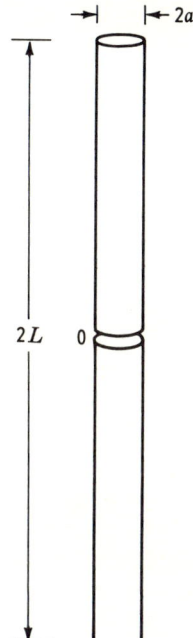

Fig. 5.2   Model cylindrical antenna.

we may presume that the electric field is known. In particular, it is zero over the whole surface except at a slot gap in which the source is located. This source region may be visualized either as a gap across which a voltage is developed by a circuit-type generator, or as a magnetic current ring on the surface of the perfectly conducting cylinder (i.e., with no gap in the metal). The electric field across the small source region E_z is related to the voltage at the input as follows

$$E_z = \begin{cases} \dfrac{V}{z_2 - z_1} & z_1 < z < z_2 \\ 0 & \text{otherwise} \end{cases}$$

where z_2 and z_1 are the values of z at the extremes of the input region. Equation (29) must apply at every point on the surface of the antenna. Thus, the vector potential at the surface of the antenna A_{zs} must satisfy the equation

$$\frac{\partial^2}{\partial z^2} A_{zs} + \beta^2 A_{zs} = \frac{j\omega\epsilon V}{(z_2 - z_1)} [U(z - z_1) - U(z - z_2)] \tag{30}$$

where the right-hand side acts like a localized source term for the differential equation. The complete solution to this equation consists of the solution to the homogeneous equation (i.e., the equation with the right-hand side equal to zero in which case the solutions are sines, cosines, or exponentials) plus the particular solution. The particular solution may be found by appropriate means, to be given below; the complete solution may then be written

$$A_{zs} = Ae^{-j\beta z} + Be^{j\beta z} + \text{Particular Solution} \tag{31}$$

On the other hand, A_{zs} must satisfy another equation, for it must be given by an integral involving the current distribution on the surface of the antenna. This integral may be expressed in alternative ways. The fundamental form is that given in Chap. 1, Eq. (17), based on the field of an elementary dipole. However, with the antenna we are studying, the current is confined to a cylindrical surface which means that it can be regarded as a superposition of a continuum of circular arrays (rings) of z-directed dipoles. We considered such circular arrays in Sec. 5.2.3. In particular, Eq. (28) gives the vector potential for one of these arrays, located at position z_0. We can make use of that solution; we can obtain the quantity A_{zs} by placing $\rho = a$ and restricting the z values to points on the antenna surface. For convenience, let us change the variable z_0 to z' (locating source points). Then let the current I be a function of z', $I(z')$. Then, by superposition of the results for several rings, each with a vector potential as given in Eq. (28), we can find the vector potential for the complete current distribution

$$\begin{aligned}A_{zs} &= \int_{-L}^{L} \frac{I(z')}{8\pi j} \int_{-\infty}^{\infty} J_0(\beta_k a) H_0^{(2)}(\beta_k a) e^{jk(z-z')} \, dk \, dz' \\ &= \frac{1}{8\pi j} \int_{-\infty}^{\infty} J_0(\beta_k a) H_0^{(2)}(\beta_k a) e^{jkz} \int_{-L}^{L} I(z') e^{-jkz'} \, dz \, dk \end{aligned} \tag{32}$$

Equations (31) and (32) are two equations for the same quantity, A_{zs}. Setting them equal gives an integral equation for the current distribution.

Before studying the integral equation in detail, let us determine a particular solution for Eq. (30). In the literature, a very popular source has been the impulse, or delta, distribution for E, $E_{zs} = V\,\delta(z)$. With this type of source distribution, the differential equation is

$$\frac{d^2 A_{zs}}{dz^2} + \beta^2 A_{zs} = j\omega\epsilon V\,\delta(z)$$

and its solution may be conveniently written

$$A_{zs} = \begin{cases} C \cos\beta z + D \sin\beta z & 0 < z < L \\ C_1 \cos\beta z + D_1 \sin\beta z & -L < z < 0 \end{cases}$$

These solutions hold everywhere except at the point $z = 0$. A solution that includes the source point may be written as follows: Consider the

INTRODUCTION TO THE MATHEMATICAL THEORY OF ANTENNAS

form

$$A_{zs} = C \cos \beta z + D \sin \beta |z| \qquad (33)$$

With this form we have

$$\frac{d}{dz} A_{zs}\bigg]_{0+} = D\beta \qquad \frac{d}{dz} A_{zs}\bigg]_{0-} = -D\beta$$

$$\frac{d^2}{dz^2} A_{zs} = -\beta^2 C \cos \beta z + 2D\beta \, \delta(z) - \beta^2 D \sin \beta |z|$$

$$= -\beta^2 A_{zs} + 2D\beta \, \delta(z)$$

Thus, Eq. (33) is a solution to the differential equation with delta-function excitation provided $2D\beta = j\omega\epsilon V$, which determines the constant D.

Thus with delta-function excitation, the integral equation for the current is

$$\frac{1}{8\pi j} \int_{-\infty}^{\infty} J_0(\beta_k a) H_0(\beta_k a) e^{jkz} \int_{-L}^{L} I(z') e^{-jkz'} \, dz' \, dk$$

$$= C \cos \beta z + \frac{j\omega\epsilon V}{2\beta} \sin \beta |z| \qquad (34)$$

At the present time, analytic solutions to this integral equation are unavailable. However, with the help of a digital computer, some good approximations can be obtained. One of the more straightforward procedures is to change the integral equation into a large number of algebraic equations. This can be done by expanding the current distribution in a Fourier series with coefficients to be determined. To facilitate the actual computations, it is also helpful to represent the z variation of the vector potential in the form of a Fourier series. We shall accomplish the latter first. The right-hand side of Eq. (34) is clearly an even function of z; this means that the left-hand side must also be even. Let us therefore expand both sides, separately, in Fourier cosine series:

$$C \cos \beta z + \frac{j\omega\epsilon V}{2\beta} \sin \beta |z| = \sum_{m=0}^{\infty} A_m \cos \frac{m\pi z}{L}$$

where

$$A_m = \frac{1}{L} \int_{-L}^{L} \left(C \cos \beta z + \frac{j\omega\epsilon V}{2\beta} \sin \beta |z| \right) \cos \frac{m\pi z}{L} \, dz \qquad m \neq 0$$

For the left-hand side to have a form

$$\text{LHS} = \sum_{n=0}^{\infty} B_m \cos \frac{m\pi z}{L}$$

the coefficients B_m must satisfy the equation

$$B_m = \frac{1}{8\pi j L} \int_{-L}^{L} \int_{-\infty}^{\infty} J_0(\beta_k a) H_0(\beta_k a) e^{jkz} \int_{-L}^{L} I(z') e^{-jkz'} dz' \, dk \, \cos\left(\frac{m\pi z}{L}\right) dz$$

$$= \frac{1}{8\pi j L} \int_{-\infty}^{\infty} J_0(\beta_k a) H_0(\beta_k a) \int_{-L}^{L} I(z') e^{jkz'} dz' \int_{-L}^{L} e^{jkz} \cos\left(\frac{m\pi z}{L}\right) dz \, dk$$

The integral on z is

$$e^{jkz}\left[\frac{jk \cos(m\pi z/L) + (m\pi/L)\sin(m\pi z/L)}{(m\pi/L)^2 + k^2}\right] = \frac{2jk \cos m\pi \cos kL}{(m\pi/L)^2 + k^2}$$

Next we can expand the unknown current distribution in a Fourier series

$$I(z') = \begin{cases} \sum_{n=0} I_n \cos \frac{(2n+1)\pi z'}{2L} & -L < z < L \\ 0 & \text{otherwise} \end{cases}$$

thus

$$\int_{-L}^{L} I(z') e^{-jkz'} dz' = \sum_{n=0} I_n \int_{-L}^{L} \cos \frac{(2n+1)\pi z'}{2L} e^{-jkz'} dz'$$

The latter integral is well known. With the notation $p_n = (2n+1)\pi/2L$ to save writing, the formula for the coefficients B_m now reads

$$B_m = \frac{1}{2\pi L} \sum_{n=0} I_n p_n \sin p_n L \cos m\pi \int_{-\infty}^{\infty} \frac{\cos^2 kL \, J_0(\beta_k a) H_0(\beta_k a) \, dk}{(p_n^2 - k^2)[(m\pi/L)^2 - k^2]}$$

$$\beta_k^2 = (\beta^2 - k^2)^{1/2}$$

For brevity, write this in the form

$$B_m = \frac{1}{2\pi L} \sum_{n=0} I_n C_{nm}$$

Then finally, because the corresponding terms in a Fourier series must be equal term by term, $B_m = A_m$, or

$$\frac{1}{2\pi L} \sum_{n=0} I_n C_{nm} = A_m \tag{35}$$

Equation (35) is a set of algebraic equations in the unknown current coefficients I_n. To find the current distribution, one decides on the number of terms required in a satisfactory approximation for the current, say 10, for example. Then, Eq. (35) can be written down for 10 different values of the parameter m. This would seem to give 10 equations in the 10 unknown current coefficients I_n. However, if the development is examined, it will be seen that the coefficients A_m contain one arbitrary constant C. Thus, for a solution, Eq. (35) should be written down 11 times, with 11 values of the parameter m, in order that all the constants are determined (in general write one more equation than the number of unknown current coefficients).

Of course, the evaluation of the coefficients C_{nm} is not trivial

$$C_{nm} = p_n \sin p_n L \cos m\pi \int_{-\infty}^{\infty} \frac{\cos^2(kL) J_0(\beta_k a) H_0^{(2)}(\beta_k a)\, dk}{(p_n^2 - k^2)((m\pi/L)^2 - k^2)}$$

but it can be evaluated numerically as a subroutine in the computer program.

In the final result, the coefficients I_n are proportional to the excitation voltage V. Thus, the input admittance is determined once the current coefficients are known

$$Y_{\text{in}} = \frac{I(0)}{V} = \frac{1}{V} \sum_{n=0}^{N} I_n$$

and V may be set to unity.

An interesting survey and summary of results pertaining to the linear antenna, including relatively long structures, has been given by King.[1]

[1] R. W. P. King, *Proc. IEEE*, **55**:2 (January, 1967).

 # APERTURE ANTENNAS

The sources for electromagnetic fields are always, apparently, electric currents. However, the electric current distribution is often unknown, and in certain structures, it may be quite a complicated function of position. This is particularly true with radiators in the forms of slits, slots, horns, reflectors, and lenses. Consequently, with these types of radiators, the theoretical work is usually not based on the primary current distributions; rather, the results are obtained with the aid of what is known as *aperture theory*.

The basic idea behind aperture theory is simple and sound. It is based upon the fact that an electromagnetic field in a source-free closed region is completely determined by the values of tangential E or tangential H on the surface of the closed region. For exterior regions, a boundary condition at infinity may be employed, in effect, to close the region. In the analysis of many antennas, bounding surfaces may be selected so that much or most of the surface coincides with conducting material. In such cases, although the current distribution on the conductors may be very complex and hard to determine, the tangential electric field is known over most of the surface—it is zero. If, in the aperture(s), the tangential electric field is known or can be easily approximated, then enough information is available to determine the fields everywhere within the surface.

In practice, the mathematical formulation of the idea expressed in the last paragraph is simplified by the use of Huygens' principle and the electromagnetic Huygens sources. In its original form, Huygens' principle simply states that each point on a wavefront acts as a new source of spherical waves. In about the year 1900, A. E. H. Love and H. M. MacDonald developed and extended the principle for application in electromagnetic theory. In effect, they defined a closed surface about

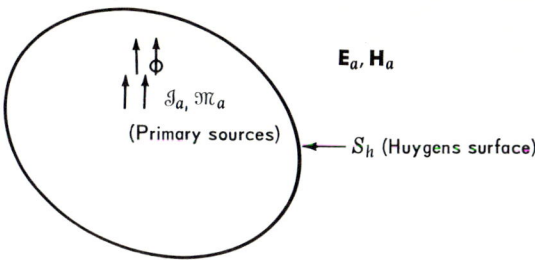

Fig. 6.1 Huygens surface around primary sources. Huygens sources consisting of surface currents on S_h generate $\mathbf{E}_a$, $\mathbf{H}_a$ outside S_h but zero field inside S_h.

the primary source (called a Huygens surface) and showed that the electromagnetic wave sources on this surface which produce the primary field outside should be surface currents specifically related to the tangential E and H on the Huygens surface. This idea was developed further in the mid-thirties by S. A. Schelkunoff[1] and called by him the *equivalence theorem*.

In essence, Huygens' principle or the equivalence theorem shows how to replace an actual source by a set of equivalent sources spread over a specified closed surface S_h. To help understand how this is possible, consider the following argument: Suppose a primary source $\mathcal{J}_a$, $\mathfrak{M}_a$ sets up a field $\mathbf{E}_a$, $\mathbf{H}_a$ everywhere (Fig. 6.1). Now enclose the primary sources in an imaginary closed surface S_h, and then remove the primary sources. Let us now see if we can find a solution to Maxwell's equations such that the field outside S_h is $\mathbf{E}_a$, $\mathbf{H}_a$ and the field inside S_h is zero. (This field is identical to the field produced by the primary sources $\mathcal{J}_a$, $\mathfrak{M}_a$ in the region outside of the surface S_h.) Such a solution would have discontinuities in tangential E and tangential H on the surface S_h. In order that the fields satisfy Maxwell's equations and the boundary conditions, it is necessary that on S_h we have a distribution of electric surface currents $\mathcal{J}_h = \hat{\mathbf{n}} \times \mathbf{H}_a(S_h)$, to account for the discontinuity in tangential H, and a distribution of magnetic surface currents, $\mathfrak{M}_h = \mathbf{E}_a(S_h) \times \hat{\mathbf{n}}$, to account for the discontinuity in tangential E. But with our hypothetical field, $\mathcal{J}_h$ and $\mathfrak{M}_h$ are the only sources; thus we conclude that they generate the field $\mathbf{E}_a$, $\mathbf{H}_a$ outside S_h, but zero field inside. The Huygens sources are then surface currents as follows:

$$\mathcal{J}_h = \hat{\mathbf{n}} \times \mathbf{H}_a(S_h) \qquad \mathfrak{M}_h = \mathbf{E}_a(S_h) \times \hat{\mathbf{n}}$$

In a field calculation, these equivalent sources are treated in the same way as any primary source distribution.

[1] S. A. Schelkunoff, *Bell System Tech. J.*, 15:92 (January, 1936).

However, because the field is zero inside a Huygens surface, a further simplification is possible. Material may be introduced into the zero-field region without changing the field anywhere. So, suppose we line the Huygens surface (inside) with perfectly conducting material. This does not change the field. On the other hand, we know that such a lining will nullify the effect of the electric currents on the Huygens surface; this is so because the current distribution $- \mathcal{J}_h$ will be induced on the conducting surface, resulting in an out-of-phase image, essentially coincident in space. It follows that the fields outside may be calculated from the magnetic currents alone, these acting in the presence of the conducting lining.

6.1 SLOT ANTENNAS

For fast-moving vehicular applications, it is usually necessary to design antennas that do not protrude appreciably from the surface of the vehicle. Slot antennas were developed to satisfy this requirement. In practice, it is usually necessary to enclose one side of a slot antenna in order to contain the fields; the enclosure often takes the form of a resonant cavity. Initially, we shall ignore this feature and simply focus our attention on the properties of the slotted surface itself.

Consider first a slot in a conducting plane. To help initially with the understanding, imagine first a transmission line consisting of a pair of conducting half-planes set up in the same plane, edge to edge, with a separation $2w$ between the edges. Such a transmission line could be excited, for example, by a current source connected from one plane to another, or, in practice, by means of a coaxial line whose shield is connected to one of the half-planes and whose center conductor is extended across to the other (Fig. 6.2). The voltage across the half-plane line would be that of an infinite (matched) transmission line. The electric field would have a similar variation in the transmission direction z and would be directed across the slot opening. If, however, the line were short-circuited at two positions equally distant from the feed point (and in such a case the gap between the short circuits might as well be filled with metal, from the short circuits on outward), a standing wave would be set up, and the voltage distribution (and E field) in the transmission direction would be those of a shorted transmission line. And although there would be the characteristic standing-wave pattern in the transmission direction, the electric field would still have the same variation in the direction across the slot as before. But with the short circuits and the remainder of the opening between the planes filled in, we have evolved the transmission structure into a rectangular slot antenna having an electric field in the slot as indicated.

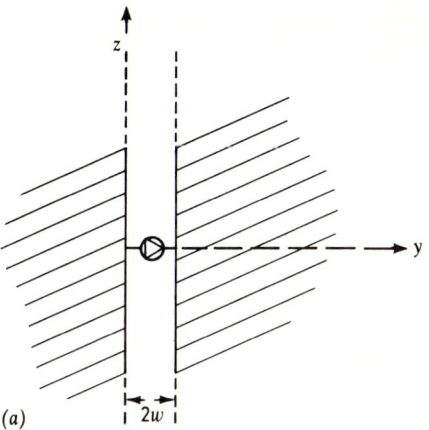

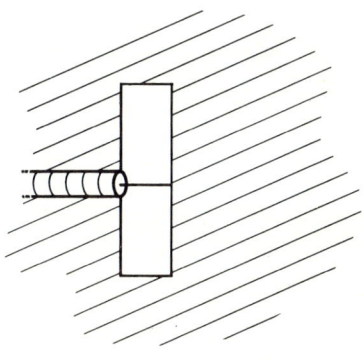

Fig. 6.2 (a) Pair of semi-infinite planes acting as a transmission line, excited by a current source. (b) The plane-transmission lines in (a) after they have been shorted, excited by coaxial system (this makes a slot antenna).

Let the length of the slot be $2L$ and its width be $2w$. Then, taking the y direction in the plane of the slot, the electric field in the opening can be written in the functional form

$$E_y = \begin{cases} V(y,z) & -w < y < w \\ 0 & \text{otherwise} \end{cases}$$

where

$$V(y,z) = \begin{cases} V_0(y) \sin \beta(L - |z|) & -L < z < L \\ 0 & \text{otherwise} \end{cases}$$

The magnetic current equivalent to this electric field is

$$\mathfrak{M} = \mathbf{E} \times \hat{\mathbf{n}} = \begin{cases} V_0 \sin \beta(L - |z|) \hat{\mathbf{y}} \times \hat{\mathbf{x}} & \begin{cases} -w < y < w \\ -L < z < L \end{cases} \\ 0 & \text{otherwise} \end{cases} \quad (1)$$

To calculate the fields on one side of the plane, say the $+x$ side, we may take the whole plane as the Huygens surface. Since it is a Huygens surface, we may line one side of the plane completely with a perfect conductor and thereafter compute the fields from the magnetic currents alone, as pointed out above. These magnetic currents [Eq. (1)] are spread on the conducting plane in the position of the aperture. By image considerations, the field on one side of the plane is then just twice the field generated by the magnetic currents acting in free space.

To find the radiation pattern, we may employ the idea of *duals*. The solution of Maxwell's equations with magnetic current sources may be obtained from the solutions which hold for the corresponding electric current sources by the appropriate interchange of E and H and ϵ and μ.[1] The radiation pattern of a narrow slot antenna is then essentially the pattern of an electric dipole antenna having a sinusoidal current distribution, except that E and H are interchanged. The exact form of the y variation of $\mathfrak{M}$ in the narrow slot is of no primary importance; it may be assumed to be uniform.

The input impedance of the slot antenna may also be deduced from the properties of dual solutions. Since the fields of the slot and dipole are dual to one another, a careful examination and comparison shows that the slot input impedance Z_s and the complementary dipole impedance Z_d are related by the equation

$$Z_s Z_d = \frac{1}{4}\frac{\mu}{\epsilon} \tag{2}$$

This equation may also be derived from Booker's extension of Babinet's principle.[2] Equation (2) is immediately useful since it may be employed to find the impedance of slot antennas, given the impedance of the complementary dipole antennas. For Eq. (2) to be highly accurate, the metallic plane sheet must be very thin, and of course the impedance of the complementary dipole must be accurately known (in this case, it is a thin-strip dipole). The effect of the sheet thickness is to decrease the slot impedance near resonance.

In practical applications, it is usually necessary to enclose the slot antenna on one side of the plane. The enclosure usually forms a resonant cavity. One type of cavity backing consists of a shorted section of rectangular waveguide. The short circuit is placed one-quarter of a TE_{10} mode wavelength from the slot (see Fig. 6.3). To a first approximation, the short circuit transforms into an open circuit at the slot; thus, since the

---

[1] W. L. Weeks, "Electromagnetic Theory for Engineering Applications," pp. 50–51, John Wiley & Sons, Inc., New York, 1966.
[2] *Ibid.*, pp. 445–462.

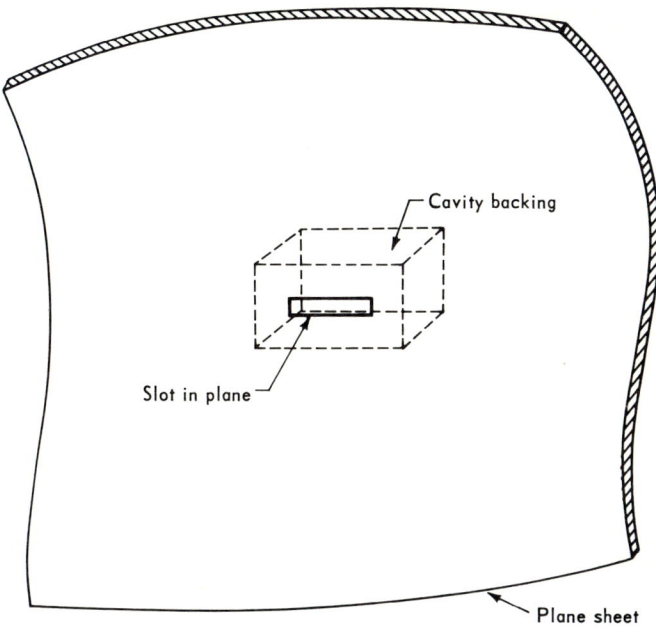

Fig. 6.3 Slot antenna with cavity backing.

radiation is confined to one side of the plane, the input admittance is one-half that of the open-plane structure. The cavity changes the input impedance variation with frequency in a complicated way.

Typical values of slot-antenna impedance are rather inconveniently high for some applications (about 400 ohms or more for the plane structures and 800 or more for the cavity-backed structures). Armed with Eq. (2) relating the impedances of a slot and its complementary dipole, and recalling the fact that a folded dipole has a *higher* impedance than the linear type (higher by a factor of about 4 or more), the engineer is led to consider a folded slot antenna, as sketched in Fig. 6.4. The detailed consideration of such a folded slot antenna is left as a problem.

In aircraft applications particularly, there is frequently a need for an antenna which is flush-mounted but whose radiation pattern is like that of a monopole antenna oriented perpendicular to the surface. One solution to this problem is the annular slot antenna. This structure resembles a large coaxial line whose conductors have been joined to an infinite ground plane in such a way as to leave the coaxial line open-circuited at the termination (Fig. 6.5). In fact, the circular slot is often excited in this way since, to obtain the desired pattern, the electric field in the slot should be radial. With such a radial electric field, the equivalent magnetic current is directed along the ring. Note that the dual electric current structure is a thin loop antenna; thus, from our knowledge of loop antennas, it follows that there is a null in the field on the axis of the ring and that the pattern is azimuthally symmetric, with the polarization such that the electric field is perpendicular to the conducting plane. If the conducting plane is infinite, the maximum in the radiation pattern lies in the plane of the slot; the effect of a finite ground plane is to elevate the maximum slightly. All the foregoing statements concerning the pattern were based on the assumption that the slot diameter is small compared with a wavelength. Unfortunately, the total radiation from small rings is small, and it increases as the size of the ring increases.

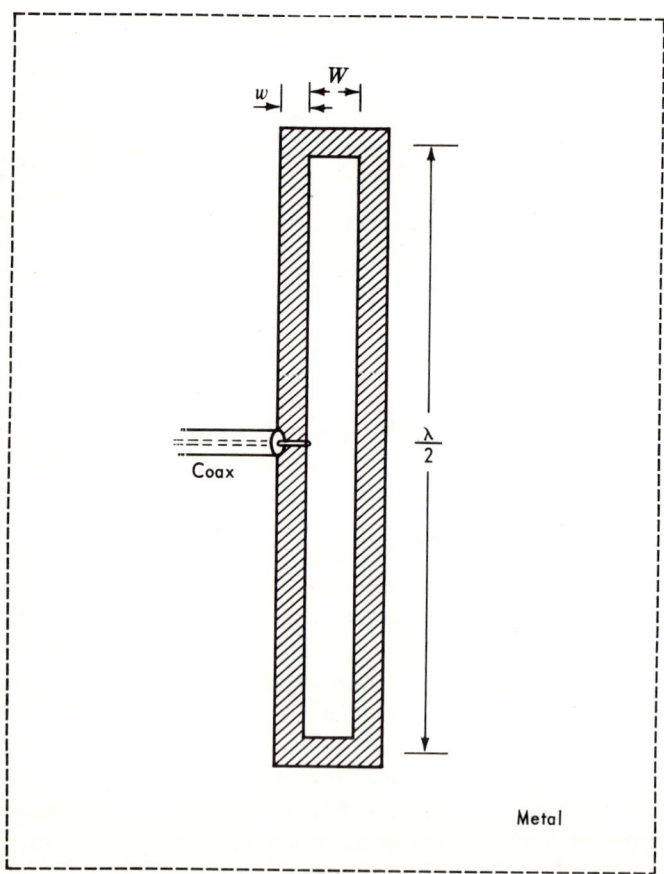

Fig. 6.4 Folded slot antenna in metal plane.

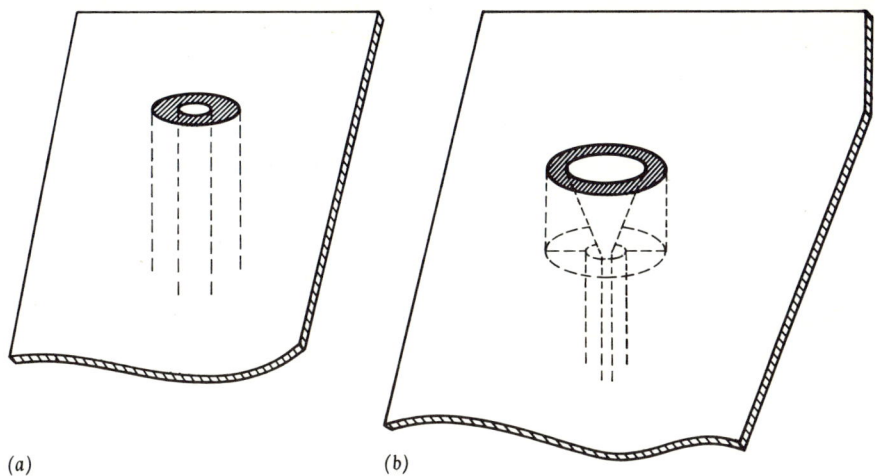

Fig. 6.5 Annular slot antennas: (*a*) Coaxial line terminated in a plane; (*b*) annular slot with cavity backing.

Problem 6.1 A slot antenna is cut in a large conducting plane as indicated in Fig. 6.4. Calculate the input impedance and the radiation pattern of the structure. Then design a cavity backing for one side of the structure, and estimate its effect on the input impedance.

6.2 SLOTTED CYLINDER ANTENNAS

Some of the vehicles on which slot antennas are employed have shapes that can be approximated by circular cylinders. Having learned about the performance of slot antennas in planes, the next question is: "What is the effect of curvature of the surface on the radiation patterns of slot antennas?" The effect of curvature of the surface in one dimension is relatively easily displayed by the consideration of a slot antenna on a conducting cylinder.

Consider a slot in an infinite conducting cylinder, and let it be excited by a transverse voltage. We shall suppose that the fields interior to the cylinder are contained in a manner that is not of concern to us at the present. Let the slot be infinitely long and narrow. Then as long as the cylinder is large enough in diameter that it does not introduce a direct short circuit across the slot, the electric field distribution in the slot is not significantly different from that of a similarly excited slot in a conducting plane. As a specific example, and as a model having some flexibility for future use, let us suppose that the electric field is a traveling wave distribution along z and is almost uniform across the slot, with the voltage such that $V = \int_{-\varphi_1}^{\varphi_1} V_0 a \, d\varphi$:

$$E_\varphi \Big]_{\rho=a} = \begin{cases} V_0 e^{-jkz} & -\varphi_1 < \varphi < \varphi_1 \\ 0 & \text{otherwise} \end{cases}$$

This equation is a boundary condition on the electric field over the complete cylindrical surface. With this boundary condition and the solutions to Maxwell's equations in cylindrical coordinates, we may find the exterior fields. Note that the magnetic current that is equivalent to the prescribed electric field in the slot is z directed. This suggests that the solutions can be obtained in terms of a "TE to z" type of field analysis.[1] It will be recalled that in this type of analysis the electric fields are related to a scalar g as follows:

$$\mathbf{E} = \nabla \times g\hat{\mathbf{z}}$$

where g satisfies

$$\nabla^2 g = \gamma^2 g \tag{3}$$

and is proportional to the z component of the "electric vector potential" (Prob. 1.2). In the cylindrical coordinate system (ρ,φ,z), the separable solutions are

$$g = R(\rho)\Phi(\varphi)Z(z)$$

where the solutions for the ρ variation are given in terms of the Bessel functions of the first and second kinds

$$R(\rho) = A_1 J_p(\beta\rho) + B_1 N_p(\beta\rho)$$

in which $\beta = (\gamma_z^2 - \gamma^2)^{1/2}$. The z and φ solutions are exponential and trigonometric functions

$$Z(z) = A_2 e^{-\gamma_z z} + B_2 e^{\gamma_z z}$$

$$\Phi(\varphi) = A_3 \cos p\varphi + B_3 \sin p\varphi$$

The A's and B's are arbitrary constants. From the curl equation relating E and g, the φ component of $\mathbf{E}$ is

$$E_\varphi = -\frac{\partial g}{\partial \rho} = -\Phi(\varphi) Z(z) \frac{\partial R}{\partial \rho}$$

The axis of the cylinder lies at $\rho = 0$. Thus, exterior to the cylinder, the ρ variation that is noninfinite at infinity and represents waves traveling away from the cylinder is

$$R(\rho) = A H_p^{(2)}(\beta\rho) = A(J_p(\beta\rho) - jN_p(\beta\rho))$$

In the exterior region, the variable φ covers the complete range, 0 to 2π.

---

[1] Weeks, *op. cit.*, pp. 298, 469, and 703.

The solutions $\Phi(\varphi)$ must therefore be periodic with period 2π, and this implies that p must be an integer n. Each order n of the functions is a solution to the equations; thus the following sum is also a solution:

$$E_\varphi = \sum_{n=0} A_n \cos n\varphi \, e^{-j\beta_z z} \frac{\partial H_n^{(2)}(\beta\rho)}{\partial \rho}$$

This solution will satisfy the boundary conditions if we force it to be equal to the stated E field at the surface of the cylinder, $\rho = a$:

$$\sum_{n=0} A_n \cos n\varphi \, e^{-j\beta_z z} \beta H_n^{(2)\prime}(\beta a) = \begin{cases} V_0 e^{-jkz} & -\varphi_1 < \varphi < \varphi_1 \\ 0 & \text{otherwise} \end{cases}$$

where

$$H_n^{(2)\prime}(\beta a) = \frac{\partial H_n^{(2)}(\beta\rho)}{\partial(\beta\rho)}\bigg]_{\rho=a}$$

The equation can be satisfied only if $\beta_z = k$. With the z variation thus canceled out, the left-hand side of the equation is basically a Fourier series. The coefficients A_n can be found in the usual way by multiplying through by $\cos m\varphi$ and integrating over a complete period. In this way, it is easily established that

$$A_0 = \frac{V_0}{\beta H_0^{(2)\prime}(\beta a)} \frac{\varphi_1}{\pi}$$

$$A_n = \frac{V_0}{\beta H_n^{(2)\prime}(\beta a)} \frac{2 \sin n\varphi_1}{n\pi}$$

We have therefore an explicit expression for the field

$$E_\varphi = \frac{V_0}{H_0^{(2)\prime}(\beta a)} \frac{\varphi_1}{\pi} e^{-jkz} H_0^{(2)\prime}(\beta\rho) + \sum_{n=1}^{\infty} \frac{2V_0 \sin n\varphi_1 \, e^{-jkz}}{H_n^{(2)\prime}(\beta a) n\pi} \cos n\varphi H_n^{(2)\prime}(\beta\rho)$$

In the distant-field region ($\beta\rho \to \infty$), the asymptotic forms for the Hankel functions may be employed:

$$H_n^{(2)}(\beta\rho) \Rightarrow \left(\frac{2}{\pi\beta\rho}\right)^{1/2} \exp\left(-j\beta\rho + j\left(\frac{2n+1}{4}\right)\pi\right)$$

thus

$$H_n^{(2)\prime}(\beta\rho) \Rightarrow -jH_n(\beta\rho)$$

The radiation field is then

$$E_\varphi = \frac{V_0}{\pi} \frac{\varphi_1}{H_0^{(2)\prime}(\beta a)} e^{-j\frac{1}{4}\pi} \left(\frac{2}{\pi\beta\rho}\right)^{1/2} e^{-j(kz+\beta\rho)}$$

$$+ V_0 2 \left(\frac{2}{\pi\beta\rho}\right)^{1/2} e^{-j(kz+\beta\rho)} \sum_{n=1}^{\infty} \frac{\sin n\varphi_1}{H_n^{(2)\prime}(\beta a)} \frac{\cos n\varphi}{n\pi} e^{j\frac{(2n-1)\pi}{4}}$$

238 ANTENNA ENGINEERING

The convergence of this series depends strongly on the functions $H_n^{(2)'}(\beta\rho)$, all of which become infinite as βa tends to zero.

The radiation patterns described by this series are indicated in Fig. 6.6. For very small βa, the first term alone predominates, which of course means that the radiation is essentially omnidirectional. As the cylinder diameter increases, the radiation concentrates on the slot side of the cylinder.

Slots may be filled with low-loss dielectric material with little change

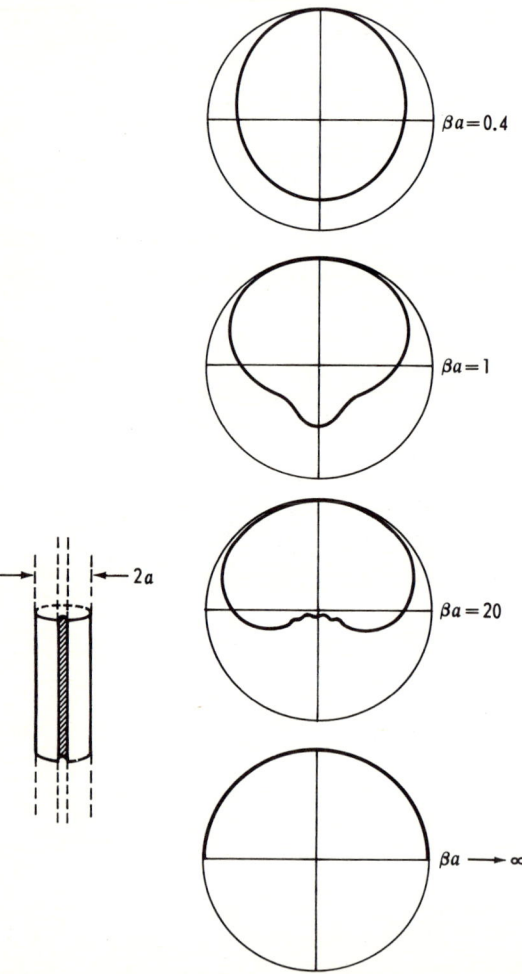

Fig. 6.6a Radiation patterns $E\varphi(\varphi)$ of a narrow slot on an infinite cylinder.

APERTURE ANTENNAS 239

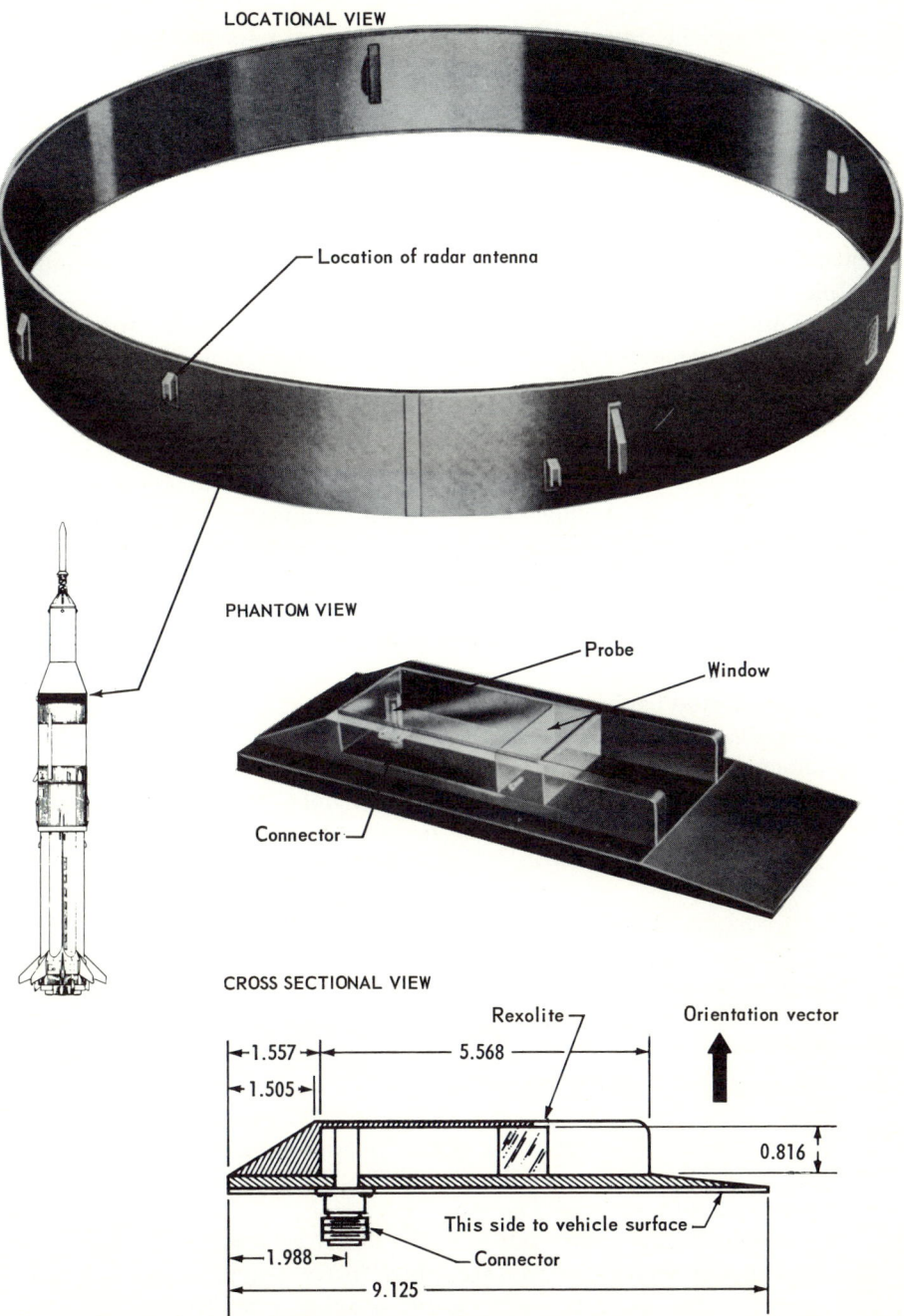

Fig. 6.6b Saturn SA-202 radar antenna; waveguide-type slot for 5,700-MHz band.

in the pattern; the impedance levels with be altered, however, and standing waves may be set up.

6.3 HORN ANTENNAS

The idea of the horn radiator for acoustic waves has been known for centuries. Loosely speaking, the flare of the horn improves the impedance match to free space and increases the radiating aperture. In radio work at microwave frequencies, metallic pipes (waveguides) are frequently employed as transmission devices to feed energy from the generator to the antenna. If the end of the waveguide is simply left open, some of the energy will leak out and be radiated. Figure 6.6b shows a practical example of this type of slot. However, it will be recalled that rectangular waveguide sizes are typically less than 1 wavelength in the largest dimension. Consequently, the magnetic current distribution that is equivalent to the electric field in the aperture resembles a relatively small magnetic dipole and therefore exhibits little directive gain. Furthermore, the abrupt ending of the waveguide constitutes a waveguide discontinuity; higher-order modes are generated at this discontinuity as required in order to satisfy the boundary conditions, and consequently some of the incident energy is reflected back down the waveguide. As is the case in acoustics, within limits, the gain of the waveguide radiator may be increased and the impedance match to free space improved by flaring the waveguide walls to make a horn. There are many types of horns for special purposes. We shall restrict discussion here to rectangular sectoral and pyramidal horns.

6.3.1 SMALL HORNS

The radiation patterns of horn antennas are usually calculated approximately by the application of Huygens' principle and equivalent sources. To illustrate the process, let us suppose that an open waveguide is set into, or flared into, an infinite, plane, perfectly conducting baffle. The radiated field may be calculated from an equivalent distribution in the aperture, with the aperture behind the magnetic currents imagined to be filled in with a perfect conductor. The magnetic currents then act in the presence of a complete conducting plane.

The magnetic current distribution is related to the electric field distribution in the horn aperture, according to the equation $\mathfrak{M}_s = \mathbf{E} \times \hat{\mathbf{n}}$. The exact electric field in the aperture is difficult to determine, but a good approximation for small horns is found easily. If the horn is connected to a rectangular aperture, the aperture field is to a first approxima-

APERTURE ANTENNAS

tion equal to the waveguide field, which, in standard waveguide notation, is

$$E_y = E_0 \sin \frac{\pi x}{a}$$

$$H_x = -\frac{E_y}{Z_{0w}} \qquad Z_{0w} = \left(\frac{\mu}{\epsilon}\right)^{1/2} \left[1 - \left(\frac{\omega_c}{\omega}\right)^2\right]^{-1/2}$$

The radiation pattern of the equivalent magnetic current is essentially given in Chap. 3 on pages 78 and 82, except that an adjustment must be made to fit the waveguide aperture coordinates into the coordinate system of Chap. 3.

To this end, introduce an alternative set of coordinates (x',y',z'), so that the magnetic current is directed along z'. The geometry is shown in Fig. 6.7. In the primed set of coordinates

$$\mathfrak{M}_s = \begin{cases} E_0 \hat{z}' \sin \frac{\pi z'}{A} & 0 < z' < A, \, 0 < x' < B \\ 0 & \text{otherwise} \end{cases}$$

A and B are the dimensions of the horn aperture. This magnetic current is the dual to the continuous electric current distribution considered in Sec. 3.1.2, Chap. 3. The fields therefore may be found from a potential function F_z as follows:

$$F_{z'} = \frac{1}{4\pi r} \iint e^{-j\beta R} \mathfrak{M}_{sz'} \, dx' \, dz'$$

where $R = r - x' \cos \phi' - z' \cos \theta'$. The explicit integral is

$$F_{z'} = \frac{e^{-j\beta r}}{4\pi r} E_0 \int_0^B e^{j\beta x' \cos \phi'} \, dx' \int_0^A e^{j\beta z' \cos \theta'} \sin \frac{\pi z'}{A} \, dz'$$

The first integral is the continuous uniform array factor, for which the result is given in Eq. (9), Chap. 3. The second integral is also of a type that was considered in Chap. 3, with the result given in Eq. (10). With these intermediate steps, the final result is

$$F_{z'} = \frac{e^{-j\beta r}}{4\pi r} E_0 e^{j(\psi_1+\psi_2)} \frac{AB \sin (\tfrac{1}{2}\beta B \cos \phi')}{\tfrac{1}{2}\beta B \cos \phi'} \left(\frac{2}{\pi}\right) \frac{\cos (\tfrac{1}{2}\beta A \cos \theta')}{1 - [(2/\pi)(\tfrac{1}{2}\beta A) \cos \theta']^2}$$

from which the fields themselves are readily determined as follows:

$$H_{\theta'} = -j2\omega\epsilon F_{z'} \sin \theta' \qquad E_{\varphi'} = \left(\frac{\mu}{\epsilon}\right)^{1/2} H_{\theta'}$$

The factor of 2 in the fields comes in because the magnetic currents are acting on a conducting plane, so that they have an in-phase image. The phase factor $e^{j(\psi_1+\psi_2)}$ arises because we chose the origin at the corner of

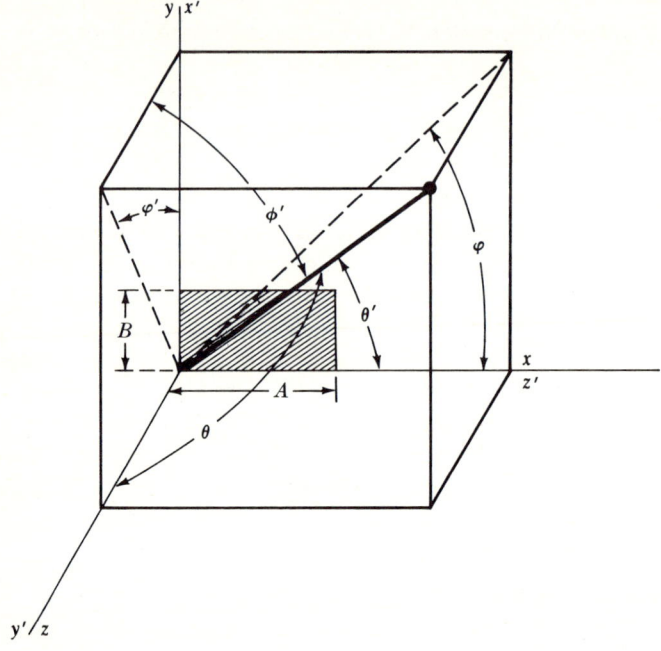

Fig. 6.7 Horn aperture in xy or $z'x'$ plane, showing the geometrical relationships between the primed and unprimed coordinate systems.

the aperture rather than at the center; this factor disappears relative to an origin at the center of the aperture.

In much of the literature concerning horns and other apertures, the coordinate system is oriented with the z axis perpendicular to the aperture; results are stated in terms of an angle variable θ measured from this z axis. With this angle and the angle φ taken in a standard spherical system (Fig. 6.7), the primed angles and the unprimed angles are related as follows:

$$\cos \theta' = \sin \theta \cos \varphi \qquad \cos \phi' = \sin \theta \sin \varphi$$

The radiation patterns usually considered are those in the "E plane" (yz plane, $\varphi = \frac{1}{2}\pi, -\frac{1}{2}\pi$) and the "$H$ plane" (xz plane, $\varphi = 0, \pi$). It should be remembered that the field components that we have calculated are stated in the primed system of coordinates $E_{\varphi'}$ and $H_{\theta'}$. The primed system is the more natural in the sense that both E and H have both θ and φ components in the unprimed system. In the E plane, the pattern is essentially a $\sin \psi / \psi$-type pattern, proportional to the aperture area, AB. In the H plane, the pattern consists of an array factor of the type given in Fig. 3.12 multiplied by a factor of cos θ. The radiation patterns of a wide variety of horns are available in the literature.[1]

In practice, horns are infrequently set in large conducting baffles. The foregoing theoretical results are deficient in this respect. Without the baffle, currents are excited on the exterior surfaces of the waveguide and/or horn, and these currents contribute to the radiation pattern. From the point of view of Huygens' principle, the patterns of horns should be calculated from the magnetic currents (equivalent to E in the aperture) acting on a perfectly conducting enclosure having the shape of the waveguide-horn system, rather than on the infinite plane. Usually, such a calculation is too difficult to be practical.

There is another useful approximation which, while it is no better than the approximation of the baffle, is frequently employed. It consists of calculating the fields arising from both the magnetic and the electric current equivalent sources in the aperture, still ignoring the currents on the exterior surfaces of the waveguide. (The tacit assumption is made that both E and H are zero over the remainder of a closed surface; this is sometimes interpreted as an aperture in an "absorbing" screen, although this does not logically follow.) The magnetic current equivalent sources are the same as those above. The equivalent electric current source density is

$$\mathcal{J}_s = \hat{n} \times H \quad \text{or} \quad \mathcal{J}_y = -\frac{E_0}{Z_{0w}} \sin \frac{\pi x}{A}$$

With this source distribution, the calculation is similar to that presented for the magnetic current equivalent sources. The integrals are essentially the same as those given above. The effects due to current or dipole orientation are different, however. Looking at the component $E_{\varphi'}$, in the E plane, we find that the sin ψ/ψ factor is multiplied by cos θ and that in the H plane, the factor cos θ that was present in the part of the field due to the magnetic current disappears from the field expression pertaining to the electric current source. The net result is that in place of the factor of 2 which comes in with magnetic currents on an infinite baffle, we find a factor of $(1 + \cos \theta)$ in both E and H planes when the field is calculated from both electric and magnetic equivalent sources without the baffle. Also, the electric current generates field components other than the θ' and φ' components characteristic of the magnetic currents. In particular, it generates the corresponding two components in a spherical coordinate system in which "z" lies along the current direction, i.e., along y, or x', in Fig. 6.7.

The gain on the axis of a small horn may be computed from the field expressions for $E_{\varphi'}$ and the integral of the Poynting vector over the

[1] D. R. Rhodes, *Proc. IRE*, **36**:1101 (September, 1948).

aperture. If the ratio of E to H in the aperture is the impedance $Z_{0w} = (\mu/\epsilon)^{1/2}(1 - (\lambda/2A)^2)^{-1/2}$, the power through the aperture is

$$P_{av} = \frac{1}{2} \text{Re} \int_0^B \int_0^A \frac{E_0^2}{Z_{0w}} \sin^2 \frac{\pi x}{A} dx\, dy = \frac{1}{2} E_0^2 \frac{AB}{2Z_{0w}}$$

The field on the axis is

$$|E_{\varphi'}| = \frac{\beta E_0}{\pi^2} \frac{AB}{r}$$

From the fundamental definition, the directivity is then

$$G_{di} = \frac{32}{\pi} \left[1 - \left(\frac{\lambda}{2A}\right)^2\right]^{-1/2} \left(\frac{AB}{\lambda^2}\right)$$

The VSWR in waveguides feeding small horns is usually less than 2:1, with a value of 1.5 being typical.

6.3.2 FIELDS AND GAIN OF LARGER HORNS

When the horn aperture is several wavelengths across, the effects of curvature of the wavefront (i.e., curvature of the surface of constant phase) become important. To understand the origin of and type of curvature, recall that as the walls of the waveguide are flared out, the fields nonetheless must still satisfy the conditions of zero tangential E and normal H at the walls. The net result of this is that the phase fronts become convex with respect to a point in the waveguide, as indicated in Fig. 6.8. To a good approximation, at the horn opening, the curvature

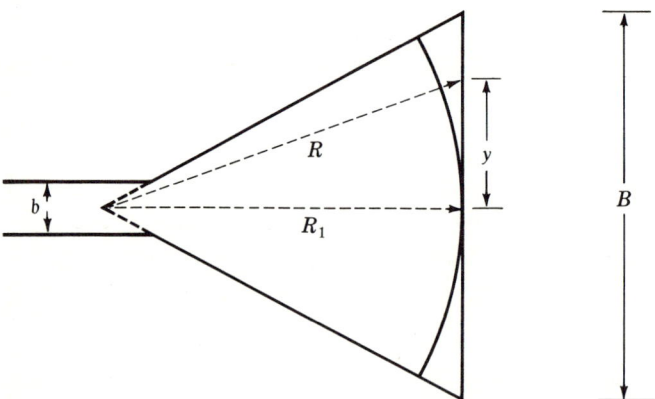

Fig. 6.8 Horn flared in E plane only. Phase difference between center of aperture and point y is represented by $e^{-j\beta(R-R_1)}$. Later, to distinguish between E-plane and H-plane flares, R_1 is called R_{1E}.

is approximately circular, with a radius equal to the distance back to a phase center. The phase center is, roughly, the point of intersection of the projection of the walls, as indicated in Fig. 6.8. For various reasons, it seems simplest to keep the Huygens surface plane. This means that the phase variation across the plane aperture is approximately quadratic, as follows.

Consider first a horn that is flared in the E plane (yz plane) only. The field in the plane aperture may then be written $E_y = E_0 (\sin \pi x/A) e^{-jf(y)}$ where $f(y)$ represents the phase variation due to the curvature of the wavefront. Taking the zero of phase at the center of the aperture, we have

$$e^{-jf(y)} = e^{-j\beta(R-R_1)} = \exp\left[-j\beta R_1 \left(\frac{R}{R_1} - 1\right)\right]$$

From the geometry,

$$R = (R_1^2 + y^2)^{\frac{1}{2}} = R_1\left[1 + \left(\frac{y}{R_1}\right)^2\right]^{\frac{1}{2}}$$

$$\doteq R_1\left[1 + \frac{1}{2}\left(\frac{y}{R_1}\right)^2\right]$$

or

$$\left(\frac{R}{R_1} - 1\right) \doteq \frac{1}{2}\left(\frac{y}{R_1}\right)^2$$

This is the origin of the quadratic phase variation:

$$e^{-jf(y)} \doteq \exp\left[-j\beta R_1\left(\frac{R}{R_1} - 1\right)\right] \doteq e^{-j\frac{1}{2}\beta y^2/R_1}$$

The equivalent magnetic current is then

$$\mathfrak{M}_x = E_0 \sin \frac{\pi x}{A} \exp \frac{-j\frac{1}{2}\beta y^2}{R_1}$$

With this magnetic current distribution acting on an infinite conducting plane, we may find the radiation pattern as before. To do so, it seems easiest to work in a primed coordinate system as in Fig. 6.7, except with the origin at the center of the aperture:

$$\mathfrak{M}_z = E_0 \cos \frac{\pi z'}{A} e^{-j\frac{1}{2}\beta x'^2/R_1}$$

Then for a point on the axis of the horn, the field strength is

$$E_{\varphi'} = -2j\beta F_z = \frac{2j\beta E_0}{4\pi r} \int_{-\frac{B}{2}}^{\frac{B}{2}} \int_{-\frac{A}{2}}^{\frac{A}{2}} e^{-j\beta R} \cos \frac{\pi z'}{A} e^{-j\frac{1}{2}\beta x'^2/R_1} \, dx' \, dz'$$

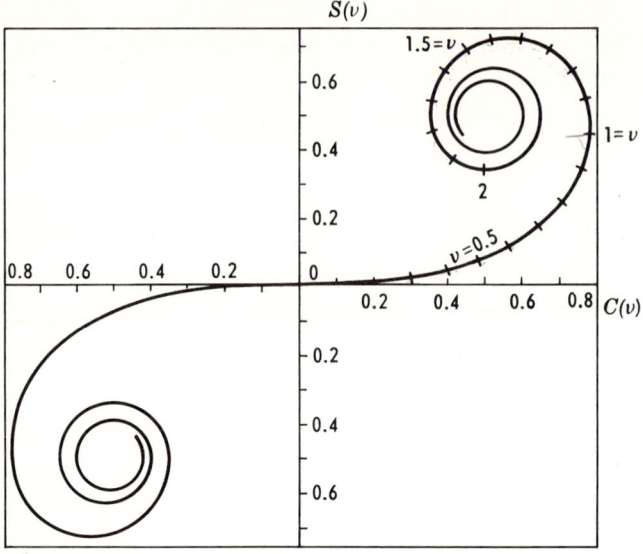

Fig. 6.9 The Cornu spiral.

For points on the axis, $R \doteq r$; thus

$$E_{\varphi'} = \frac{2j\beta A E_0}{\pi^2 r} e^{-j\beta r} \int_0^{\frac{B}{2}} e^{-j\frac{1}{2}\beta x'^2/R_1} \, dx'$$

The integral in the foregoing expression is an integral related to the error integral. It is most easily expressed in terms of a pair of integrals that were studied long ago in connection with the theory of the diffraction of light, and that are known as the Fresnel integrals (see, for example, Jahnke and Emde, pages 35 to 37). The Fresnel integrals, $C(v)$ and $S(v)$, are defined such that

$$C(v) - jS(v) = \int_0^v e^{-j\frac{1}{2}\pi u^2} \, du$$
$$= \int_0^v \cos \tfrac{1}{2}\pi u^2 \, du - j \int_0^v \sin \tfrac{1}{2}\pi u^2 \, du$$

The function, $C(v) - jS(v)$, plotted in the complex plane, gives the curve known as the Cornu spiral, Fig. 6.9. A glance back at the expression for the field on the axis of the horn shows that this field strength is proportional to the exponential form of the Fresnel integrals; it is therefore proportional to the magnitude of the radius vector from the origin to a point on the Cornu spiral. To make use of the standard tabulations of the Fresnel integrals, we may use the substitution $\tfrac{1}{2}\beta x'^2/R_1 = \tfrac{1}{2}\pi u^2$; thus $dx' = (\tfrac{1}{2}\lambda R_1)^{\frac{1}{2}} \, du$. Then the field strength on the axis of the

E-plane sectoral horn is

$$E_{\varphi'}\bigg]_{\text{axis}} = \frac{2j\beta A E_0}{\pi^2 r} e^{-j\beta r} (\tfrac{1}{2}\lambda R_1)^{\frac{1}{2}} \int_0^{\frac{B}{2}\sqrt{\frac{2}{\lambda R_1}}} e^{-j\frac{1}{2}\pi u^2}\, du$$

Examination of the Cornu spiral curve shows that the magnitude of the integral has a maximum when the variable v has a value of about 1.23. This suggests that there should be a particular combination of horn apertures and axial lengths of horns that give maximum gain (since $v = (\tfrac{1}{2}B^2/\lambda R_1)^{\frac{1}{2}}$). The power through the aperture has the same form as the expression for small horns

$$P_{\text{av}} = \tfrac{1}{2}E_0^2 \frac{AB}{2Z_{0w}}$$

where in this case $Z_{0w} \doteq (\mu/\epsilon)^{\frac{1}{2}}$. From the field strength on the axis and the power radiated, we may compute the directivity of a sectoral horn flared in the E plane only:

$$G_{dE} = \frac{AB}{\lambda^2}\frac{32}{\pi}\frac{2\lambda R_1}{B^2}\left\{C^2\left[\left(\frac{\tfrac{1}{2}B^2}{\lambda R_1}\right)^{\frac{1}{2}}\right] + S^2\left[\left(\frac{\tfrac{1}{2}B^2}{\lambda R_1}\right)^{\frac{1}{2}}\right]\right\}$$

The aperture-axial length combinations that give maximum gain may be displayed by plotting this directivity function. A plot of the function $G_{dE}\lambda/A$ is shown in Fig. 6.10. The curves show that there is an optimum

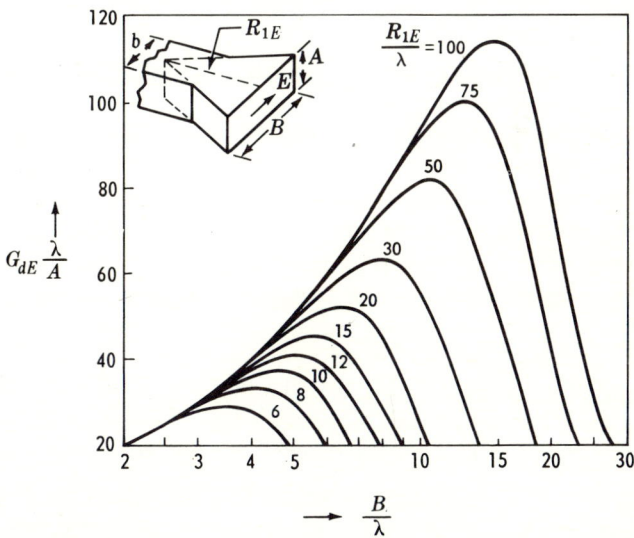

Fig. 6.10 Directivity of sectoral horn flared in the E plane only. Each curve represents a given R_{1E}/λ as indicated on the curves.

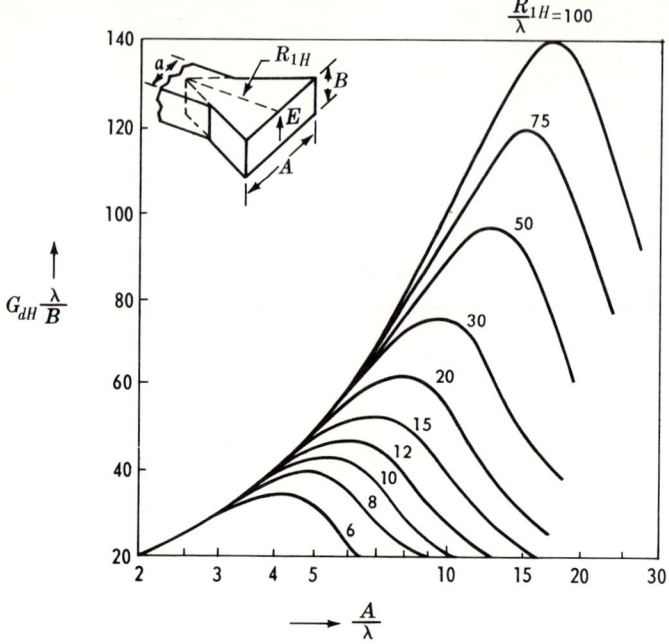

Fig. 6.11 Directivity of sectoral horn flared in H plane only. Each curve represents a given R_{1H}/λ, as indicated on the curves.

aperture for a given slant length. Recall that this is due to a quadratic phase variation across the aperture.

If the horn is flared in the H plane alone, the same types of results apply. The results are quantitatively different, however, because the integral for the field strength on the axis includes the cosine term times the quadratic phase factor. To handle the integral, the cosine function is expressed in terms of exponentials. Again, the results may be expressed in terms of the Fresnel integrals, although the change in variables that is required makes the formulas a little more complicated. The gain curves for a horn flared in the H plane only are shown in Fig. 6.11.

From the data in Figs. 6.10 and 6.11, pyramidal horns (i.e., horns with a flare in both E and H planes) may be designed for a prescribed gain. The results agree very well with experiment. When the horn is flared in both planes, it is easily shown that the directivity is given by

$$G_{di} = \frac{\pi}{32}\left(G_{dE}\frac{\lambda}{A}\right)\left(G_{dH}\frac{\lambda}{B}\right)$$

where the latter two factors are given in Figs. 6.10 and 6.11. Of course in the actual design of a horn, the walls of the horn must be cut so that they will fit together and will match the dimensions of the feeding waveguide. This means that their slant lengths and axis projections must be the same.

Problem 6.2 Design a pyramidal horn for use with 0.4 in. $\times$ 0.9 in. ID rectangular waveguide. The horn should have a gain of 20 db at 10,000 MHz, and should be optimum in the sense that it has maximum gain for a given axial length.

6.4 OTHER WAVEGUIDE RADIATORS

For some applications, a radiator consisting of an open-ended waveguide or horn is not acceptable. For example, there may be severe flush mounting requirements, or the required pattern may be that typical of a long line source. For such reasons, radiators consisting of slots or holes in the walls of waveguides have been developed. The detailed performance depends on slot type and location. We shall discuss a few of the more popular types.

6.4.1 SHUNT SLOTS

A slot in a waveguide wall will radiate if there is an appreciable tangential component of electric field in the slot. Such a field generally develops when the current flow path in the waveguide wall is interrupted. It will be recalled that in a rectangular waveguide, the current lines in the broad wall do not cross the centerline (dominant mode). Thus, a narrow slot along the centerline of the broad wall radiates very little; the small tangential component of electric field that does exist in the narrow gap is oppositely directed on either side of the centerline. As the slot width is increased, more radiation takes place, as will be discussed below. Also, the radiation is increased substantially either as the slot is moved away from the centerline or as it is turned so that its axis is not lined up with the waveguide axis. Consider first the displaced slot in the broad wall, as shown in Fig. 6.12a. With the slot in such a position, the tangential component of electric field is not symmetric about the slot axis; as a result there is a net equivalent longitudinal magnetic current. Thus, such an individual slot radiates almost broadside, with a polarization that is perpendicular to the waveguide axis.

The effect of the slot on the dominant mode fields in the waveguide can be accounted for by representing the slot as a shunt element on an equivalent transmission-line representation of the dominant mode. Detailed formulas for the impedance values are available in a paper by Oliner.[1] Experimental values are given in Jasik. The general results are that the slot is resonant when its length is about 0.49 wavelength. The shunt

---

[1] A. A. Oliner, *IRE Trans.*, **AP-5**:4 (January, 1957).

250 ANTENNA ENGINEERING

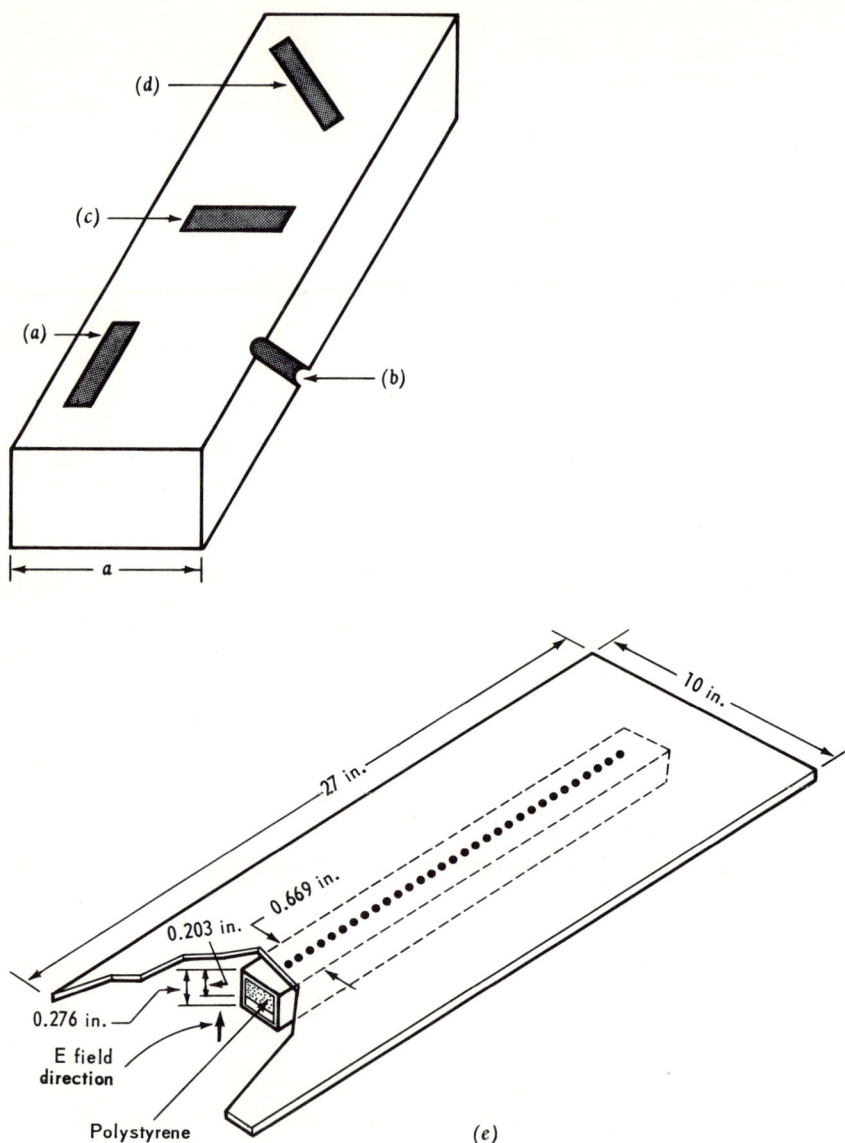

Fig. 6.12 (a) to (d) Different slot configurations in rectangular waveguide. (e) X-band Holey-waveguide radiator with dielectric loaded waveguide.

conductance at resonance depends on both the displacement from the centerline and the slot width; the normalized shunt conductance for displacements from the centerline of 20 percent of the guide width (with slot width of 10 percent of the guide width) is about 0.7. With a frequency change of ± 10 percent from the resonant frequency, the conductance falls to about one-third of its value at resonance, while the normalized susceptance could be expected to vary within the range of approximately ± 0.4.

The shunt admittance values may be employed in a transmission-line model of the TE_{10} mode in the rectangular waveguide. If the waveguide is terminated in a short circuit, located a quarter wavelength from the slot, the reflection coefficient for H is

$$\rho = \frac{Y_s/Y_0 - 1}{Y_s/Y_0 + 1}$$

This means that if the slot is cut for resonance and has a normalized conductance of, say, 0.7, the reflection is relatively small.

When these slots are to be employed as the individual elements in an array (so as to produce some desired pattern), the slot displacements and/or the slot widths must be adjusted to give the desired illumination in each slot opening. The ratio of the electric field in the slot opening to that in the center of the waveguide has been found to be

$$\left|\frac{E_{\text{slot}}}{E_{TE_{10}}}\right| = \frac{AY_s}{\cos(\pi x/a)}$$

where A is a constant, Y_s is the slot admittance, and x is the displacement from the centerline. A standard but rather involved transmission-line analysis is required in order to proceed from a specified set of slot field amplitudes to a set of slot admittances which will produce the required set of slot field amplitudes.

Note that slots spaced a half wavelength apart radiate in phase if they are located on opposite sides of the centerline.

Problem 6.3 Find the shunt admittances required in a three-element resonant slot array in order to radiate broadside with a binomial amplitude distribution (1:2:1). The waveguide is terminated with a short circuit a distance of a quarter wavelength from the last slot position.

Radiation with polarization parallel to the waveguide axis may be obtained with shunt slots cut in the narrow wall (edge slots), as shown in Fig. 6.12b. Admittance data for both slot types are given in the chapter by M. J. Ehrlick in Jasik's "Antenna Engineering Handbook."

6.4.2 SERIES SLOTS

If a slot is cut in a waveguide, as shown in position c of Fig. 6.12, the effect on the dominant mode in the waveguide corresponds to the effect of a series element inserted into the equivalent transmission line. Impedance data are available in the references cited above. Arrays of series slots have been employed less in practice than shunt slots; one reason for this is that there is a strong cross-polarized component in the radiation patterns of series slots.

Rotated series slots in the broad wall (Fig. 6.12d) have some interesting properties. A pair of them cut perpendicular to each other and set at a 45° angle with respect to the waveguide axis can be arranged to radiate circular polarization.[1] The sense of the circular polarization depends on the direction of the incident wave in the waveguide. By reciprocity, it follows that the device behaves as a polarization splitter, or filter, when it is employed as a receiving antenna. If detectors are placed at both ends of the waveguide, the signal having one sense of circular polarization will appear at one detector, and the opposite sense of polarization will appear at the other detector.

6.4.3 LONG SLOTS AND TRAVELING WAVE SLOT ARRAYS

When the requirement is for an end-fire or near end-fire flush-mounted line source, either a long slot or an array of closely spaced nonresonant slots or holes in a waveguide is worthy of consideration. Such structures constitute, respectively, continuous or discrete linear arrays. The phase of the tangential electric field components (and therefore that of the equivalent magnetic current sources) corresponds to the phase velocity of the mode in the perturbed waveguide. As will be recalled (see Chap. 3, Secs. 3.1.1 and 3.1.2), the main lobe appears at an angle which depends on the phase shift per unit length along the line of the sources. In ordinary waveguide, the phase velocity is always greater than c, which means a phase shift per unit length greater than that of a plane wave in space. This means that the beam is pulled away from the end-fire direction, since the conical angle between the beam and the line of the array is given by

$$\phi = \cos^{-1} \frac{c}{v_p} = \cos^{-1} \frac{\lambda}{\lambda_g}$$

where v_p is the phase velocity, and λ_g is the guide wavelength. End-fire beams may be obtained by slowing down the wave in the waveguide, either by dielectric loading or by some form of metallic slow-wave structure. Clearly, the phase velocity in the perturbed structure is a quantity

---

[1] A. Simmons, *IRE Trans.*, **AP-5**:31 (January, 1957).

of great interest. A good discussion of the calculation of the propagation constants may be found in papers by Goldstone and Oliner[1] and Honey.[2] When the phase velocity in such traveling wave arrays is greater than c, the exciting wave exhibits attenuation consistent with the power radiated per unit length. It follows that in a long uniform slot or in an array of identical slots or holes, the amplitudes of the equivalent sources show an exponential taper. Except for slow-wave structures, the power leaving the waveguide per unit length is proportional to the square of the tangential component of the electric field in the slot, which in turn is proportional to the electric field in the waveguide, that is,

$$\frac{dP(z)}{dz} = -k|E_{\tan}(z)|^2 = -k_1 P(z)$$

where k_1 is a constant if the structure is uniform, and $P(z)$ is the power traveling down the waveguide. It follows that

$$E(z) = E(0)e^{-\frac{1}{2}k_1 z}$$

Now suppose the requirement is for an array having, say, a uniform amplitude distribution or a sine-tapered amplitude distribution. Clearly, in this case it is necessary to force the radiation per unit length to vary along the structure. Such a variation may be obtained by an appropriate design of the width of the long slot or the size of the holes. In such a nonuniform structure, the power loss per unit length is given by

$$\frac{dP(z)}{dz} = -k(z)|E_{\tan}(z)|^2 = -k_1(z)P(z)$$

There are indeed many varieties of these so-called "leaky-wave" antennas. The subject of traveling wave slot antennas was discussed early by Rumsey[3] and Harrington.[4] The design of these and many other types is discussed in a book by C. H. Walter.[5] As a specific example, some of the details of an X-band "Holey waveguide," a 1955 (University of Illinois) design by the author, are presented. The design is for an essentially end-fire beam having moderate directivity (half-power beamwidth about 14°) at a center frequency of 9.8 GHz, with polarization perpendicular to the ground plane. The holes were $11\frac{1}{32}$ in. on center, of varying sizes, drilled with standard drill bits. The result is an approximation to a specified tapered amplitude distribution, giving low side

---

[1] O. L. Goldstone and A. A. Oliner, *IRE Trans.*, **AP-7**:307 (October, 1959).
[2] R. C. Honey, *IRE Trans.*, **AP-7**:335 (October, 1959).
[3] V. H. Rumsey, *J. Appl. Phys.*, **24**:1358 (1953).
[4] R. F. Harrington, *J. Appl. Phys.*, **24**:1366 (1953).
[5] C. H. Walter, "Traveling Wave Antennas," McGraw-Hill Book Company, New York 1965.

Table 6.1 Hole Sizes for "Holey-waveguide" Design for X-band Dielectric Loaded Waveguide (Fig. 6.12e)

Hole No.	Diam., in.	Hole No.	Diam., in.	Hole No.	Diam., in.
1	0.125	11	0.281	21	0.313
2	0.125	12	0.290	22	0.297
3	0.141	13	0.295	23	0.281
4	0.147	14	0.302	24	0.266
5	0.152	15	0.313	25	0.250
6	0.171	16	0.313	26	0.234
7	0.187	17	0.313	27	0.219
8	0.219	18	0.313	28	0.203
9	0.250	19	0.313	29	0.187
10	0.266	20	0.313	30	0.187

lobes. The hole sizes are indicated in Table 6.1, and a cutaway view of the structure is shown in Fig. 6.12e. A significant fraction of the power is dissipated in the termination.

6.5 CIRCULAR APERTURE ANTENNAS

Many practical microwave radiators have fields which, in one plane, are pretty much confined to a circular opening. Examples of such structures are paraboloidal reflectors, circular lenses, and conical horns, to name a few. Thus, the theory describing the radiation from a circular aperture is of interest and importance. In some cases, it is possible to arrange the feed system so that the fields are almost uniform across the aperture. Of course they cannot be exactly uniform on an aperture in a conducting plane, because the electric field lines must bend in order to satisfy the boundary conditions at the circular opening. Nevertheless, the theory of the uniformly illuminated circular aperture provides a starting point for circular aperture theory. Other aperture distributions are considered in the book by Silver.[1]

For simplicity, let us consider the circular aperture set in an infinite conducting baffle occupying the xy plane. Then, if the electric fields in the aperture are uniform and directed along y, the magnetic currents that are equivalent sources for this distribution are uniform and directed along x. For variety, let us compute the radiation patterns with the aid of the reciprocity theorem. Since the magnetic currents are solely along x, the radiated electric field is axial about the x axis. In the yz plane, we can pick out this sole component of E by orienting a hypo-

[1] S. Silver, "Microwave Antenna Theory and Design," M.I.T. Radiation Laboratory Series, Vol. 12, McGraw-Hill Book Company, New York, 1949.

thetical point-test dipole along $\hat{\theta}$ in the distant field (yz plane). The reciprocity theorem applied between this test dipole $\mathcal{J}_t$ and the equivalent magnetic current in the aperture is

$$\int \mathbf{E}_a \cdot \mathcal{J}_t \, dV = -\int \mathfrak{M}_a \cdot \mathbf{H}_t \, dS$$

Since $\mathcal{J}_t$ is a point source oriented along $\hat{\theta}$ (in the yz plane), the left-hand side of this equation gives just $E_{a\theta}$; the magnetic field of the test source is $H_t = H_x = (-j\beta/4\pi R)e^{-j\beta R}$, or really twice this value at the perfectly conducting plane:

$$E_{a\theta} = 2\int_0^{2\pi}\int_{\rho=0}^a \mathfrak{M}_a \cdot \frac{j\beta}{4\pi R} e^{-j\beta R} \rho' \, d\rho' \, d\varphi'$$

where a is the radius of the circular aperture. As we have seen many times before, the distance between the observation point and a point in the aperture is

$$R = r - y' \sin\theta \sin\varphi - x' \sin\theta \cos\varphi$$

In the yz plane, $\varphi = \frac{1}{2}\pi$; therefore $R = r - y' \sin\theta$. In the aperture itself, $y' = \rho' \sin\varphi'$; therefore the integral for the field strength is

$$E_{a\theta}(\theta)\bigg]_{\varphi=\frac{\pi}{2}} = \frac{j\beta E_0}{2\pi r} e^{-j\beta r} \int_0^{2\pi}\int_0^a e^{j\beta\rho' \sin\varphi' \sin\theta} \rho' \, d\rho' \, d\varphi'$$

To evaluate this integral, put $u = \beta\rho' \sin\theta$; the integral on φ' is then

$$\int_0^{2\pi} e^{ju \sin\varphi'} \, d\varphi'$$

Although this is a nonelementary integral, it is one that is well known. The general form of this integral can be found for example on pages 148 and 149 of Jahnke and Emde, from which it is clear that

$$\int_0^{2\pi} e^{ju \sin\varphi'} \, d\varphi' = 2\pi J_0(u)$$

where $J_0(u)$ is the zero-order Bessel function of the first kind. The field strength is then

$$E_{a\theta}\bigg]_{\varphi=\frac{\pi}{2}} = \frac{j\beta}{r} E_0 e^{-j\beta r} \int_0^a J_0(\beta\rho' \sin\theta)\rho' \, d\rho'$$

The integral of the Bessel function may be found on page 145 of Jahnke and Emde

$$\int J_0(x) x \, dx = x J_1(x)$$

$$E_{a\theta}\bigg]_{\varphi=\frac{\pi}{2}} = \frac{j\beta}{r} E_0 e^{-j\beta r} a^2 \frac{J_1(\beta a \sin\theta)}{\beta a \sin\theta}$$

That is, the pattern for a uniformly illuminated circular aperture varies as $J_1(x)/x$, in contrast to the $\sin x/x$ variation characteristic of rectangular apertures. The function $\Lambda(x) = 2J_1(x)/x$ is tabulated in Jahnke and Emde on page 181. A sketch of the function is shown in Fig. 6.13, along with a construction to yield the pattern.[1]

With the explicit radiation pattern, it is an easy matter to find the directive gain. The limit of $J_1(x)/x$ as x tends to zero is one-half. Thus the field strength on the axis is

$$|E_{\text{axis}}| = \frac{\beta a^2}{2r} E_0$$

The power coming through the uniformly illuminated aperture of radius a is

$$P_{\text{av}} = \tfrac{1}{2} E_0^2 \left(\frac{\epsilon}{\mu}\right)^{\frac{1}{2}} \pi a^2$$

The directivity is then

$$G_d = \beta^2 a^2$$

That is, the directivity is $4\pi/\lambda^2$ times the area of the aperture.

Let us consider the radiation pattern in the H plane. To do this, we keep the hypothetical test dipole oriented along y and move it about in the xz plane ($\varphi = 0$ plane). In this case, the component of the magnetic field that is parallel to the magnetic currents in the aperture is $-2j\beta/4\pi R e^{-j\beta R} \cos \theta$. The expression for the electric field is then

$$E_{ay}(\theta)\bigg]_{\varphi=0} = 2j\beta \cos \theta \int_0^{2\pi} \int_0^a \mathfrak{M}_a \frac{e^{-j\beta R}}{4\pi R} \rho' \, d\rho' \, d\varphi'$$

Clearly the integral is the same as that in the expression for the field in the E plane. The H-plane pattern for a uniformly illuminated aperture is thus the E-plane pattern multiplied by a factor of $\cos \theta$.

With practical antennas that radiate from a circular aperture, it is difficult to obtain a uniform illumination. Moreover, such a uniform illumination may not even be acceptable for some applications, since the side-lobe level is rather high, about 17.5 db according to Fig. 6.13. As a consequence, many paraboloidal reflectors, for example, are deliberately designed with an amplitude taper. Fortunately, in a circular aperture, aperture distributions of quite a general form can be integrated. In particular, if the aperture distribution can be approximated by a finite

---

[1] Numerical data may also be found in the "Handbook of Mathematical Functions," Nat. Bur. Standards Applied Mathematics Series, vol. 55.

APERTURE ANTENNAS 257

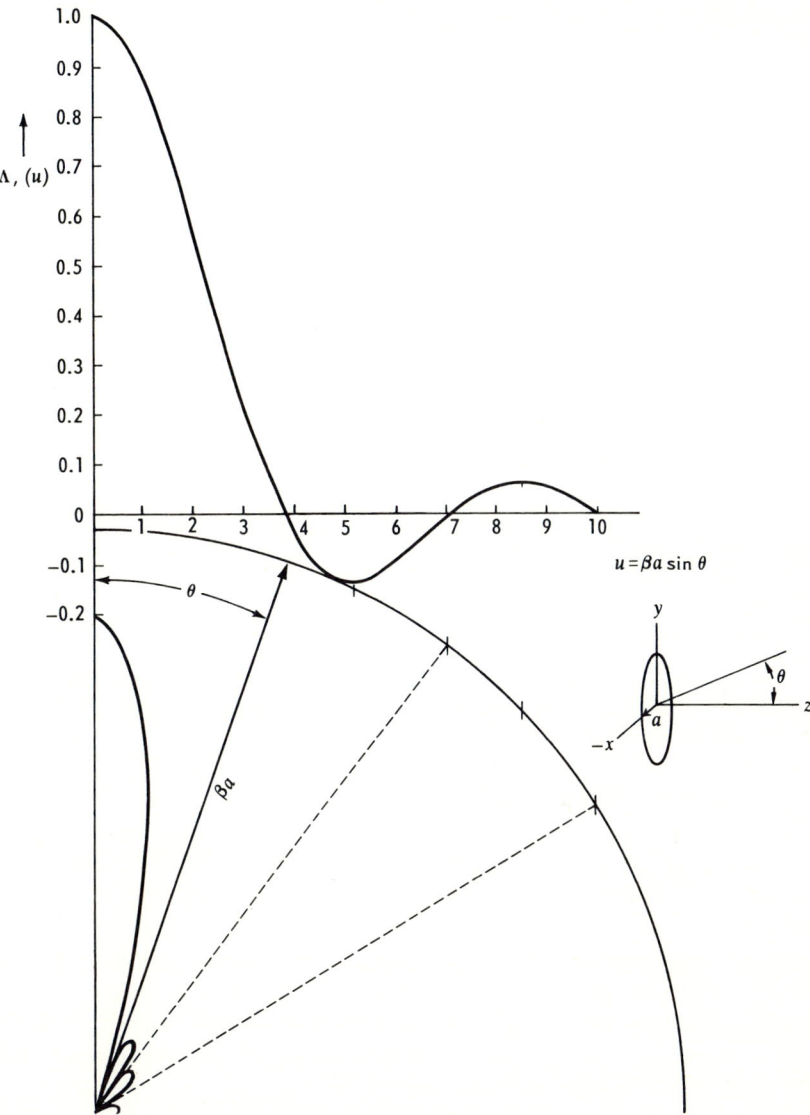

Fig. 6.13 Field-variation characteristic of uniformly illuminated circular aperture ($\Lambda(\mu) = 2J_1(\mu)/\mu$) with pattern construction.

number of terms from the series

$$\mathfrak{M}_a = \sum_{n=-\infty}^{\infty} A_n \left[1 - \left(\frac{\rho'}{a}\right)^n\right] e^{jn\varphi'}$$

then the field in the $\varphi = \frac{1}{2}\pi$ plane has a form,

$$E_{a\theta}\bigg]_{\varphi=\frac{\pi}{2}} = \frac{j\beta E_0}{2\pi r} e^{-j\beta r} \int_0^{2\pi} \int_0^a \sum_{n=-\infty}^{\infty} A_n \left[1 - \left(\frac{\rho'}{a}\right)^n\right] e^{jn\varphi'} e^{ju\sin\varphi'} \rho' \, d\rho' \, d\varphi'$$

($u = \beta\rho' \sin\theta$) which can be integrated, with the help of the integral formulas on the bottom of page 145 in Jahnke and Emde, after the φ' integration is carried out to obtain the nth-order Bessel function. The integral formulas referred to are as follows:

$$\int x^{p+1} Z_p(x) \, dx = x^{p+1} Z_{p+1}(x)$$
$$\int x^{-p+1} Z_{p(x)} \, dx = -x^{-p+1} Z_{p-1}(x)$$

Typically, the gain for practical paraboloidal reflector antennas is 50 to 60 percent of that of a uniformly illuminated aperture.

Problem 6.4 Design a paraboloidal reflector for a directive gain of 36 db at a frequency of 10 GHz. Determine the width of the beam between the first nulls in one of the principal planes.

6.5.1 EXCITATION OF CIRCULAR APERTURES

In the excitation of circular apertures, the object, usually, is to produce a field distribution across the aperture which has a constant phase and some prescribed amplitude distribution. The geometry of the parabolic reflector is such as to provide the desired constant phase, if it is illuminated by a point source located at the focus. To demonstrate this, let us look for the locus of points on a curve such that the total distance from a point to a plane by way of the surface is a constant, since this condition will produce a constant phase delay in a homogeneous medium. Let the point lie on the z axis at a distance f from the origin (see Fig. 6.14). The distance, which is to be common for all points on the curve, is then $f + z_0$, where z_0 locates the plane. Let the desired curve pass through the origin and locate points on the surface by the coordinates (r,z) where r measures the distance from the z axis. A wave traveling from the point to the plane via the arbitrary point (r,z) will then travel a distance $((f-z)^2 + r^2)^{1/2} + (z_0 - z)$. The locus of points having the constant distance will then be given by the equation

$$[(f-z)^2 + r^2]^{1/2} + (z_0 - z) = f + z_0$$

APERTURE ANTENNAS 259

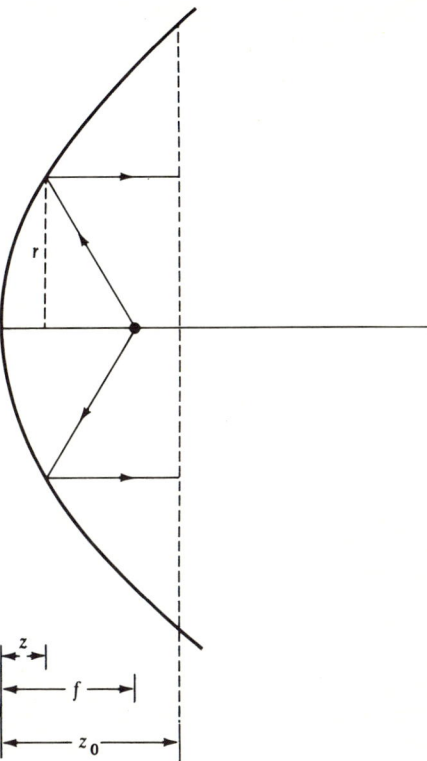

Fig. 6.14 Geometry of the parabolic reflector.

or

$$r^2 = 4fz$$

This is the equation of a parabola or a paraboloid of revolution

$$x^2 + y^2 = 4fz$$

with vertex at the origin, as stated. These results show that to produce a constant phase across the circular aperture, the phase center of the primary feed should be located at the focus of the paraboloid.

Assuming that the current distribution on the primary feed is not altered by the proximity of the parabolic reflector, the radiation pattern of the system is the superposition of the field of the primary feed, by itself, and that of the currents on the reflector surface in the presence of the feed structure. Alternatively, the effect of the reflector (or even the pattern of the complete system) can be represented by Huygens sources located on some convenient surface outside the reflector surface. Usually, the plane circular aperture discussed above is a part of the Huygens surface, and it makes the most important contribution to the pattern.

A practical paraboloidal reflector, being finite in size, ends in a circle which is the intersection of the paraboloid with the "aperture plane." This circle has a diameter, $D = 2a$, and the shapes of various paraboloids are described by the f/D ratio, or alternatively, by the angular aperture χ, the angle subtended at the focus by the radius of the aperture. This parameter is important to the gain and pattern characteristics of the system. For example, suppose a paraboloid having a very small f/D ratio is illuminated by a small dipole at the focus. It is clear that, because of the pattern of this primary feed, the illumination of the aperture will not be the same in the two principal planes. The H-plane illumination will be much more nearly uniform than the E-plane illumination. On the other hand, this effect will tend to disappear with parabolas having large f/D ratios, in which the feeding dipole is a relatively large distance from the reflector and the angular aperture subtended is small. But of course in the latter case, only a small portion of the feed radiation will be intercepted by the reflector.

In general, the practice will be to select a primary feed which directs its field into the solid angle subtended by the reflector; clearly, this feature will result in the minimum energy wasted in the form of "spillover." It will also minimize interference in the forward direction between the primary pattern and the pattern of the reflector. Within these limits, the primary-feed pattern can be selected so as to produce in the circular aperture an amplitude distribution which leads to some desired characteristics (such as low side lobes or a "shaped beam") in the distant field of the circular aperture. However, the phase fronts associated with the primary-feed pattern at the position of the parabola must be spherical, or the advantage of the paraboloid in producing the uniform phase at the aperture is lost.

The primary-feed structure almost inevitably blocks off a portion of the aperture of the paraboloid. The general effect of this "aperture blocking" is to reduce the gain somewhat and to increase the side-lobe levels.

Problem 6.5 Suppose that the center portion of a uniformly illuminated circular aperture of radius a is blocked by a circular disk of radius a_1, $a_1 < a$. Assume that the disk does not influence the field in the other portions of the aperture and that it simply reduces the field to zero in the range $0 \leq a_1$. This type of effect can be handled by considering the blocked aperture as the superposition of the original aperture and a smaller aperture of radius a_1, whose fields are 180° out of phase with the original. With these assumptions, plot the pattern of the blocked aperture in the case that $a_1 = 0.1a$.

A circular lens may be employed as another form of circular aperture antenna. Since it is fed from behind, the use of a lens eliminates the

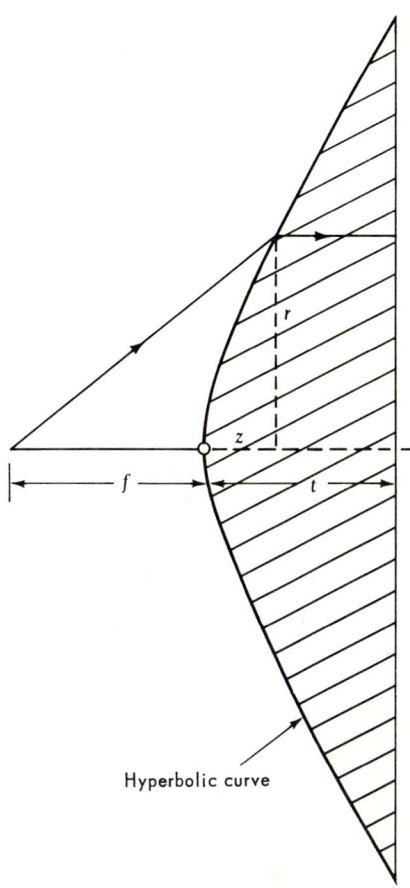

Fig. 6.15 Geometry of plane-surface dielectric lens.

aperture blocking effect; it has other disadvantages, however. As an example, consider a dielectric lens having a plane face which coincides with the circular aperture. If this lens is illuminated by a point source, the back face of the lens should be a hyperboloid. That this is so may be seen as follows: To produce a constant phase across the plane front face, the curvature of the lens surface should be such that all parts of the wavefront from the point source experience the same phase shift in reaching the front face. Along the axis of the lens (Fig. 6.15), the phase shift is $\beta_0 f + \beta_1 t$, where f is the distance between the point source and the back face, t is the lens thickness, and $\beta_1 = \beta_0(\mu_r \epsilon_r)^{1/2}$, where μ_r, ϵ_r are the relative values for the lens. Along any other path, the phase shift is

$$\beta_0[(f + z)^2 + r^2]^{1/2} + \beta_1(t - z)$$

For constant phase by all paths, then we must have the equation

$$\beta_0 f + \beta_1 t = \beta_0[(f+z)^2 + r^2]^{1/2} + \beta_1(t-z)$$

or

$$f = [(f+z)^2 + r^2]^{1/2} - (\mu_r \epsilon_r)^{1/2} z$$

or

$$r^2 = (\mu_r \epsilon_r - 1)z^2 + 2f[(\mu_r \epsilon_r)^{1/2} - 1]z$$

This equation may be recognized as the equation of a hyperboloid of revolution (in rectangular coordinates $r^2 = x^2 + y^2$). These results show that a lens with a plane surface and a hyperbolic surface may be designed to convert spherical wavefronts to plane wavefronts in the aperture.

Lenses have a few special problems. In general, there will be a reflection from both interfaces of a lens. Let us consider the reflection from the plane surface first. If the system is employed as a transmitter, the energy reflected from the plane face will pass back through the lens surface and be focused at the feed. Thus, in addition to the loss in efficiency, the feed system will be mismatched. Some type of nonreflecting surface coating is clearly desirable. A quarter-wave thickness having the geometric mean intrinsic impedance will eliminate this reflection. Corrugations or other loadings may also be employed as matching layers.

The effect of the reflection at the curved surface is more complicated. The amount of the reflection depends on the angle of incidence to the surface and the polarization with respect to the plane of incidence (good approximate results can be obtained with the usual Fresnel formulas). The net result is that amplitude distribution on the plane face may depend significantly on these reflections from the back face; for example, the effect of the Brewster angle may be seen. Consequently, it is usually desirable to eliminate the reflections from the curved lens surface as well as the plane one. A quarter-wave matching section added to the curved surface will work satisfactorily.

The inherent loss and structural instability of some dielectric lenses can be eliminated by a different approach to lens construction. In particular, the lens may be constructed from metal instead of dielectric. With this approach, the required phase delays have been obtained from "waveguide lenses" as well as with artificial dielectrics made of metal disks, rods, and spheres. See, for example, the chapter in Jasik[1] by Seymour Cohn.

[1] H. Jasik, "Antenna Engineering Handbook," chap. 14, pp. 14-21 to 14-41, McGraw-Hill Book Company, New York, 1961.

7. ANTENNAS FOR MULTIPLE FREQUENCIES

In practice, there are usually rather stringent space and cost limitations on antenna systems; moreover, for flexibility, it is usually necessary to provide for operation at several frequencies. This implies that antennas must be designed so that they will operate, often with almost identical performance, at several frequencies. However, as we have seen throughout this book, antenna performance, with regard to both input impedance and radiation pattern, is characterized by the antenna dimensions measured in wavelengths; the performance for a simple structure is then inherently frequency-dependent. In the applications, there are two broad classes of requirements: sometimes, an antenna must operate only in two or, at most, a relatively small number of discrete narrow bands; on the other hand, some antennas must provide an almost continuous frequency coverage over a frequency range that may cover from 2 to 10 octaves. For the former applications, it is appropriate to employ what might be called "spot-band" antennas. For the latter, a more elaborate "frequency-independent" antenna is required.

7.1 SPOT-BAND ANTENNAS

Spot-band antennas are designed to operate only within a small number of narrow-frequency bands. The performance at other frequencies is usually erratic and widely variable but is of no concern in the specific application. The guiding principle in the design is usually the incorporation of structural features that serve to enhance or suppress the excitation of different frequencies. Perhaps the simplest example of this technique is the design of a simple dipole for operation at two frequencies. The idea is to incorporate a high impedance to suppress currents on a part of the structure. Thus, for example, a dipole may be cut with a length

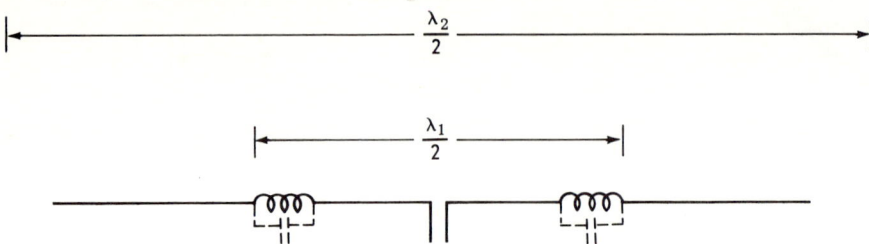

Fig. 7.1 Idea of simple two-band antenna. Impedance of L is very large (choke) at frequency f_1, but much less at frequency f_2 (providing only a loading effect).

slightly less than a half wavelength at the higher frequency. This is separated from an additional length of wire or rod by a choke (see Fig. 7.1). The choke inductance is selected so that the impedance is very high at the high frequency, but low at the low frequency. Alternatively, a capacitor may be added so that the high impedance is obtained by a parallel-resonant circuit. Thus, at the low frequency, the structure also closely approximates a half-wave dipole, differing only in the small inductive loading. As a result, the performance can be nearly the same at two well-separated frequency bands; of course, the scheme will not operate satisfactorily at two arbitrarily related frequencies.

Another approach is to include two or more radiators but to employ transmission-line sections so as to diplex the signals into the appropriate radiator. For example, suppose the application is suited to a two-element Yagi antenna but calls for operation at two frequency bands. In this case, the main transmission line might be branched into lines leading to the two radiating elements (Fig. 7.2). Then, transmission-line tuning stubs are introduced into each of the branched lines, so as to provide, at the branch point, an open circuit at one frequency and a match at the other frequency (looking into that particular line). This provides for the excitation of one or the other of the dipoles. Also, a short circuit at the feed point of that dipole attached to the unexcited branch may be provided by selecting the length of line to the tuning stubs or with another set of stubs. The net result is a structure which at one frequency is a driven dipole with a parasitic director and at the other frequency a driven dipole with a parasitic reflector. This technique is appropriate as long as the two frequency bands are not too widely separated.

Exercise Work out the details of the design for the two-band antenna suggested in Fig. 7.2. Assume a 50-ohm line and take the frequencies of operation to be 60 MHz and 80 MHz.

A somewhat more complicated antenna design may be illustrated with the description of a dual-channel Yagi-type antenna for a pair of tele-

ANTENNAS FOR MULTIPLE FREQUENCIES

vision channels. Consider for example a Yagi antenna having four directors and a reflector, as shown in Fig. 7.3. The directors are cut so as to give best operation at the higher frequency band, and the reflector is cut to give best operation at the lower frequency band. The driven element is a folded dipole, basically cut for operation at the higher frequency. However, on the unfed side of the folded dipole, a tuning-stub arrangement is connected which somewhat resembles a large hairpin. The length of the open-circuited stub is adjusted to be a quarter wavelength at the upper frequency so that the open transfers into a short circuit at the junction. The curved (short-circuited) side of the hairpin provides sufficient length so that, with allowances for the capacitive effect of the open stub, the folded dipole is also resonant at the lower frequency.

As an example of another type of broadbanding, consider an antenna type described by J. T. Bolljahn in U.S. Patent No. 2,505,751. This structure was developed as an aircraft antenna, aimed to cover almost an octave in the lower part of the UHF band. The antenna consists of a "partial sleeve" monopole, made up of a driven element and two short

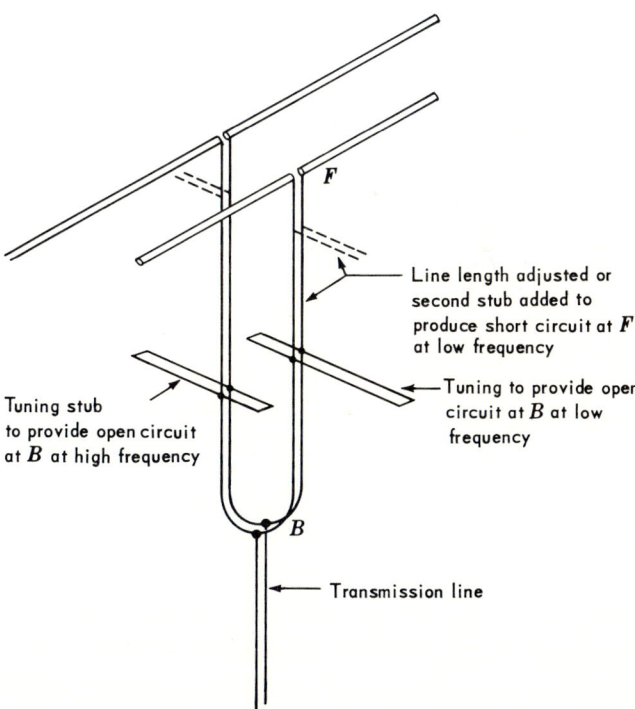

Fig. 7.2 A two-band Yagi array with tuning-stub diplexer.

266 ANTENNA ENGINEERING

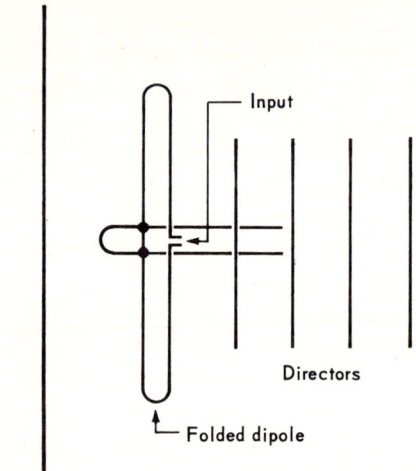

Fig. 7.3 Two-channel Yagi array, with hairpin-tuning stubs on folded dipole.

parasites, as shown in Fig. 7.4. With the particular dimensions shown, the structure provides a VSWR of less than 1.8 on a 50-ohm line over the frequency band from 310 to 510 MHz.

The other main technique for designing spot-band antennas is to attempt to fit antenna systems for the different frequencies into the same

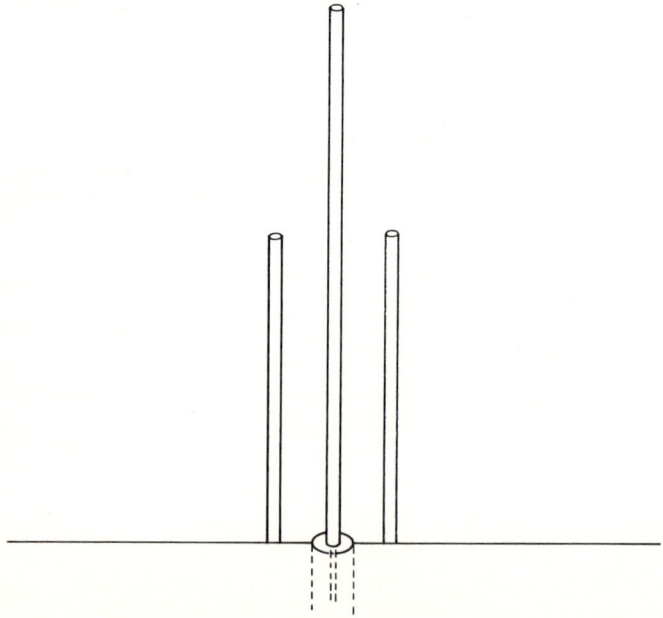

Fig. 7.4 Open-sleeve antenna for operation between 310 and 520 MHz. Longer element is $8^{11}/_{16}$ in.; shorter ones are $5^{1}/_{8}$ in.

space. This process of interleaving and interlacing usually requires a large measure of ingenuity along with compromises in order to prevent one antenna system from drastically influencing the performance of another.

A large step forward in antenna engineering came with the development of "frequency-independent" antennas, the subject of the next section. This section includes somewhat more than the usual amount of informal history; this detail is presented as a case history of a not-so-straightforward antenna development. It is also presented in the belief that this information aids in the understanding of the great variety of antennas of this type now employed.

7.2 FREQUENCY-INDEPENDENT ANTENNAS[1]

The research work which led to the development of antennas whose performance is almost independent of frequency was carried out mainly at the University of Illinois in the period from 1955 to 1958. The work, along with several other projects, was sponsored by the Air Force in order to relieve the problems associated with the increasing numbers of different electromagnetic systems and equipment being carried on high-speed military aircraft. So many different antennas were required that the finding of locations for the antennas was a very serious problem. It was recognized that the problem would be relieved if a given antenna could serve several systems and frequencies, and consequently the Air Force sponsored a research program on the general subject of broadband antennas.

Out of this research work came many different novel, unconventional structures, few of which are such as to permit a tractable mathematical description. It seems that the best way to provide some insight into the operation and design of these novel broadband structures is to trace their historical development. This is done in the next paragraphs.

In connection with the sponsored research work on broadband antennas, Professor V. H. Rumsey, then antenna laboratory director at the University of Illinois, asked himself the question: "What is it that makes an antenna sensitive to frequency?" In thinking about this, he noticed that the features which introduce the frequency dependence are the *characteristic lengths* of the structure. Antenna performance is generally a function of length/wavelength. On the other hand, by the principle of modeling or scaling, to ensure that a given type of structure has the same performance at different frequencies, it is only necessary to scale the size of the structure in the ratio of the frequencies. Thus, Rumsey concluded that the structural feature required for frequency-independent operation is the absence of characteristic lengths. With this feature, a structure could be self-scaling.

[1] E. C. Jordan, G. A. Deschamps, J. D. Dyson, and P. E. Mayes, *IEEE Spectrum*, **1**:58 (April, 1964).

But what kind of physical structure is there that has no physical lengths? Rumsey's answer was that the structure should be completely described by angles. Thus, he put forward the "angle concept," which said, essentially, that a structure whose shape is defined by angles alone, with no characteristic lengths, should be a frequency-independent structure. In looking for structures that can be defined by angles alone, the first that come to mind are the infinite biconical antenna (or transmission-line) and the infinite bifin (bow-tie) antenna. However, practical versions of these structures are obviously finite in size, and although these structures do have comparatively broadband tendencies, the truncation to a finite size introduces a characteristic length, and this destroys the frequency-independent behavior. In thinking about other structures, Rumsey recalled the logarithmic or equiangular spiral, whose curve in a plane is given by the equation $\rho = \rho_0 e^{a\phi}$; or, if the curve is developed on a cone of angle $\theta = \theta_0$, its equation is $r = r_0 e^{a\phi}$. He noted that, except for a rotation in space about the axis of the spiral, this structure, if infinite, should look the same at any frequency. Although the infinite structure could not be built, Rumsey reasoned that if the spiral were excited at the origin, the currents on the arms might fall off rapidly enough that, at least through a wide band of frequencies, the fact that the structure must be finite in size would not matter. Moreover, earlier Air Force experiments reported by E. Turner had indicated that an archimedean spiral had promise as a broadband antenna. Rumsey therefore asked J. Dyson, then a graduate student employed by the laboratory, to build and test an equiangular spiral antenna.

At the same time, Rumsey recalled another fact that he had seen earlier. Namely, he recalled from Booker's work the relation between the impedance of a slot antenna and the complementary dipole antenna (page 258), $Z_{\text{slot}} \cdot Z_{\text{dipole}} = (1/4)(\mu/\epsilon)$. This equation implies that if the slot and the complementary dipole could be made to look the same, then Z_{slot} should equal Z_{dipole}, and therefore the input impedance of either should be a constant, independent of frequency. If the slot and the complementary dipole are the same, it is appropriate to call the structures *self-complementary*. Examples of self-complementary structures are shown in Fig. 7.5. Self-complementary structures of one class may be constructed by drawing an arbitrary (nonoverlapping) curve from the origin and letting it extend to infinity. This curve is then rotated 90° about the origin and redrawn; it is rotated another 90° and redrawn, and so on until it is back to its original position. The plane is thereby split into four parts. If the alternate parts are filled in with metal, the resulting structure is self-complementary.

Rumsey noticed that the equiangular spiral could be made from sheet metal in such a way that it would be self-complementary, and he asked

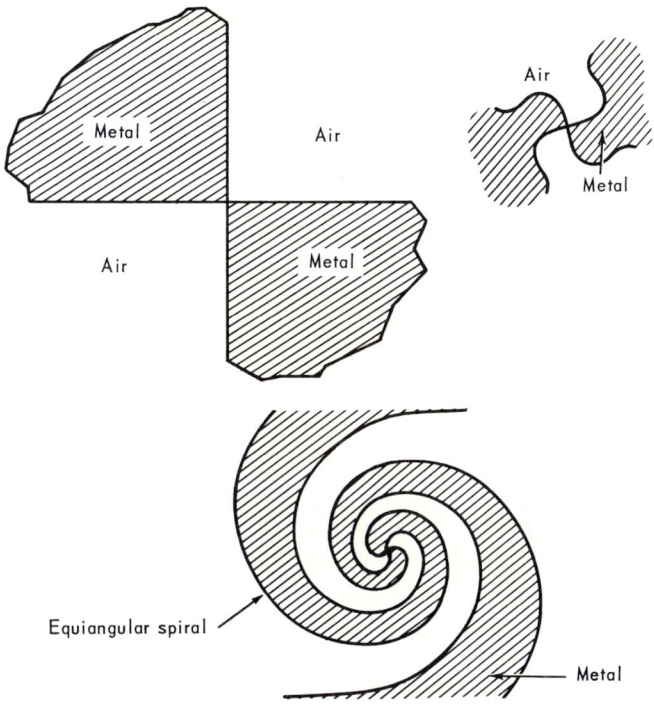

Fig. 7.5 Self-complementary structures.

Dyson to build this feature into the first model for testing. The results were striking. The first models had bandwidths of several octaves in the sense that they had almost constant input impedance and a nearly constant radiation pattern (constant except for a rotation about the axis of the spiral). Moreover, the patterns were observed to be circularly polarized. The current distribution along the arms of the spiral was measured, and it was found that (1) the currents fall off faster than $1/r$, and (2) at all frequencies within the band over which the performance was observed to be constant, the current amplitude was substantially zero at or before the point of truncation. At frequencies such that the diameter of the truncated spiral is approximately equal to a wavelength, the currents at the point of truncation begin to be significant and the performance begins to deteriorate. On the other hand, the upper limit on the frequency-independent operation is determined by the accuracy with which the feed region is (or can be) constructed.

When the spiral is operating within its frequency-independent band, it is self-scaling. Thus, its input impedance should be independent of frequency, even without the self-complementary feature discussed above.

Consequently, this feature was eliminated on later experimental models, and it was found that the bandwidth was indeed maintained, as expected (although the VSWR is more nearly constant with the self-complementary structures).

The spiral antenna is typically excited by running a small coaxial line from the outer extremities along one of the arms of the spiral, into the origin. There the center conductor is led out and joined to the other arm of the spiral. With this type of feed, the structure provides its own balun. Usually, a dummy cable is soldered on the second arm in order to preserve the symmetry. Further details on the design of spiral antennas can be found in the publications of Dyson.[1]

The work on spirals was a significant advance in antenna art, but it certainly did not solve all broadband-antenna problems. The radiation patterns of the plane spirals are broad and bidirectional with the maxima along the axis perpendicular to the sheet metal, and are circularly polarized. Consequently, R. H. DuHamel (then a research assistant professor employed by the University of Illinois Antenna Laboratory) addressed himself to the problem of designing a broadband antenna with linear polarization. He realized that the bifin or bow-tie antenna could be constructed in a self-complementary fashion and of course that it radiates linear polarization. But he also realized that the bandwidth of the bifin was limited because of the truncation, or more particularly, because the currents were not negligible at the point of truncation. Consequently, the problem as DuHamel saw it was to somehow alter the bow-tie structure in such a way as to cause the currents to fall off with distance from the feed point more rapidly than usual. His method for accomplishing this was to introduce discontinuities, for example, teeth, into the fins in an attempt to increase the radiation and speed up the decay of current. But the question he had to ask himself was, "How should the teeth be designed?" DuHamel decided that he should adhere to Rumsey's angle concept as far as possible. Consequently, he decided to cut the teeth along circular arcs and let the length of the arcs be determined by an angle (see Fig. 7.6). However, this did not fix the tooth spacing, since the latter could not be specified by angles alone.

In trying to decide what spacing to use on the teeth, DuHamel noticed that on the equiangular spiral (a successful structure), along a line drawn from the center outward, the spacings from one conductor to the next were in a constant ratio (since the defining curve was $r = r_0 e^{a\phi}$). He therefore considered spacing the teeth in the bifin such that the spacings were in a constant ratio. He accomplished this by choosing the radii of

[1] J. D. Dyson, *IRE Trans.*, **AP-7**:181–187 (April, 1959). *IRE Trans.*, **AP-7**:329–334 (October, 1959). *IEEE Trans.*, **AP-13**:488–499 (July, 1965).

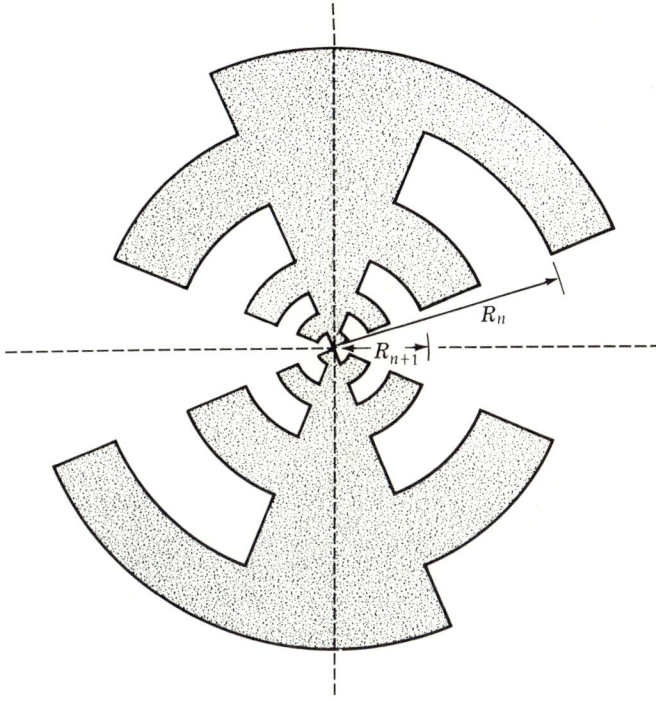

Fig. 7.6 Log-periodic toothed structure (self-complementary).

the circular arcs forming the corresponding parts of the successive teeth such that they were in a constant ratio, $R_{n+1}/R_n = \tau$. He recognized that the structure would not necessarily be frequency-independent but that, on the other hand, the performance on an infinite structure would be identical at a discrete number of frequencies. (For example, consider a frequency at which the pth tooth is, say, one-tenth of a wavelength in size. Then, at a lower frequency, the $(p-1)$st tooth will be one-tenth of a wavelength in size, and all the rest of the structure will be scaled accordingly; therefore the performance at the lower frequency will be the same as that at the first frequency.) In fact, if the structure has a performance (E_1, Z_1) at frequency f_1, the performance should be identical at frequencies τf_1, $\tau^2 f_1$, $\tau^3 f_1$, and so on as long as the structure is modeled accurately at the feed point and is effectively infinite in size (i.e., current zero at the point of truncation). Again, τ is the common ratio of distances. The frequencies at which the performance should be identical are related by the equation $f_n = f_{n+1}\tau$, or $\log f_{n+1} = \log f_n + \log(1/\tau)$. Inspection of this latter equation shows that the performance is a periodic function of the logarithm of the frequency (i.e., the frequencies at which the performance is the same are spaced equally when plotted on log paper). Thus, these types of structures were subsequently named *log-periodic* antennas.

Although initially it was by no means clear what the performance would be as a continuous function of frequency, DuHamel decided that, since the percentage bandwidth would also be independent of frequency, it should be well worth the effort to investigate the structures experimentally. Thus, toothed structures of the type indicated in Fig. 7.6 were built and tested. Initially, the self-complementary feature was built into all models, to guarantee a constant input impedance on the infinite structure. The experiments showed that on structures with large enough teeth, the current decay was rapid and that the log-periodic frequency behavior was verified. But perhaps more important, on some of the structures, it was found that the change in the radiation pattern with frequency at frequencies between the log-periodic frequencies was relatively minor. That is, the structures were self-scaling with only minor variations between periods. Eventually it was learned that the self-complementary feature is not essential to acceptable impedance bandwidth.

Many structures of this type were built and tested. Some were less successful (i.e., frequency-independent) than others, and it was noticed that some of the more successful had a polarization whose strongest linear component was perpendicular to that normally radiated by a bow-tie

Fig. 7.7 Nonplanar log-periodic antenna.

Fig. 7.8 Trapezoidal toothed log-periodic structure.

antenna. This indicated that significant currents were flowing along the teeth.

One of the objectives, namely a broadband linearly polarized antenna, was attained with the invention of the toothed log-periodic structures. However, the patterns were still broad and bidirectional, normal to the plane of the sheet metal from which the antennas were constructed. The next step forward came when D. E. Isbell (then a laboratory assistant working with DuHamel), in an attempt to produce a unidirectional pattern, decided to ignore the planar restrictions suggested by Babinet's principle. He modified one of the successful planar structures by folding it into a wedge. To the surprise of almost everyone in the laboratory except Isbell, the experiments with the wedge structure (Fig. 7.7) showed that it radiated unidirectionally, with the maximum in the radiation pattern off the tip (feed point) of the structure. Moreover, the bandwidth was found to be practically the same as that of the planar structure (although the impedance levels were lower). When oriented as in Fig. 7.7, the polarization was predominantly in the x direction. It was found subsequently that the teeth could be cut straight as well as curved (Fig. 7.8) to give what was called a trapezoidal toothed structure.

About this time it began to be evident that most of the current excitation and radiation were associated with teeth whose lengths were in the vicinity of a quarter wavelength. Still later, DuHamel recognized that the sheet-metal structures could be simulated with wires or tubes which outline the periphery of the sheet structures. The tube type of construction was later developed by DuHamel and coworkers at Collins Radio Company into a class of commercial antennas for application in the HF band, the type Collins 237A-1 (Fig. 7.9). Antennas of this type, made of both sheet metal and wires, were also employed as feeds for

274   ANTENNA ENGINEERING

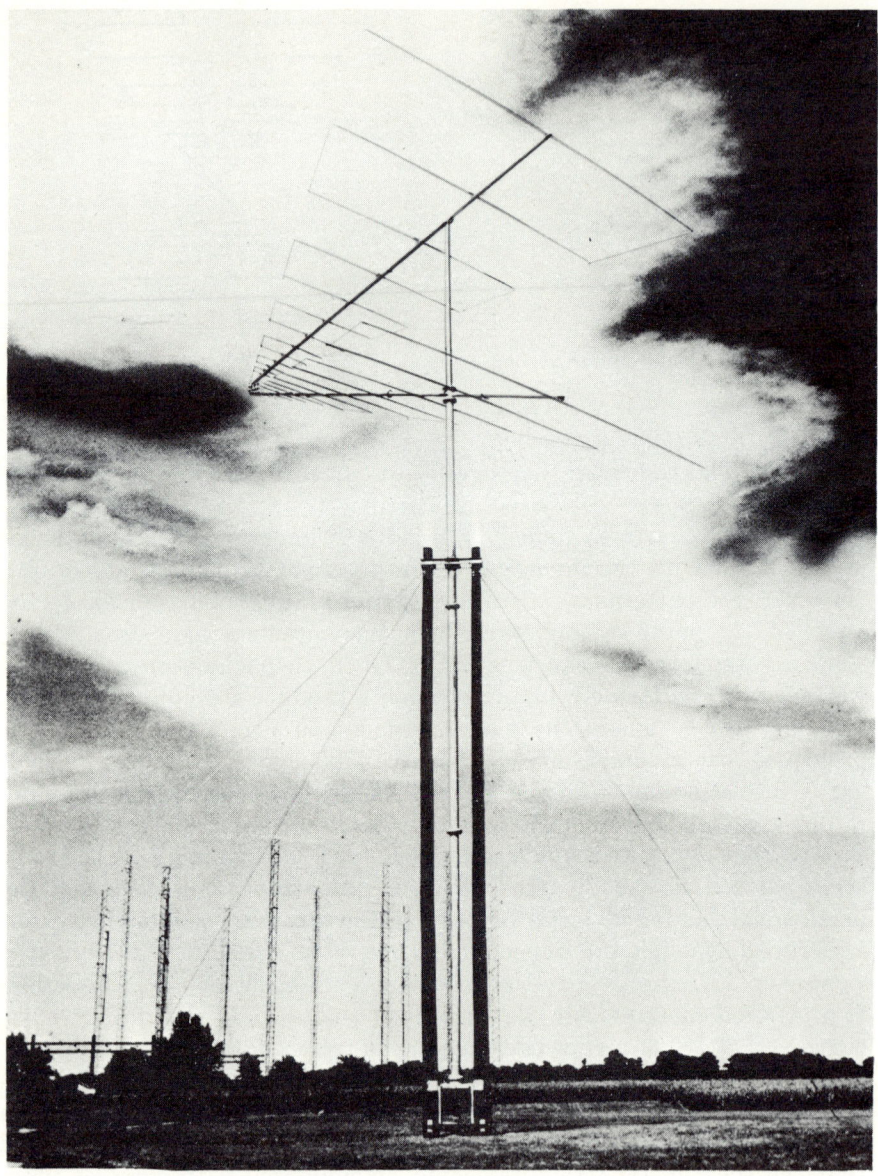

Fig. 7.9 A rotatable nonplanar log-periodic antenna system for 6.5 to 60 MHz. (Collins Radio Company 237A-1.)

parabolic reflectors. Typical gain for log-periodic antennas of the type in Figs. 7.8 and 7.9 is about 8 to 10 db, with a front-to-back ratio of 15 and an H plane beamwidth of 80 to 110°. Further details are available in the literature.[1]

The next major step came with Isbell's invention of the log periodic dipole array. His work was motivated by the desire to develop broadband arrays of more conventional construction. Thus he decided to build and test an antenna array constructed of conventional wirelike elements; however, the lengths of the elements were determined by an angle α as before, and the spacings were such as to give the log-periodic type of behavior; that is, successive distances between the apex and the elements were in a constant ratio, $R_{n+1}/R_n = \tau$. With this general structural type in mind, the big question was how to excite the elements in this array. Taking a hard look at a successful trapezoidal tooth antenna (Fig. 7.8), Isbell saw that when this structure was bent over into a wedge to give the unidirectional pattern, the center portion on each arm could be regarded as a transmission line feeding the teeth as radiators; the radiating teeth were simply connected in shunt on this line. For the new array, Isbell decided to ignore the fact that the feed line should, strictly speaking, be conical and expanding in radius and to try instead an ordinary parallel-wire line as the feeder transmission line for the teeth. Taking another hard look at the trapezoidal toothed structure, he noted that if the two arms were folded back so as to be almost parallel, then pairs of equal-length elements would almost be lined up as dipoles. However, he noticed also that on a given arm of the structure, successive teeth came out from the centerline in opposite directions. Consequently, he decided that in order to simulate the toothed structure, the dipoles should be connected to the parallel-wire line in such a way that the successive elements came out from the line in opposite directions. Isbell accomplished this with the type of construction indicated in Fig. 7.10a, in which the two halves of the dipoles are offset slightly. Note also that the parallel-wire line was itself excited by means of a coaxial line which was led through one of the rods making up the parallel-wire line. The center conductor of the coax is led over to the other member of the parallel-wire pair. This method of feeding provides a built-in broadband balun. Isbell also recognized immediately that his method of connecting the dipoles to the line is equivalent to crisscrossing the wire line between the elements as indicated in Fig. 7.10b, which clearly introduces a 180° phase shift between elements. The experiments with the structure demon-

[1] R. H. DuHamel and D. E. Isbell, *IRE Natl. Conv. Record*, **pt. 1:**119–128 (1957).
R. H. DuHamel and F. R. Ore, *IRE Natl. Conv. Record*, **pt. 1:**139–151 (1958).
R. H. DuHamel and D. G. Berry, *IRE WESCON Conv. Record*, **pt. 1:**161-174 (1958).
Ibid., *IRE Natl. Conv. Record*, **pt. 1:**42 (1959).

276 ANTENNA ENGINEERING

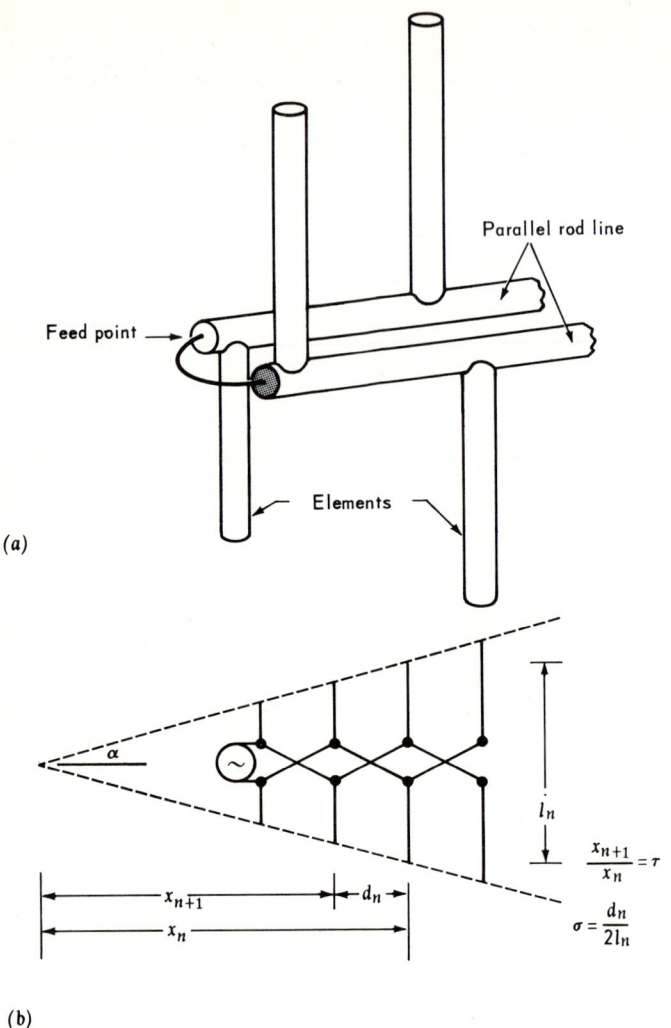

Fig. 7.10 Log-periodic dipole construction.

strated that in a certain range of values for τ and α, the structure was indeed a broadband log-periodic structure with a unidirectional pattern. Isbell also demonstrated experimentally that most of the radiation was coming from those dipole elements which were in the vicinity of a half wavelength long and that the currents and voltages at the large end of the structure were negligible within the operating band of frequencies. Finally, it was shown once again that the operating band of frequencies was bounded on the high side by frequencies corresponding to the size of the smallest elements (and the feed-line size) and on the low side by the frequencies at which the largest dipole element is about a half wavelength long.[1]

A careful and extremely valuable analysis of the log-periodic dipole array was made by R. L. Carrel in a doctoral dissertation. The physical makeup of the log-periodic array, unlike its predecessors, is such that an analysis of it may be based on more or less conventional theory of linear antennas and transmission lines. The main difficulty is the inherent complication. Carrel's analysis consisted of breaking the overall problem into parts, each of which was programmed for the digital computer. First, making the assumption that the element currents were sinusoidally distributed, he computed in the conventional way the mutual impedances between the dipole elements and the self-impedance of each element. In the second part of the problem, Carrel focused his attention on the parallel-wire transmission line, fed at one end and shunt-loaded with impedances corresponding to the dipole antenna elements having sizes and spacings characteristic of log-periodic dipole arrays; of course, the impedance values came from the first part of his computer program. He carried out (on the digital computer) a circuit type of analysis to find the input impedance, voltages, and currents on the loaded transmission line, together with the base (i.e., input) currents at each antenna element. As the last part of the problem, with the specific values for the magnitude and phase of the currents in the antenna elements, he calculated the radiation patterns. Having developed a systematic computer program, Carrel completed calculations on more than 100 different log-periodic dipole designs. He then compared the results of several of these with corresponding experimental models. The measurements included not only impedances and radiation patterns but also the voltage and current distributions in the structure. The agreement between the computer output and the experimental results was excellent. Carrel's work includes two important features: (1) he presents enough detail concerning the voltage and current distribution on the structures to provide insight into the operation of the antenna; (2) he presents a set of design curves which show how to adjust the dimensions of a structure in order to meet specified design objectives.

First, to better understand the operation of the antenna, let us examine the typical results for voltages and currents on a log-periodic dipole array. Figure 7.11, from Carrel's work,[2] shows the amplitude and phase of the transmission-line voltage (voltage at the base of the elements) as a function of distance along the line (from the apex). In this particular

---

[1] D. E. Isbell, *IRE Trans.*, **AP-8**:260–267 (May, 1960).
[2] R. L. Carrel, University of Illinois Antenna Lab. Tech Rept. 52, "Analysis and Design of the Log-periodic Dipole Antenna," Contract AF 33(616)-6079.

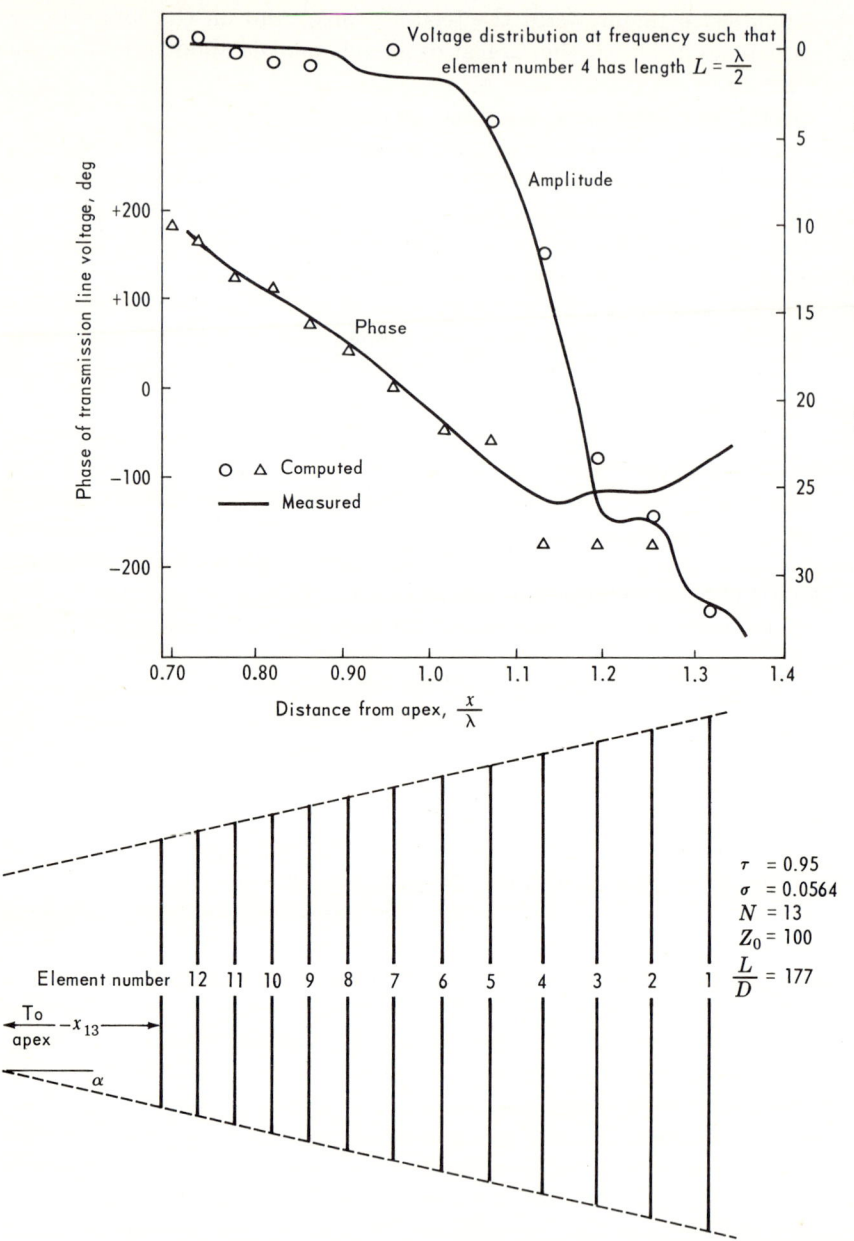

Fig. 7.11 Feeder-line voltage on log-periodic dipole array. Lower figure shows relative position of the elements.

figure, the frequency is such that the element numbered 4 is a half wavelength long. Note in the plot that the voltage is almost constant in amplitude and uniform in phase progression from the point of feed down to about the element numbered 6 (the largest element is numbered 1). This region of constant voltage is called the transmission region at this frequency, since the voltage distribution is almost like that on a matched transmission line. Note, however, that the phase changes by about 150° along a length of line whose length is about a quarter of a free-space wavelength. This means that the phase velocity on the line is only about 0.6 of that of plane waves in space. The decrease in phase velocity is caused by the shunt-capacitive loading of the line by the smaller antenna elements. This loading turns out to be almost constant per unit length, because the larger elements are more widely spaced. Immediately after the element numbered 6, the amplitude of the voltage falls off rapidly; the linear progressive phase variation is also disturbed. The voltage falloff is associated with a relatively strong current excitation in the antenna elements, as will be discussed immediately below. Note finally that the voltage at the largest element is very small, being lower by some 30 db than the voltage at the smaller elements.

Next (Fig. 7.12), consider the currents in the antenna elements. Note that the amplitudes of the currents in elements 7, 6, 5, and 4 are some 5 to 10 db greater than the current amplitudes in the other elements; in fact, the current amplitude in the largest element is too small to plot. The region of relatively high-current amplitudes is termed the "active" region. The fields at the large end of the structure are so small that the fact that the structure does not extend indefinitely is of no consequence at this frequency. The smaller elements are too small to be excited effectively. Next note that the relative phase of the element currents, particularly in the active region, is such that there seems to be a linear progressive phase shift in the direction opposite to that which would occur on an unloaded line. The phasing of the currents in the active elements is in fact suggestive of a wave traveling back toward the feed point. This phasing also accounts for the "backfire" characteristic of the radiation pattern.

As the frequency of operation is changed, the general pattern of voltage and current distribution remains the same, but the active region moves to different elements. In particular, as the frequency is increased, the active region moves in the direction of the shorter elements. When, as a result of the frequency increase, the active region has moved up to the smallest element, the performance begins to change with frequency. Roughly speaking, the high-frequency limit is the frequency at which the shortest element is about a half wavelength long, provided the feed-line separation is small compared to a wavelength. As the frequency is

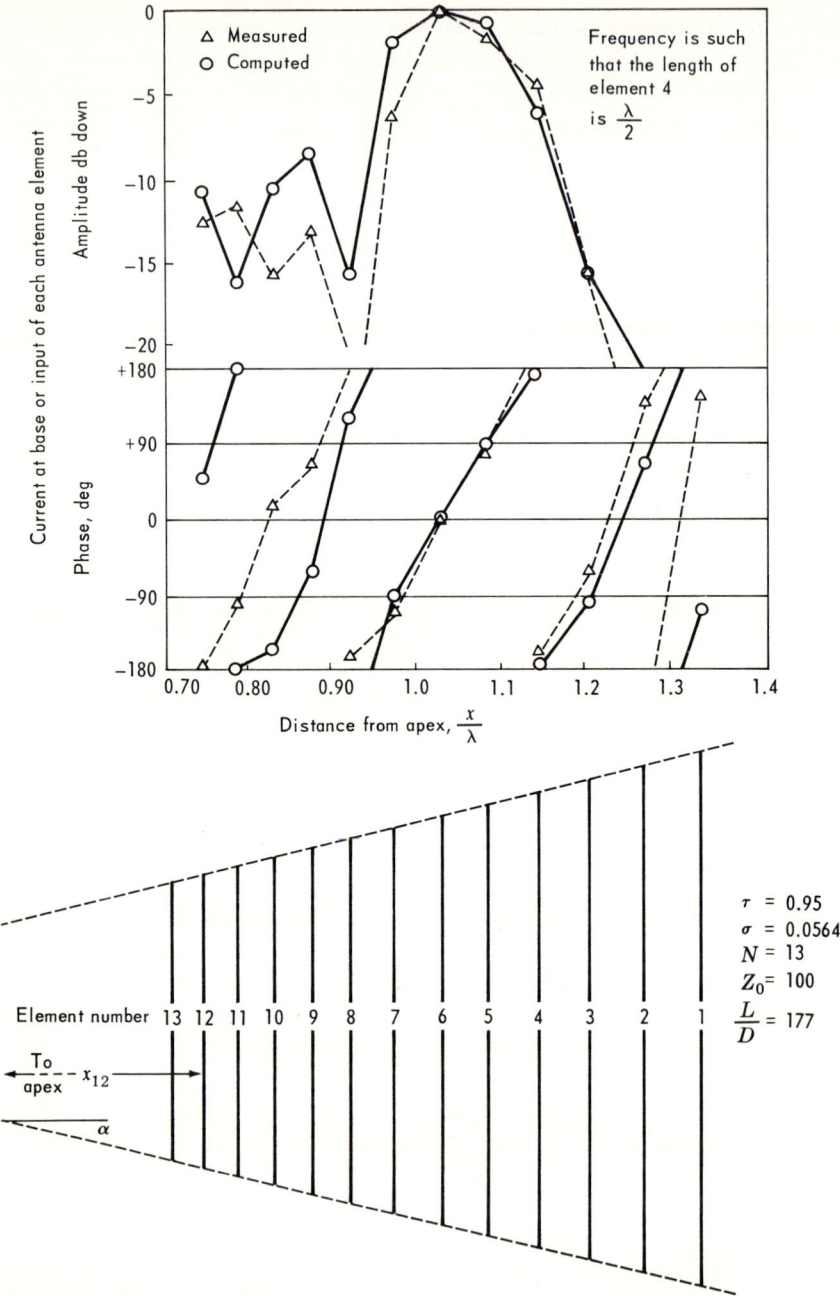

Fig. 7.12 Element input currents on log-periodic dipole array. Lower figure shows relative position of the elements.

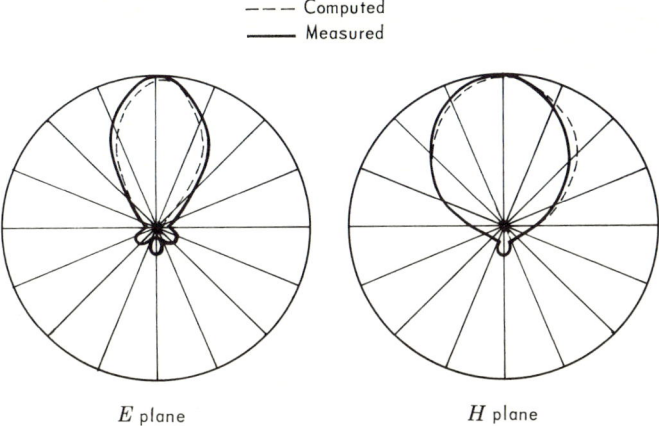

Fig. 7.13 Log-periodic dipole radiation patterns.

decreased, the active region moves toward the longer elements. The performance is satisfactorily constant until the active region reaches the largest element. Roughly speaking, the low-frequency limit is the frequency at which the largest element is a half wavelength long.

Given the element currents and positions, it is a relatively easy matter to compute the radiation pattern. Figure 7.13 shows the pattern of a typical log-periodic antenna which has parameters similar to those for which the voltage and current data are presented. As mentioned above, the direction of the maximum in the radiation pattern is off the small end of the structure, which is opposite to the direction in which the wave starts down the parallel-wire feeder.

As the frequency is varied within the operating bandwidth of the structure, the input impedance values at the feed point, when plotted on a Smith chart, lie on and within a small circle centered on the real axis of the chart. The mean input-resistance level for the structure is designated R_0. This mean resistance level is essentially the characteristic impedance of the equivalent line comprising the transmission-line region. It is found that the actual value, although it is a function of several variables, is only a slowly varying function of the parameter τ. The mean resistance level as a function of σ is shown in Fig. 7.14. Next, Fig. 7.15 shows a typical variation of R_0 with Z_0, the characteristic impedance of the unloaded parallel-rod feeder line. The input-resistance level also depends on the length-to-diameter ratio of the dipole elements, as indicated in Fig. 7.16. In general, it is found that if Z_0 is less than about 75 ohms, then an appreciable amount of power remains on the feeder at the large end of the antenna.

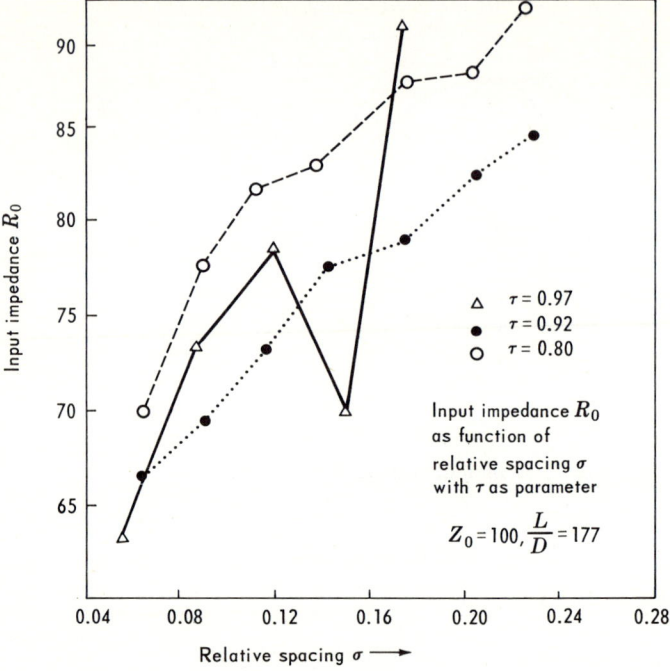

Fig. 7.14 Mean input impedance level of log-periodic dipole array as a function of relative spacing σ (σ is ratio of element spacing to twice the length of the next larger element).

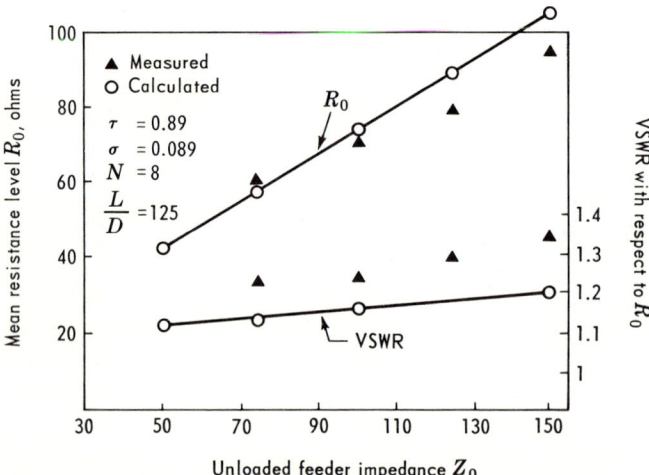

Fig. 7.15 Input impedance of log-periodic dipole array as a function of the characteristic impedance of the unloaded feeder line.

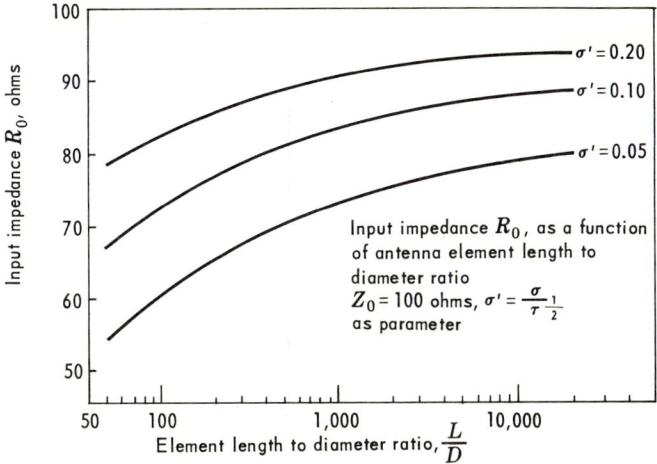

Fig. 7.16 Mean input impedance level of log-periodic dipole arrays as function of element length-to-diameter ratio.

Carrel has prepared curves that are invaluable in the design of log-periodic dipole antennas. As pointed out earlier, the design parameters are τ (the ratio of distances of successive elements from the apex), α (the half-angle subtended at the apex), and σ (the ratio of the element spacing to twice the next-larger element length). Of course, only two of these variables are independent. Figure 7.17 shows how the directive gain varies with σ and τ. The data are displayed in the form of directivity contours. It will be noted that gains in the range from 7.5 to 12 db over isotropic can be obtained with the proper choice of parameters. For a given σ, τ, and element length-to-diameter ratio, the input impedance depends on the characteristic impedance of the parallel-rod feeder line. Thus an antenna may be designed so as to give a specified directivity, and thereafter the input impedance can be almost independently adjusted to a required level. The bandwidth and the specific frequency range required determine the absolute size and arrangement of the structure.

As an example, let us illustrate the design of a log-periodic dipole array for use between 20 and 60 MHz. The antenna is to have a gain of 10 db and an input impedance of 75 ohms. From Fig. 7.17, along the optimum σ line, we find $\tau = 0.917$, and $\sigma = 0.172$. From the geometry, the angle α is

$$\alpha = \tan^{-1} \frac{1 - \tau}{4\sigma}$$

or about 7°. The bandwidth required is 3:1, and roughly speaking the

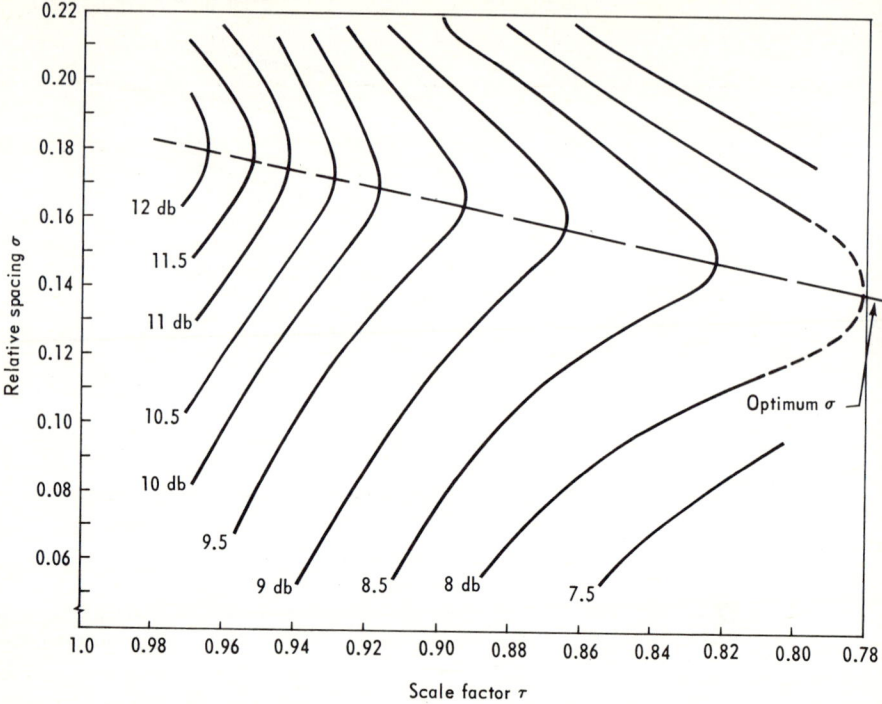

Fig. 7.17 Contours of constant directivity versus τ and σ, for log-periodic dipole arrays.

longest element should be a half wavelength at the lowest frequency. However, the size of the active region depends on the specific design. To incorporate this feature into the design planning, Carrel has introduced and calculated a factor which in essence gives the required structure size to achieve a desired bandwidth. Essentially, one designs for a slightly larger bandwidth than would be required if the active region were of negligible length along the structure. This larger bandwidth B_s is related to the actual required bandwidth B by a relation $B_s = BB_{ar}$, where the factor B_{ar} is tabulated as a function of σ and τ. Carrel calls this quantity the bandwidth of the active region. A nomograph for B_{ar} is presented in Fig. 7.18. In our example, since α is about 7°, B_{ar} is 1.54; therefore $B_s = 3 \cdot 1.54 = 4.56$. Taking the longest element to be a half wavelength at the low-frequency end of the band B_s, it follows from the geometry of the structure that the distance L between the shortest element and the longest element is

$$\frac{L}{\lambda_{max}} = \frac{1}{4}\left(1 - \frac{1}{B_s}\right) \cot \alpha \quad (\doteq 1.6)$$

Again, from the geometry and the relation $x_N/x_1 = \tau^{N-1}$, the number of elements is given by the equation

$$N = 1 + \frac{\log B_s}{\log (1/\tau)} \quad (= 18)$$

The directivity is a slight function of length-to-diameter ratio. In

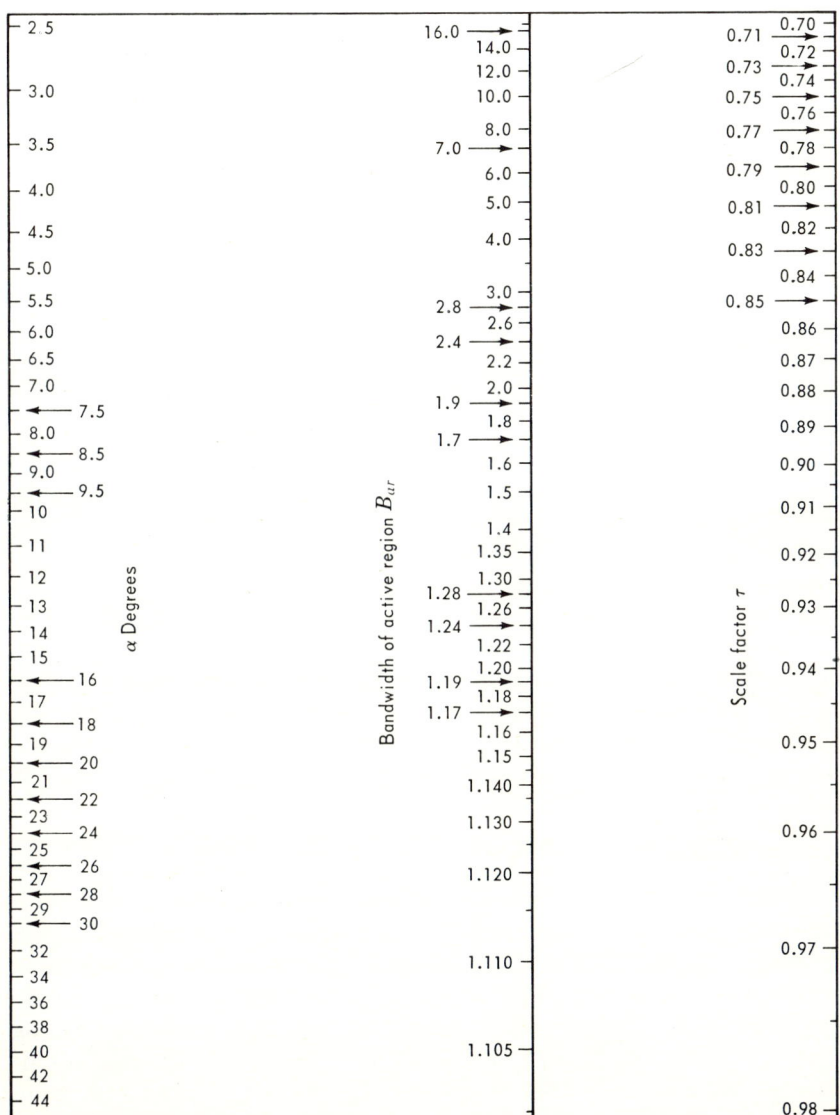

Fig. 7.18 Nomograph, $B_{ar} = 1.1 + 7.7(1 - \tau)^2 \cot \alpha$.

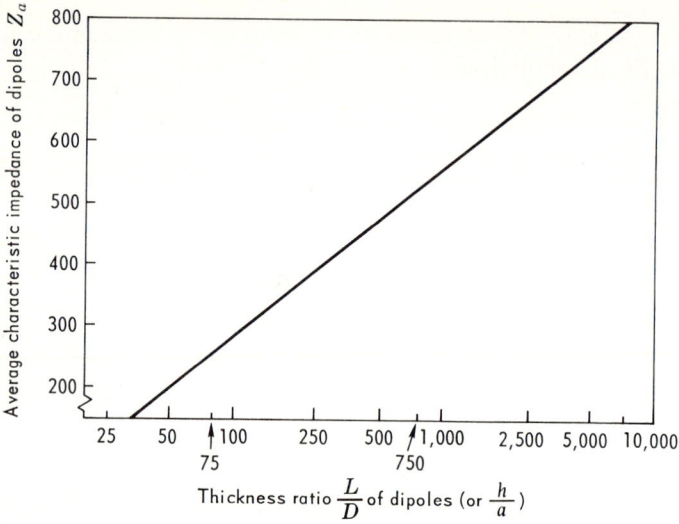

Fig. 7.19 Average characteristic impedance of dipoles Z_{al}, as a function of length-to-diameter ratio.

our particular case, to reach the lower end of the frequency band, the largest elements would have a half-length of about 14 ft. For strength, the elements might be constructed of 1½-in. aluminum tubing. If this were the case, the length-to-diameter ratio would be about 200; therefore no correction to Carrel's main curves would be required.

Next we must design for the required input impedance R_0. This quantity depends on Z_0, σ, and the length-to-diameter ratio of the elements. To bring in the effect of the latter, it is convenient to work in terms of an average characteristic impedance for the elements. An approximate formula for the average characteristic impedance of the elements is

$$Z_a = 120 \left(\ln \frac{h}{a} - 2.25 \right)$$

where h/a is the half-length-to-radius ratio. A graph for Z_a is given in Fig. 7.19. For our example, the average characteristic impedance is something like 360 ohms. The loading produced by such elements depends on the spacing. The effect is indicated quantitatively in the graph, Fig. 7.20. Therein, the quantity σ' is a mean relative spacing, $\sigma' = \sigma/(\tau)^{1/2} = 0.18$. Thus, since $Z_a/R_0 = 360/75 = 4.8$, the graph indicates that $Z_0/R_0 = 1.1$; that is, we should design a parallel-rod feeder such that if unloaded it would have a characteristic impedance of $Z_0 = 1.1 \times 75 = 83$ ohms. It will be recalled that the characteristic impedance of a parallel-rod line is given by

$$Z_0 = 120 \cosh^{-1} \frac{s}{D}$$

where s is the center-to-center spacing, and D is the diameter of the rod. This completes the design, except for the details of picking the specific lengths and tube sizes.

If the diameter of the largest element were selected to be 1½ in., then this implies that the feeder-rod diameter should also be at least 1½ in. in diameter. This sets the center-to-center spacing on the feeder line as follows:

$$s = D \cosh \frac{Z_0}{120} = 1.87 \text{ in.}$$

The length of the parallel-rod line between the longest and the shortest elements is $1.6\lambda_{max}$, or about 90 ft, a rather long structure. The lengths and positions of the other elements follow from τ and σ. An effort should be made to select tubing for the various elements in such a way that the length-to-diameter ratio is nearly the same for all elements. Sometimes it is necessary or desirable to scale down the size of the parallel-rod feeder line in the region of the smaller elements.

The design presented in the foregoing example is by no means optimum. In practice, several such designs would be worked out in detail and the results studied, in order to minimize the antenna length, or the

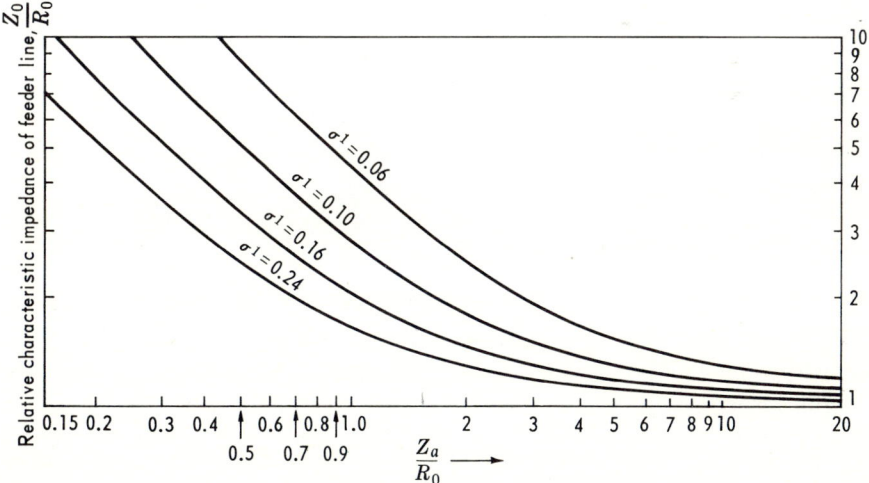

Fig. 7.20 Relative characteristic impedance of feeder line as function of relative characteristic impedance of dipole elements loading the line, Z_a/R_0.

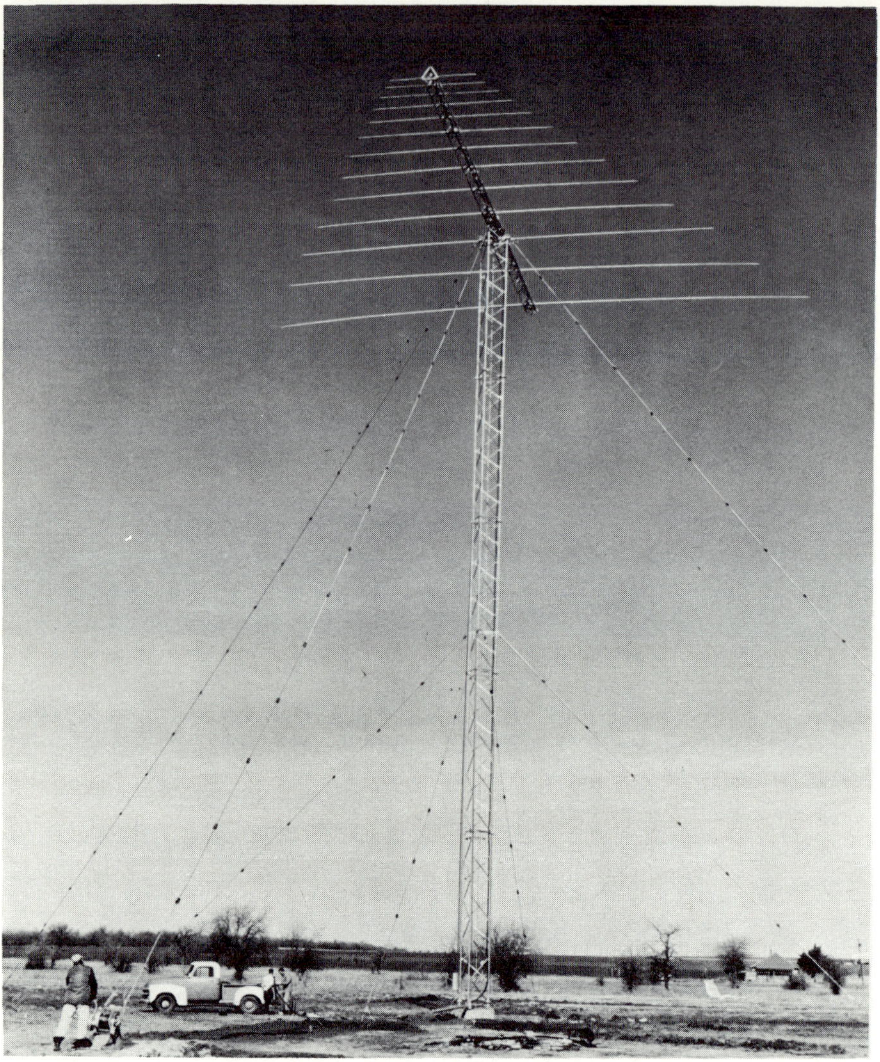

Fig. 7.21 Collins Radio Company 237B series log-periodic dipole array.

number of elements, or to optimize the design in some other respect. As might be expected, the higher gains require the longer structures. A photograph of a commercial version of a log-periodic dipole array, for use in the HF band, is shown in Fig. 7.21.

We have now traced the development of the log-periodic concept and have studied in some detail the log-periodic dipole array as a concrete example of this concept. But a little thought reveals that there is an infinite variety of log-periodic structures. It is therefore worthwhile to summarize by discussing the log-periodic concept in general, without regard to any specific structure. The basic idea is that of electromagnetic modeling or scaling. If all dimensions of a lossless antenna are changed by a factor of τ, then the antenna performance is identical at a frequency $1/\tau$. Now consider the class of antennas made up of an infinite number of "cells," which differ from each other only in size; in particular, each differs from its neighbor only by a constant scale factor τ, as indicated in Fig. 7.22. Structures of this class may be called self-similar, because they possess the unique property of transforming into themselves under a uniform expansion by τ or an integral power of τ. This means that the performance is identical for all frequencies related by $f = f_0 \tau^p$, $p = \pm 1, \pm 2, \ldots$. If voltages and currents are defined at corresponding points in the cells of the structure, these must satisfy the relationships

$$V_n(f) = V_{n+1}(\tau f)$$
$$I_n(f) = I_{n+1}(\tau f)$$

There are essentially no restrictions on the internal composition of the cells as long as they scale with frequency; they may include scaled sources and sinks. However, most work to date has been confined to a single

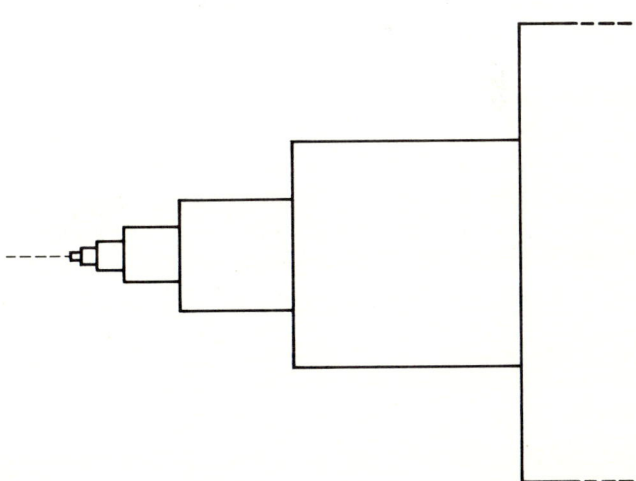

Fig. 7.22 A generalized set of self-similar cells in a structure, selected so that the dimensions of one are τ times the size of the next larger. This is a generalized log-periodic structure.

290 ANTENNA ENGINEERING

Fig. 7.23 Rotatable log-periodic antenna for HF band. Shunt-fed elements. (Collins Radio Company 637B.)

generator at a single location, which must be at the small end in a finite structure.

The practical structure must be finite in size; it must start with some smallest cell and end with some largest cell. Yet at the frequencies within the operating band, the structure must "look" infinite. This implies that at the lower frequencies the electrically small cells must behave as transmission networks. If the cells are electrically small enough, the structure is a *circuit*, which is to say that whether the generator is located in one of these small cells or another is of no significance. It also implies that the larger cells must not be excited, so that their presence or absence will make no difference in the performance within the operating band of frequencies. If these restrictions can be met, the result will be a structure of the log-periodic type, independent of the details of the individual cells. The further art in making a broadband antenna is to choose the cells and the scaling parameters so that the performance between frequencies $f_0 \tau^p$ is satisfactorily constant. Figure

7.23 is a photograph of a log-periodic antenna having a different type of cell structure from that discussed earlier. In this structure, the radiators are shunt-fed through circuit elements. The results of near-field measurements performed on a structure of this type, excited, however, from the small end in a series fashion, are available in the literature.[1]

Problem 7.1 Design a log-periodic dipole antenna to operate over the frequency band from 50 to 250 MHz. To have a specific goal in mind, suppose that the antenna is to cover the VHF TV bands plus the FM bands.

After carrying out the design in a straightforward way, consider what might be done to make a shorter structure, in view of the fact that continuous coverage is not required all the way from 50 to 250 MHz in the specific application.

[1] R. L. Bell, C. T. Elfving, and R. Franks, *IRE Trans.*, **AP-8:**559 (November, 1960).

8. RECEIVING ANTENNAS

The relative signal at the antenna terminals as a function of the angle of arrival of a plane wave is known as the receiving pattern of the antenna. It follows from the reciprocity theorem that the receiving pattern is the same as the radiation pattern of the same antenna when excited as a transmitter. In reciprocal systems, therefore, this question requires no further discussion; in nonreciprocal systems, the voltage at the antenna terminals should be calculated directly.

In the considerations of receiving antennas, the question of vital concern is the ratio of the signal to the noise that they deliver. In this chapter we shall describe methods for calculating the amount of power delivered by receiving antennas; we shall also describe a system for estimating the amount of noise delivered by antennas. The noise delivered by a receiving antenna can in general originate either within the antenna itself or from sources external to it. By definition, a receiving antenna must be located in an open, nonshielded, noisy, and often hostile environment. At the present state of the art at least, antenna structures do not reduce the noise originating in the environment without at the same time reducing the signal, unless the noise is directional in character. If the noise is nondirectional, any signal-to-noise improvement comes therefore only by reducing the ohmic resistance of the antenna system. If the noise is directional, the signal-to-noise ratio may be improved by increasing the directivity of the antenna.

8.1 POWER DELIVERED BY A RECEIVING ANTENNA

In most applications, the signals received by an antenna are delivered to an electronic circuit. This implies that the antenna must somehow lead to a pair of circuit terminals. We can find the power delivered to the circuit by standard circuit analysis, provided we have the current in the circuit elements or the voltage across the circuit. That is, as far as the receiving circuit is concerned, we can find the power delivered if we can determine either the Thévenin equivalent or the Norton equivalent of the antenna as seen from the receiver terminals. We shall now assume that the receiver terminals and the antenna terminals are the same since any transmission-line effects between these may be accounted for easily in a separate analysis. To find the Thévenin equivalent, we must find (1) the open-circuit voltage at the antenna terminals and (2) the impedance looking into the antenna terminals with the primary source (distant transmitter) switched off. To find the Norton equivalent, we must determine (1) the short-circuit current at the antenna terminals and (2) the admittance looking into the antenna terminals with the primary source switched off. *Note* that the immittance quantities in the equivalent circuits are the input impedance or input admittance of the antenna when it is acting as a *transmitter*.

In principle, the required short-circuit currents or open-circuit voltages can be calculated exactly. To do so, it is necessary to treat the receiving-antenna structure as an electromagnetic scatterer and solve for the scattered field. The short-circuit current can then be found from the tangential component of the total magnetic field at the input. Or, the open-circuit voltage can be found from an integral of the electric field from one terminal to the other. However, in practice, the determination of the required scattered fields is very difficult, except in a few special cases (for example, with a spherical structure it is fairly easy). Consequently, the required quantities are commonly determined by means of approximations and experiment.

First let us develop a formula from which the short-circuit current can be obtained. In this case, the receiving-antenna terminals are joined by a perfectly conducting wire. Let us designate the primary source at the transmitter by a, and let us presume that this primary source is a current source $\mathcal{J}_a$. It will be recalled that, with the aid of the reciprocity theorem, the magnetic field anywhere may be found by placing a magnetic current source at the observation point. In this particular case, let us place a uniform magnetic current ring $\mathfrak{M}_r$ about the perfectly conducting wire connected between the antenna terminals. The reciprocity theorem applied between $\mathcal{J}_a$ and $\mathfrak{M}_r$ is then

$$-\int \mathfrak{M}_r \cdot \mathbf{H}_a \, dv = \int \mathcal{J}_a \cdot \mathbf{E}_r \, dv \tag{1}$$

where $\mathbf{H}_a$ is the magnetic field produced by $\mathcal{J}_a$ in the presence of the receiving-antenna structure, and $\mathbf{E}_r$ is the field produced by $\mathfrak{M}_r$ when it is connected around the shorted receiving-antenna terminals. Taking the form of $\mathfrak{M}_r$ as follows

$$\mathfrak{M}_r = -\hat{\phi}\,\delta(\rho - a)\,\delta(z)V_g$$

where a is the radius of the wire, and V_g is a constant equal to 1 volt/m², the integral on the left of Eq. (1) is

$$-\int \mathfrak{M}_r \cdot \mathbf{H}_a\, dv = \int_0^{2\pi} H_{\varphi a}(a,\varphi,0) a\, d\varphi = I_{sc}(0)V_g \tag{2}$$

where I_{sc} is the required short-circuit current; combining Eqs. (1) and (2), we obtain the result

$$I_{sc} = \frac{1}{V_g} \int \mathfrak{J}_a \cdot \mathbf{E}_r\, dv \tag{3}$$

This formula is exact, but it is frequently inconvenient and/or impractical, since to find the short-circuit current it is necessary to know the field produced by $\mathfrak{M}_r$ (with the antenna structure present) at the location of the primary transmitting source, $\mathfrak{J}_a$. Consequently it is worthwhile to obtain an alternative form.

An alternative form of Eq. (3) has the substantial advantages that (1) it is based only on the field at the receiver, therefore the transmitting current distribution $\mathfrak{J}_a$ is immaterial, and (2) in the alternative equation, the field quantity at the receiving position is that in the absence of the receiving structure. To obtain the alternative form, consider the fields produced by the receiving-antenna structure when it is excited as a transmitter by the source $\mathfrak{M}_r$. The fields are set up by the primary source $\mathfrak{M}_r$ plus the induced currents on the structure. Suppose we know the antenna current distribution (as a transmitter). Then if we were to set up in free space this same distribution of currents, the fields that would be produced by these currents in free space would be the same as those that are set up by the currents on the actual antenna. Then consider the reciprocity theorem as it applies between $\mathfrak{J}_a$ and the hypothetical distribution of currents in space. Calling the hypothetical currents $\mathfrak{J}_h$ we have

$$\int \mathbf{E}_r \cdot \mathfrak{J}_a\, dv = \int \mathbf{E}_{a0} \cdot \mathfrak{J}_h\, dv - \int \mathbf{H}_{a0} \cdot \mathfrak{M}_r\, dv \tag{4}$$

since $\mathfrak{J}_h + \mathfrak{M}_r$ generates $\mathbf{E}_r$. The subscript 0 has been added to $\mathbf{E}_{a0}$ and $\mathbf{H}_{a0}$ to emphasize the fact that these are the fields produced by the distant transmitter $\mathfrak{J}_a$ in the *absence* of the receiving-antenna structure, since we are here applying the reciprocity theorem between $\mathfrak{J}_a$ and a hypothetical current distribution in free space. This means that $\mathbf{H}_{a0}$ is almost constant over the extent of the magnetic current ring $\mathfrak{M}_r$, which in turn implies that the integral is essentially zero. Combining Eqs. (3) and (4) we can obtain an alternative formula for the short-circuit current:

$$V_g I_{sc} = \int \mathbf{E}_{a0} \cdot \mathfrak{J}_h\, dv \tag{5}$$

This formula is almost exact. To evaluate it exactly, however, we need to know the exact distribution of currents on the receiving antenna structure when it is excited as a transmitter by the magnetic current ring. This exact distribution is not often known; however, as we have seen in Chaps. 4 and 5, a good approximation for the current distribution is usually available. Consequently, Eq. (5) is useful and provides a good approximation to the short-circuit current.

Let us rewrite Eq. (5) in a form that should promote its correct application. The quantity $\mathbf{E}_{a0}$ is the field produced by the distant source in the absence of the antenna structure. The nature of the source a is irrelevant; therefore let us relabel $\mathbf{E}_{a0}$ as $\mathbf{E}_{\text{inc}}$, where the subscript inc is to suggest *incident*. Also, since $\mathfrak{J}_h$ is the equivalent of the transmitting current distribution, let us relabel $\mathfrak{J}_h$ as $\mathbf{J}_t$. Then recall that in Eq. (5), this current distribution $\mathbf{J}_t$ is that produced by a particular primary source, the magnetic current ring $\mathfrak{M}_r = -V_g \hat{\phi}\, \delta(\rho - a)\, \delta(z)$; this source is the equivalent of a unit constant-voltage source at the input. If we have a voltage source V_g at the input, the reciprocity theorem gives

$$V_g I_{sc} = \int \mathbf{E}_{\text{inc}} \cdot \mathbf{J}_t \, dv$$

where $\mathbf{J}_t$ is the current generated by V_g. The current at the input is related to V_g by the input admittance, $I_{\text{in}} = V_g Y_{\text{in}}$. With this relation, we can replace V_g and obtain the most useful and suggestive form of Eq. (5),

$$I_{sc} = \frac{Y_{\text{in}}}{I_{\text{in}}} \int \mathbf{E}_{\text{inc}} \cdot \mathbf{J}_t \, dv \tag{5a}$$

With this result, we are in a position to identify a Norton-equivalent circuit for the receiving antenna, as indicated in Fig. 8.1a. In this circuit, Y_{in} is the admittance as seen looking into the antenna input terminals, which can be measured or calculated as indicated in Chaps. 4 and 5;

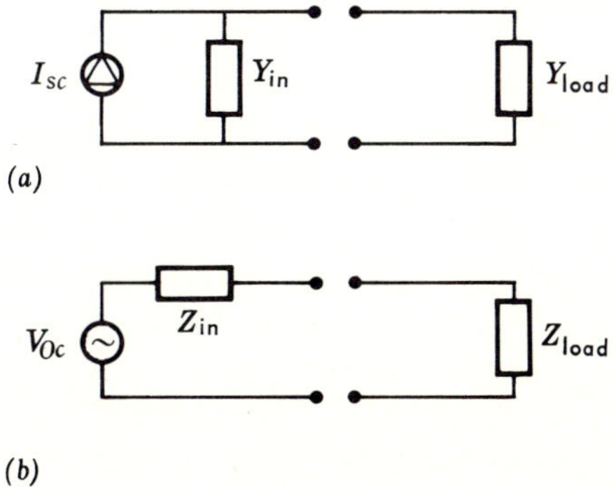

Fig. 8.1 Equivalent circuits for receiving antennas: (a) Norton-equivalent circuit for receiving antenna $I_{sc} = \left(\dfrac{Y_{\text{in}}}{I_{\text{in}}}\right) \int \mathbf{E}_{\text{inc}} \cdot \mathbf{J}_t \, dv$ (Y_{in} is input admittance of the receiving antenna if it were operating as a transmitter); (b) Thévenin-equivalent circuit for receiving antenna $V_{oc} = 1/I_{\text{in}} \int \mathbf{E}_{\text{inc}} \cdot \mathbf{J}_t \, dv$ (Z_{in} is input impedance of receiving antenna if it were operating as a transmitter).

Y_{load} represents the impedance presented by the receiving electronic system at the input terminals of the receiving antenna. The power delivered to the load is then readily obtained with conventional circuit analysis.

Equation (5a) provides a convenient way to determine the short-circuit current for wirelike antenna structures. The determination of an equation that is readily applicable to aperture-type antennas is suggested as a problem.

Problem 8.1 Show that the short-circuit current at the terminals of an aperture type of antenna is given by

$$V_g I_{sc} = \int \mathbf{E}_{\text{inc}} \cdot \mathcal{J}_h \, dv - \int \mathbf{H}_{\text{inc}} \cdot \mathcal{M}_h \, dv \tag{6}$$

where $\mathcal{J}_h$ and $\mathcal{M}_h$ are the equivalent electric and magnetic current sources for the fields in the aperture, assuming that the antenna is excited with a voltage generator V_g at a pair of input terminals.

The open-circuit voltage for the Thévenin-equivalent circuit may be obtained by a procedure similar to that by which the Norton equivalent was obtained above. The only real difference in the analysis is that the terminals of the receiving antenna are open-circuited and that an electric current source density $\mathcal{J}_r$, connected between the terminals, takes the place of the magnetic current ring $\mathcal{M}_r$. The details of the development are left as a problem for the reader.

Problem 8.2 Show by a field analysis that the open-circuit voltage at a receiving antenna can be calculated from the relationships

$$I_g V_{oc} = \int \mathbf{E}_r \cdot \mathcal{J}_a \, dv \tag{7}$$

(notation same as the development above) and

$$V_{oc} = \frac{1}{I_{\text{in}}} \int \mathbf{E}_{\text{inc}} \cdot \mathbf{J}_t \, dv \tag{8}$$

where again $\mathbf{E}_{\text{inc}}$ is the field at the position of the receiving antenna but with the antenna structure removed, and $\mathbf{J}_t$ is the current distribution which would result if the receiving antenna were excited as a transmitter by the point current $\mathcal{J}_r$.

The Thévenin-equivalent circuit, shown in Fig. 8.1b, may also be obtained from the Norton equivalent in Fig. 8.1a by the standard procedure of circuit analysis. The open-circuit voltage at the terminals of the Norton equivalent is clearly $V_{oc} = I_{sc} Z_{\text{in}}$; thus Eq. (8) together with the equivalent circuit in Fig. 8.1b follows immediately.

In the event that the receiving antenna is an array of discrete elements, the quantity I_{in} is the current at the common terminal pair when the system is excited there so as to act as a transmitter.

8.2 EFFECTIVE AREA OF A RECEIVING ANTENNA

System design and analysis is greatly aided by the introduction of the concept of effective area A_e, sometimes called effective aperture, of a receiving antenna. The utility of the concept should be clear from its definition:

$$A_e \equiv \frac{\text{power delivered by receiving antenna to a conjugate-matched load}}{\text{power density incident}}$$

The power delivered to the load can be found from either of the equivalent circuits in Fig. 8.1. In terms of rms values, it is

$$P_{\text{del}} = \left|\frac{(1/I_{\text{in}})\int \mathbf{E}_{\text{inc}} \cdot \mathbf{J}_t \, dv}{Z_{\text{in}} + Z_{\text{load}}}\right|^2 \cdot R_{\text{load}} \tag{9}$$

so that if the antenna and load are conjugate-matched

$$P_{\text{del}} = \frac{|(1/I_{\text{in}})\int \mathbf{E}_{\text{inc}} \cdot \mathbf{J}_t \, dv|^2}{4R_a} \tag{10}$$

where R_a is the input resistance of the antenna. The effective area is then

$$A_e = \frac{|\int \mathbf{E}_{\text{inc}} \cdot \mathbf{J}_t \, dv|^2 \left(\frac{\mu}{\epsilon}\right)^{1/2}}{4|I_{\text{in}}|^2 R_a |E_{\text{inc}}|^2} \tag{11}$$

The effective area of a receiving antenna is related to the gain of the same antenna employed as a transmitter. To establish the connection, imagine that the plane wave which is incident on the receiving antenna is set up by a distant point dipole, $\mathcal{J}_a = \hat{\mathbf{p}} \, \delta(x) \, \delta(y) \, \delta(z)$. Then recall the definition of system gain

$$G \equiv \frac{\text{power density at some location}}{\text{power input}} \times 4\pi r^2$$

The power input as a transmitter is, of course, $|I_{\text{in}}|^2 R_a$. The power density at the position of the point dipole is $|E_t|^2/(\mu/\epsilon)^{1/2}$. The field at the point dipole, E_t, produced by the antenna when excited as a transmitter, may be obtained from the reciprocity theorem

$$\int \mathbf{E}_t \cdot \mathcal{J}_a \, dv = \int \mathbf{E}_a \cdot \mathbf{J}_t \, dv$$

or since, as we have said, $\mathcal{J}_a$ produces the incident field, $\mathbf{E}_a$ is $\mathbf{E}_{\text{inc}}$,

$$E_t = \int \mathbf{E}_{\text{inc}} \cdot \mathbf{J}_t \, dv$$

where the last step follows because $\mathcal{J}_a$ is a unit amplitude point dipole. It follows from the definition that the gain is

$$G = \frac{|\int \mathbf{E}_{\text{inc}} \cdot \mathbf{J}_t \, dv|^2}{(\mu/\epsilon)^{1/2} |I_{\text{in}}|^2 R_a} 4\pi r^2 \tag{12}$$

where R_a is the input resistance of the antenna. We may now find the connection between A_e and G by forming the ratio of Eqs. (11) and (12):

$$\frac{A_e}{G} = \frac{1}{4}\frac{\mu}{\epsilon}\frac{1}{|E_{\text{inc}}|^2 4\pi r^2}$$

Here, the quantity E_{inc} is the distant field of a point dipole with unit current moment. If the antennas are in free space, this field is given by

$$|E_{\text{inc}}|^2 = \frac{\omega^2 \mu^2}{(4\pi r)^2} \qquad (13)$$

Recognizing the quantity $\omega^2 \mu\epsilon$ to be equal to $(2\pi/\lambda)^2$, we obtain the result

$$A_e = \frac{G\lambda^2}{4\pi} \qquad (14)$$

This is the connection between effective area and gain that we were seeking. With this equation in hand, it is clear that compilations of antenna characteristics need only contain numbers representing either their gains or their effective apertures, not both. Table 8.1 gives the directivities of common antenna structures. Efficiencies must be known or estimated.

Exercise Identical antennas are employed for transmission and reception in a communication system and are located far from ground. The directivity of the antennas is 20 db, the frequency is 6000 MHz, the range is 50 km. Find the ratio in db of the power input to the power delivered to a conjugate-matched load.

Table 8.1 Directivities of Common Antennas

Antenna Type	Directivity
Small antenna	1.5
Half-wave dipole	1.64
Full-wave dipole	2.41
Long continuous uniform line source	$4L/\lambda$
Half-wave dipole $\frac{1}{4}\lambda$ from a large reflector	5.5
Large broadside rectangular array	$2\pi A\dagger/\lambda^2$
Large broadside rectangular array with reflector	$4\pi A/\lambda^2$
N turn, axial-mode helix, with circumference C, turn-to-turn spacing S ($C \sim 0.8$ to 1.3λ, $S \sim 0.15$ to 0.3λ)	$15(C/\lambda)^2 (S/\lambda)$
Optimum horn	$10A/\lambda^2$
Uniformly illuminated circular aperture	$4\pi A/\lambda^2$
Parabolic reflector	6 to $7.5A/\lambda^2$

† A is area.

Although Eq. (14) is very simple and useful, it must be used with great caution when the antenna is situated in or near a lossy environment. The basic difficulty is revealed in the derivation, in that the result involves the field due to a distant-point dipole. In general, this incident field is not given by Eq. (13); therefore Eq. (14) does not follow directly. To establish the corresponding connection, one must find the field of a point dipole in the actual environment; this is often complicated. For systems work, to be sure of a correct, consistent result, when the antenna is located in or near anything other than a free-space environment, the effective area should be computed from Eq. (11); it should be noted that the quantity E_{inc} in that equation is the field produced *in the prescribed environment* by a distant-point dipole.

8.3 POWER TRANSFER BETWEEN ELLIPTICALLY POLARIZED ANTENNAS

To simplify the considerations of power transfer between elliptically polarized antennas, let the antennas be facing each other along the z axis. Let the transmitter be facing along positive z, which means of course that the receiver is facing negative z. The state of polarization of the transmitted signal can be represented by an E field of the form $\mathbf{E} = E_x\hat{\mathbf{x}} + E_y\hat{\mathbf{y}}$, where E_x and E_y differ in amplitude and phase so as to give the required state of polarization. To find the power delivered to the receiver, it will be recalled [Fig. 8.1 and Eq. (11)] that we need to know the current distribution $\mathbf{J}_t$ on the receiving antenna when it is excited as a transmitter. For an elliptically polarized antenna, this current distribution may be written in the form $\mathbf{J}_t = J_x\hat{\mathbf{x}} + J_y\hat{\mathbf{y}}$, where J_x and J_y are the components required to give a prescribed state of polarization. Then, for example, it follows from Eq. (8) that the open-circuit voltage at the receiving antenna is

$$V_{oc} = \frac{1}{I_{\text{in}}} \int \mathbf{E}_{\text{inc}} \cdot \mathbf{J}_t \, dv$$
$$= \frac{1}{I_{\text{in}}} \int (E_x J_x + E_y J_y) \, dv$$

Then, by Eq. (11), the effective area of the receiving antenna is

$$\frac{|\int (E_x J_x + E_y J_y) \, dv|^2}{4|I_{\text{in}}|^2 R_{\text{in}} (E_x{}^2 + E_y{}^2)} \left(\frac{\mu}{\epsilon}\right)^{\frac{1}{2}}$$

where again I_{in} and R_{in} refer to the values at the common input-terminal pair of the receiving antenna.

Let us define orthogonal polarizations to be those for which the integrand of the reaction integral vanishes:

$$E_x J_x + E_y J_y = 0$$

Note then that there are several kinds of orthogonal polarization. The obvious orthogonal polarizations are those in which $E_x = 0$ when $J_y = 0$, or $E_y = 0$ when $J_x = 0$. But there are others. For example, suppose the transmitted signal is right-hand circularly polarized, $\mathbf{E}_{\text{inc}} = E_x(\hat{\mathbf{x}} - j\hat{\mathbf{y}})$. Suppose further that the receiving antenna is phased for *left*-hand circular polarization: $\mathbf{J}_t = J_x(\hat{\mathbf{x}} - j\hat{\mathbf{y}})$. (The sign on the y component is negative, because the receiving antenna when acting as a transmitter is facing negative z.) In such a case it is clear that the reaction integral vanishes: $\mathbf{E}_{\text{inc}} \cdot \mathbf{J}_t = E_x J_x - E_x J_x = 0$; therefore the receiving antenna is in a state of orthogonal polarization to the transmitting antenna. On the other hand, if $\mathbf{J}_t$ is phased to give right-hand circular polarization, $\mathbf{J}_t = J_x(\hat{\mathbf{x}} + j\hat{\mathbf{y}})$, then the reaction integrand is

$$\mathbf{E}_{\text{inc}} \cdot \mathbf{J}_t = 2 E_x J_x$$

It is interesting to examine the result for the case of, for example, right-handed elliptical polarization being received by a left elliptically polarized antenna. To examine this case, we may set $\mathbf{E}_{\text{inc}} = E_0(a\hat{\mathbf{x}} - jb\hat{\mathbf{y}})$, and $\mathbf{J}_t = J_0(c\hat{\mathbf{x}} - j\,d\hat{\mathbf{y}})$ where a, b, c, and d are real and positive. To see if these states of polarization are orthogonal, we examine the integrand

$$\mathbf{E}_{\text{inc}} \cdot \mathbf{J}_t = E_0 J_0 (ac - bd)$$

Thus, one condition for orthogonality is $ac = bd$, or $a/b = d/c$. This shows that the left elliptically polarized receiving antenna will not receive a right elliptically polarized signal if the polarization ellipses have the same axial ratio but have major axes that are perpendicular. Alternatively, the elliptical polarizations may be represented in terms of their left and right circular components.

Finally, as an explicit special case, let us find the effective area of a linearly polarized antenna receiving a circularly polarized signal. With the current $\mathbf{J}_t$ directed along $\hat{\mathbf{x}}$, Eq. (11) gives for the effective area

$$A_e = \frac{|\int E_x J_x \, dv|^2 (\mu/\epsilon)^{1/2}}{4|I_{\text{in}}|^2 R_{\text{in}} 2 E_x^2}$$

Let us compare this to the effective area of a circularly polarized receiving antenna for circular polarization. The quantity $|I_{\text{in}}|^2 R_{\text{in}}$ for the linear antenna is one-half that of the matched circularly polarized pair; the integral in the numerator is also only one-half as large. Thus, in view of the squaring of the integral, we find that the effective area of a linearly polarized antenna in a circularly polarized field is only half that of the matched circularly polarized antenna made up of the same currents at right angles.

8.4 NOISE CONSIDERATIONS

In the preceding sections we have learned how to calculate the signal power delivered to a load by a receiving antenna. The remaining task is to determine the noise power delivered to the load by the antenna, so that we can find the signal-to-noise ratio presented to the receiving system.

From fundamental physical considerations, it is known that any material body at temperature T radiates electromagnetic energy; the fields are generated by the thermal motions of its constituents. If a body has a temperature greater than its surroundings, this radiation results in cooling. If a body is at the same temperature as its surroundings, its radiated energy is balanced by energy absorbed from the radiations of its surroundings. Because of the randomness of the thermal motions, the radiated electromagnetic energy is *noise*. The noise power S_n radiated by a blackbody, at absolute temperature T per unit solid angle, in the frequency range df, is given by the famous Planck distribution law:

$$S_n = \frac{2hf}{\lambda^2} \frac{df}{\exp(hf/kT) - 1}$$

where k is Boltzmann's constant, 1.38×10^{-23} joules/°K.

For ordinary temperatures and frequencies in the usual radio-frequency spectrum, to a good approximation we may write $\exp(hf/kT) = 1 + hf/kT$; this means that a good approximation for S_n is

$$S_n = \frac{2kT}{\lambda^2} df$$

which is the Raleigh-Jeans distribution law.

To begin to understand the significance of this to receiving antennas, imagine that a simple perfectly conducting antenna is set into a perfectly absorbing (black) enclosure at temperature T. Let the antenna be conjugate-matched, in frequency band df, to an impedance whose temperature is also T. With the system in equilibrium, the antenna acts simultaneously as a receiver and as a transmitter; the load resistor delivers a power $kT\,df$ to the antenna to be radiated, and to maintain the equilibrium, the antenna must also be delivering a power $kT\,df$ to the load resistor, which it is receiving from the surroundings. The mean square open-circuit noise voltage delivered by the antenna is $|v_{oc,n}|^2 = 4R_a kT\,df$. From these considerations, we conclude that the noise power available in a frequency band df from an absorbing environment at temperature T is $P_n = kT\,df$. It is convenient in this case to assign to the antenna the absolute temperature of the environment, $T_a = T$, so that the noise delivered by the antenna to the load can be calculated by the simple formula $P_n = kT_a\,df$.

It is interesting to note at this point that with a uniformly noisy environment, the gain or effective area of the antenna does not influence the noise power delivered by the antenna. To demonstrate this explicitly, let the incident power density per unit solid angle be denoted by S_{in}. Then, the power delivered by the antenna is

$$P_{\text{del},n} = \iint S_{in}(\theta,\varphi) A_{e,n}(\theta,\varphi) \sin\theta\,d\theta\,d\varphi$$

Now for a single frequency and matched polarizations, the connection between the effective area and the gain is

$$A_e(\theta,\varphi) = G(\theta,\varphi) \frac{\lambda^2}{4\pi}$$

But since all polarizations are represented in the incoming noise band, the effective area for noise is only one-half of this. Thus, if S_{in} is uniform for all directions, as stated, the power delivered can be written

$$P_{\text{del},n} = \frac{S_{in}\lambda^2}{8\pi} \int G(\theta,\varphi) \sin\theta\,d\theta\,d\varphi$$

From the definition of (directive) gain (PD is power density)

$$G(\theta,\varphi) = \frac{\text{PD}(\theta,\varphi)}{P_{\text{rad}}} 4\pi r^2$$

and from the fact that

$$\int \text{PD}(\theta,\varphi) 4\pi r^2 \sin\theta\,d\theta\,d\varphi = 4\pi P_{\text{rad}}$$

we find for the noise power delivered

$$P_{\text{del},n} = S_{in} \frac{\lambda^2}{2}$$

which is indeed independent of the gain or effective area. Also, from the Raleigh-Jeans law, the power radiated by a blackbody at absolute temperature T per unit solid angle is

$$S_{in} = 2kT \frac{df}{\lambda^2}$$

Thus we find again that the power delivered by any antenna in a uniformly black environment at temperature T is

$$P_{\text{del},n} = kT\,df$$

independent of its gain or effective area, as stated earlier.

In most actual environments, the temperature of the surroundings is not uniform and the environment is not perfectly absorbing. Thus, the noise power incident per unit solid angle $S_{\text{in}}(\theta,\varphi)$ is not uniform. Assuming that the radiation from a given solid angle is still determined by the Raleigh-Jeans law, the noise power from that direction correlates with an environmental temperature, often called the brightness temperature, $T_b(\theta,\varphi)$

$$S_{\text{in}}(\theta,\varphi) = 2kT_b(\theta,\varphi)\frac{df}{\lambda^2}$$

The brightness temperature for a given direction is an effective temperature that includes the effects of any departures of the environment from the ideal of a blackbody radiating according to the Raleigh-Jeans law. The power delivered in a nonuniformly noisy environment is then

$$\begin{aligned}P_{\text{del},n} &= \iint \tfrac{1}{2}A_e(\theta,\varphi)S_n(\theta,\varphi)\sin\theta\,d\theta\,d\varphi \\ &= \frac{\lambda^2}{4\pi}\iint \tfrac{1}{2}G(\theta,\varphi)\frac{2k}{\lambda^2}T_b(\theta,\varphi)\,df\sin\theta\,d\theta\,d\varphi \\ &= \frac{k\,df}{4\pi}\iint G(\theta,\varphi)T_b(\theta,\varphi)\sin\theta\,d\theta\,d\varphi\end{aligned}$$

In this case, if we wish to characterize the antenna by an effective temperature T_a such that the noise power delivered can be calculated from $P_{\text{del},n} = kT_a\,df$, then the effective antenna temperature is

$$T_a = \frac{1}{4\pi}\iint G(\theta,\varphi)T_b(\theta,\varphi)\sin\theta\,d\theta\,d\varphi$$

In order to determine this effective temperature of the antenna, we must know the gain function and the brightness-temperature function of the environment.

In most practical cases, the brightness temperature of the environment is an extremely complicated function. It depends on frequency, time of day and season, and antenna location, as well as the direction in space. At the present time, there is no good way to calculate it. It must be measured experimentally, and the data for it are accurate only in a statistical sense.

8.5 EXTERNAL NOISE

The external noise that deteriorates communication-systems performance may be classified as extraterrestrial (cosmic), atmospheric, or manmade. The relative importance of these types of noises depends on frequency and location. At frequencies of the order of 1 MHz and lower,

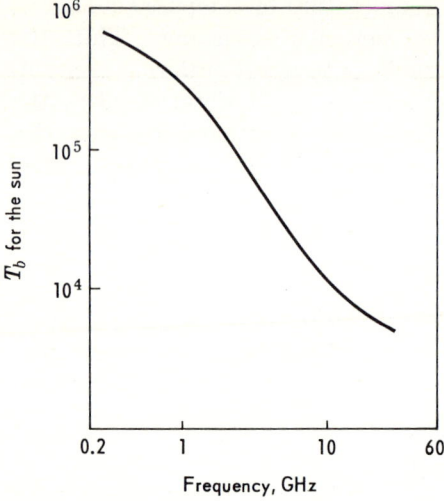

Fig. 8.2 Brightness temperature of the sun as a function of frequency.

atmospheric noise originating from lightning is dominant. Precipitation noise due to charged-particle bombardment may also be important in some cases. In heavily populated areas, man-made noise usually dominates in the frequency range from 10 to 200 MHz. At higher frequencies, in a system with low-noise amplifiers, the noise resulting from the natural

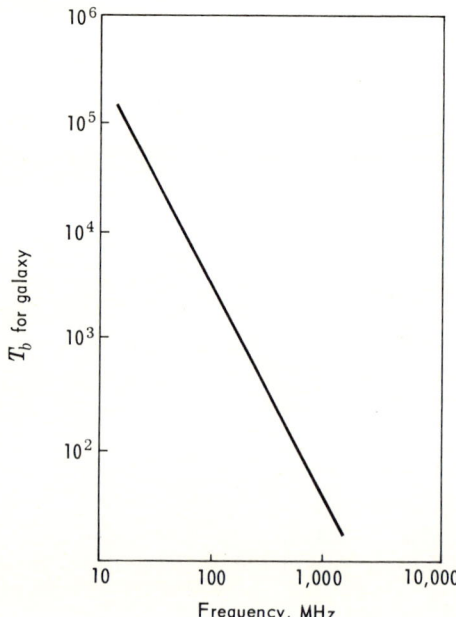

Fig. 8.3 Brightness temperature for the galaxy.

resonant frequencies of the atmospheric gases, from the earth, from the sun, and from the radio stars becomes important.

It should be clearly understood that the noise from all of these sources is time-varying. The time variations are in part systematic and in part random. The detailed character of the fluctuations must also be determined experimentally, for modulation-systems considerations. The difficulty of an adequate representation of man-made noise as a time-varying brightness temperature of the environment should be obvious.

Brightness temperatures for the noise sources that are important at the higher frequencies are indicated in Figs. 8.2 to 8.4. By far the noisiest of the extraterrestrial noise sources is the sun. Figure 8.2 shows the brightness temperature of the quiet sun as a function of frequency, as measured by S. Matt and O. J. Jacomimi and reported by Hogg and Mumford.[1] From that figure, it can be seen that the brightness temperature of the sun is in the range between about 10^4 and $10^{6}°K$, depending on the frequency range. From a point on the earth, the sun subtends a plane angle of approximately a half degree. The next most important noise source is the center of our galaxy. Measured bright-

[1] D. C. Hogg and W. W. Mumford, *Microwave Journal*, p. 60 (March, 1960).

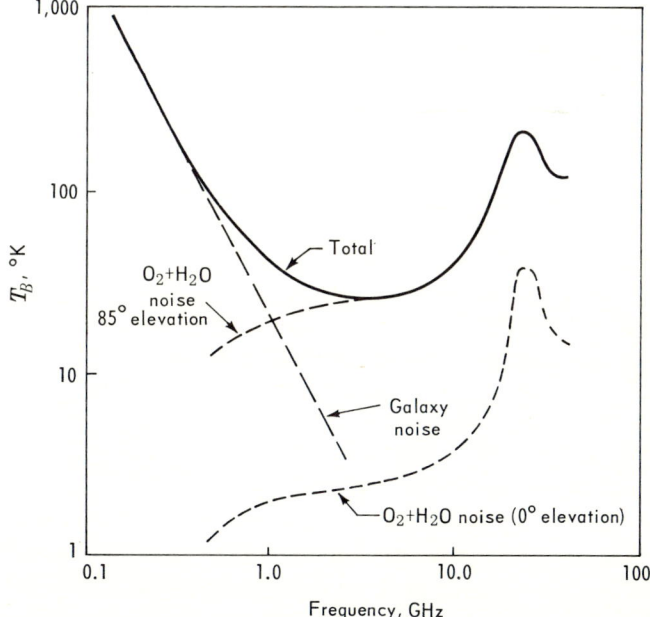

Fig. 8.4 Composite of sky-brightness temperature of galaxy noise plus noise from oxygen and water vapor in atmosphere.

ness temperatures for our galaxy are indicated in Fig. 8.3. Also, the noise associated with the resonance absorption frequencies in water vapor and oxygen have been calculated and checked experimentally. Figure 8.4 is a composite of galactic noise and sky noise due to water vapor and oxygen.

Problem 8.3 A parabolic-reflector receiving antenna is designed to have a gain of 40 db. It is pointed toward a satellite at a distance of 200 km, which is radiating 1 watt at 1 GHz from an antenna having a gain of 1.5. Estimate the signal-to-noise at the output of the receiving antenna at a time during which the bearing to the satellite and the bearing to the sun differ by only 5°. Assume a bandwidth of 1 MHz.

As was pointed out earlier, at frequencies below about 10 MHz, the noise generated in the atmosphere, chiefly from thunderstorms, predominates. Since electromagnetic waves at this frequency and lower usually propagate great distances around the earth, the noise as seen by antennas near the earth's surface is nearly omnidirectional. The noise levels have been studied extensively at many places on earth. Results of these studies are reported in the publications of the National Bureau of Standards and CCIR.[1] The results of the atmospheric-noise measurements are typically given in terms of a quantity F_a stated in decibels, which is defined as follows:

$$F_a = 10 \log_{10} \frac{P_{\text{av},n}}{kT_0B}$$

where $P_{\text{av},n}$ is the noise power available from a perfect, short, vertical monopole, T_0 is 288°K, and B is the bandwidth. Since we have shown earlier that the noise power available from a lossless antenna in a uniformly noisy environment is independent of directive gain and is equal to $kT_b \, df$, where T_b is the effective brightness temperature of the environment, it is clear that the quantity F_a can be interpreted as ten times the log of the ratio of the equivalent brightness temperature of the atmospheric noise to the reference temperature, 288°K:

$$F_a = 10 \log_{10} \frac{T_b}{288}$$

The data given in CCIR No. 332 for F_a can be plotted in the form of effective brightness temperature for atmospheric noise. This is done, for a typical set of data, in Fig. 8.5. Since the noise seems to arrive from all directions, the brightness temperature is also the effective antenna temperature for any antenna above and near the surface of the earth.

[1] "World Distribution and Characteristics of Atmospheric Radio Noise," *CCIR Rept. no.* 332, International Telecommunication Union, Geneva, 1964.

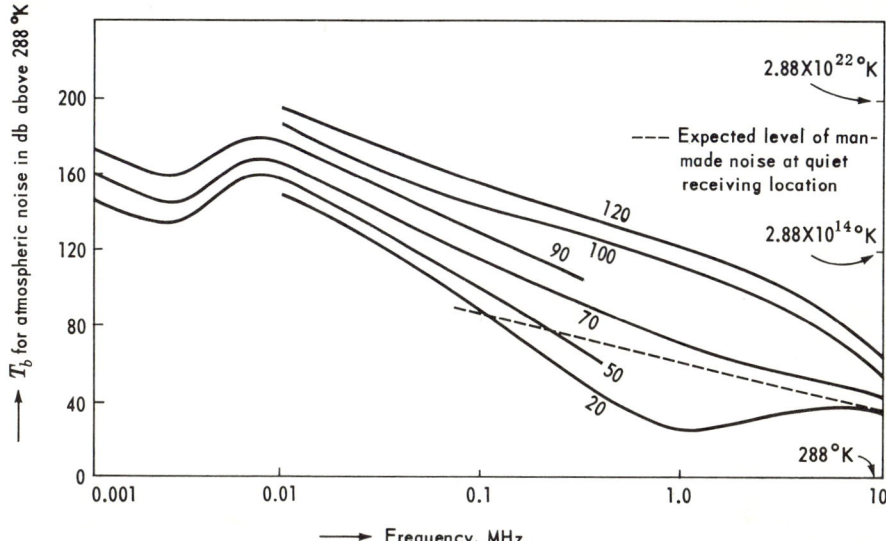

Fig. 8.5 Atmospheric-noise equivalent brightness temperature. The numbers on the curves are different CCIR noise grades.

The different curves in Fig. 8.5 correspond to different *noise grades;* the noise-grade designation is simply a system for categorizing the wide variety of noise levels encountered from time to time and from place to place on the surface of the earth. Typical median noise grades for the United States range from 70 to 100.

(It is worthy of note that the noise levels at these lower frequencies are vastly lower, both at points above the ionosphere and at depths beneath the earth's surface.)

Problem 8.4 An 800-ft vertical monopole antenna radiates at 100 kHz with an efficiency of 70 percent. Find the input power required in order that, at a distance of 600 miles over good earth, the signal level in a 3-kHz band is 20 db above the average noise level.

8.6 INTERNAL NOISE

So far we have been concerned only with the noise originating outside the receiving-antenna system. Actual antenna systems always have some ohmic resistance, and this in itself adds noise to any incoming signal.

To see how to account for the effect of the ohmic resistance in a structure, let us first find the noise added by a lossy transmission line at a temperature T. For example, suppose that the antenna itself is lossless but that it is connected to the load through a lossy transmission line having a power gain, G (G is less than 1; note that G could be defined as the transmission efficiency $\eta_t = G =$ power out/power in). To find the noise power added by the lossy line, consider a hypothetical situation in which the line is matched at both ends with resistors at temperature T. At the receiving end, if we look back toward the generating end we see a resistance equal to the characteristic resistance of the line; from such a resistance, the noise power available is $kT\,df$. The actual resistor at the generating end is putting out a noise power $kT\,df$ into the line, of which $GkT\,df\ (= \eta_t kT\,df)$ arrives at the load end. The difference, $kT\,df - GkT\,df$, must be added by the lossy line; that is, the noise added is

$$N_{\text{added}} = kT\,df\,(1 - G) = kT\,df\,(1 - \eta_t)$$

The noise contributed by the ohmic resistance of the antenna may be determined in a similar fashion. Let us define the receiving efficiency η as the ratio of the power delivered by the lossy antenna to the power delivered by a lossless antenna of the same type. [The efficiency is then also the ratio of the effective areas and, by Eq. (11), assuming the same current distribution, also the ratio of the input resistances.] Again suppose that the antenna is matched with a load resistor at temperature T and that it is in equilibrium with the environment represented by an antenna temperature $T_a = T$. The load resistor is receiving a noise power $P_{\text{del},n} = kT\,df$. The environment supplies a noise power $kT\,df$ to the antenna, of which a part $\eta kT\,df$ is delivered to the load. Thus the difference, $kT\,df - \eta kT\,df$, between the power actually reaching the load and that supplied by the environment must be supplied by the ohmic resistance in the antenna:

$$N_{\text{added}} = kT\,df\,(1 - \eta)$$

Finally, the total noise output from a lossy antenna in an environment that would provide, for a perfect antenna of that type, an antenna temperature T_a is

$$P_{\text{del},n} = kT_a\,df\,\eta + kT\,df\,(1 - \eta)$$

If the antenna is represented by a Thévenin-equivalent circuit with $R_{\text{in}} = R_l + R_r$ (loss resistance plus radiation resistance), the efficiency is $\eta = R_r/(R_l + R_r)$, and the total noise power can be written in the form

$$P_{\text{del},n} = kT_a\,df\,\eta\left(1 + \frac{T}{T_a}\frac{R_l}{R_r}\right)$$

From this form it may be seen that if the antenna temperature is very large compared to the physical temperature of the antenna, then unless R_l is large compared to R_r (i.e., a very inefficient structure) the second term in the parentheses is insignificant. As we have seen, at frequencies below 10 MHz, the antenna temperature is 10^6 to 10^{10} times that of the usual physical temperatures; this means that the antenna efficiency must lie in the range, say, of 10^{-6} to 10^{-10} for the term TR_l/T_aR_r to be significant compared with 1.

Alternative forms for calculating the noise power delivered by a lossy antenna at temperature T, with external noise sources represented by an antenna temperature T_a, are

$$P_{\text{del},n} = kT_a\, df\, \eta \left(1 - \frac{T}{T_a} + \frac{T}{\eta T_a}\right)$$

or

$$P_{\text{del},n} = kT\, df\, \eta \left(\frac{T_a}{T} + \frac{1-\eta}{\eta}\right)$$

In summary, when the external noise levels (antenna temperatures) are high, the incoming noise is decreased in the same ratio as the signals, so that poor antenna efficiency may not significantly degrade the signal-to-noise ratio. On the other hand, in low-noise environments, low antenna efficiency means an increase in the noise level proportional to the ambient temperature.

It is also interesting to note that when the antenna temperature is very high, a bad mismatch between a receiving antenna and its line or load may not significantly degrade the signal-to-noise ratio. It is only necessary to ensure that the signal level reaching the first amplifier is high enough to override the receiver noise.

Example Relative performance as receiving antennas of a quarter-wave monopole and a small ferrite loop antenna at 1 MHz.

As a concrete example, consider the performance of a small ferrite loop antenna, say 1 in. diameter $\times$ 10 in. long, compared with a perfect quarter-wave monopole (about 225 ft in height) as receiving antennas in a location in which the noise grade is 70. In this case, the effective brightness temperature of the atmospheric noise compared to the ambient temperature of the antennas is about $T_b/T = 10^7$. To have some other figures in mind, recall that the directivity of the monopole over perfect ground is 2×1.64, and the directivity of a small lossless loop is 2×1.5; also, it is shown below, the efficiency of the small ferrite loop antenna might be about 10^{-5}. From the discussions above, it is clear that the signal power delivered to a matched load by the monopole is

$$S = \frac{P_{\text{inc}}G_m\lambda^2}{4\pi}$$

and the noise power delivered is

$$N = kT_a\, df$$

where T_a is the brightness temperature. The signal power delivered by the loop is

$$S = \frac{P_{inc} G_{pL} \eta_L \lambda^2}{4\pi}$$

where η_L is the loop efficiency and G_{pL} is the directivity of the lossless loop. The noise delivered by the ferrite loop is

$$N = kT_a\, df\, \eta_L \left(1 + \frac{T}{T_a} \frac{1 - \eta_L}{\eta_L}\right)$$

The signal-to-noise delivered by the ferrite loop is then

$$\left(\frac{S}{N}\right)_{\text{loop}} = \frac{P_{inc} G_{pL} \lambda^2 / 4\pi}{kT_a\, df\, [1 + T(1 - \eta_L)/T_a \eta_L]}$$

Putting all these equations together, we may compute the ratio of the signal-to-noise ratios

$$\frac{(S/N)_{\text{monopole}}}{(S/N)_{\text{loop}}} = \frac{G_m}{G_{pL}} \left[1 + \frac{T(1 - \eta_L)}{T_a \eta_L}\right]$$

Under the conditions stated, the quantity $T(1 - \eta_L)/T_a \eta_L \approx 10^{-2}$; thus the ratio of signal-to-noise ratios is approximately 1.64:1.5. The slight improvement provided by the 225-ft quarter-wave monopole is certainly not worth the cost. On the other hand, the ratio of the signals is

$$\frac{S_{\text{monopole}}}{S_{\text{loop}}} = \frac{1.64}{1.5 \eta_L} = 1.1 \times 10^5$$

The practical question is then, how much effective area is required or is useful? The answer depends on the noise levels in the receiver. If the noise power delivered by the antenna overrides the receiver noise, then only that amount of effective area is useful; that is, if $T_a \eta \gg T_e$, where T_e is the effective noise temperature of the receiver, then the receiver noise will not degrade the signal-to-noise ratio significantly. Then, in situations of high external noise, the signal level must also be high, and if the signal level is M db above the atmospheric noise it will remain so all the way to the final readout.

8.6.1 EFFICIENCY OF FERRITE LOOP ANTENNAS

As was pointed out above, the efficiency of a small ferrite loop antenna can be about 10^{-5}. As an interesting sidelight on receiving antennas, we shall now demonstrate this, and in the process show that a ferrite loading of the coil may substantially increase the efficiency of a given small loop. It was shown in Chap. 2 that the radiation resistance of a small loop is

$$R_r = 20N^2\beta^4A^2$$

where A is the loop area. If the turns are distributed in solenoidal fashion over a length l, the radiation resistance is increased by a factor of the square of μ_e, the effective permeability; the effective permeability is a function of core shape, in particular the length-to-diameter ratio. To a good approximation, the effective permeability is

$$\mu_e = \frac{\mu_r}{1 + D(\mu_r - 1)}$$

where μ_r is the relative permeability and D is the static demagnetization factor for the particular shape (for example, for long needles, $D = 0$; for spheres, $D = \frac{1}{3}$; for thin disks, $D = 1$). The radiation resistance of a ferrite-filled loop is then

$$R_r = 20|\mu_e|^2N^2\beta^4A^2 \tag{15}$$

(At this point in the text, the easiest way to obtain this result for the radiation resistance is to note that for small perfect antennas the gain should always be 1.5 maximum, so the maximum effective area of a given loop should be constant at a given frequency, with or without ferrite loading. But the power delivered to a matched load is $|V_{oc}|^2/4R_r$, so this ratio must be a constant independent of ferrite loading. With an incident magnetic field H_{inc} when the loop is oriented for maximum reception, the field inside the ferrite is $\mu_e H_{\text{inc}}$; thus the open-circuit voltage is

$$V_{oc} \doteq \mu_0 \mu_e H_{\text{inc}} A N$$

This is a factor of μ_e greater than the open-circuit voltage that would be induced in an air-core loop. Thus, the term R_r in the denominator must increase by a factor of $|\mu_e|^2$ in order that the ratio $|V_{oc}|^2/4R_r$ be a constant. The same result may be obtained with a more direct approach to the calculation.)

The input resistance of a ferrite loop antenna arises from (1) the radiation resistance, (2) the ohmic resistance of the wires, and (3) the series resistance introduced by the core losses. The wire resistance may be obtained with standard skin-effect formulas (with proximity effect included if this is significant). A good approximation for the inductance and core-loss resistance may be obtained from application of Maxwell's equation in a situation in which the ferrite loop is energized by a constant-current generator, I_g;

$$-\oint \mathbf{E} \cdot \mathbf{dl} = V = Nj\omega \iint \mu \mathbf{H} \cdot \mathbf{dA}$$

To a first approximation, if we neglect radiation and wire loss, the quantity $\mu H \doteq \mu_0 \mu_e N I_g C / l$, where μ_e is the effective permeability, and C is Nagaoka's constant ($C \doteq 1 - D$); thus the input impedance is

$$Z_{in} = \frac{V}{I_g} = j\omega N^2 \mu_0 \mu_e \frac{AC}{l}$$

We may introduce the effects of core losses by making use of a complex permeability $\mu = (\mu' - j\mu'')\mu_0$, which implies an effective permeability as follows

$$\mu_e = \frac{\mu_r}{1 + D(\mu_r - 1)} = \frac{\mu' - j\mu''}{1 + D(\mu' - 1) - jD\mu''}$$

Momentarily introducing a symbol $F = 1 + D(\mu' - 1)$ to save writing, we have

$$\mu_e = \frac{\mu'F + D\mu''^2}{F^2 + D^2\mu''^2} - \frac{j\mu''(1 - D)}{F^2 + D^2\mu''^2}$$

In most ferrites of practical interest, the quantity $D\mu''$ is very much less than F; therefore to a good approximation

$$\mu_e = \frac{\mu'}{1 + D(\mu' - 1)} - \frac{j\mu''(1 - D)}{[1 + D(\mu' - 1)]^2}$$

or

$$Z_{in} = \frac{j\omega\mu' N^2 A C \mu_0}{[1 + D(\mu' - 1)]l} + \frac{\omega\mu'' N^2 A C (1 - D) \mu_0}{[1 + D(\mu' - 1)]^2 l} \tag{16}$$

The real part of this impedance is the series resistance introduced by the ferrite core losses.

In a typical design, the parameters would be selected so that the ohmic loss in the wire and the core loss are about equal; then, a good approximation to the input impedance would be

$$R_{in} = R_r + 2R_c$$

where R_c is the real part of the impedance given by Eq. (16). The receiving efficiency (of course equal to the radiation efficiency) can be defined as the power actually delivered over the power that could be delivered if the losses were zero:

$$\eta = \frac{R_r}{R_r + 2R_c} = \frac{1}{2} \frac{R_r}{R_c} \frac{1}{1 + \frac{1}{2} R_r / R_c}$$

We know from the last part of Chap. 2 that the efficiency will be small; therefore a good approximation will be

$$\eta = \frac{R_r}{2R_c}$$

With the aid of Eqs. (15) and (16), we may find an explicit result for the efficiency. For simplicity, we will assume a low-loss ferrite so that $\mu'' \ll \mu'$, as would usually be the case; this in turn implies the approximation $|\mu_e| = \mu'/[1 + D(\mu' - 1)]$. With these approximations, the efficiency is

$$\eta \doteq \frac{20|\mu_e|^2 N^2 \beta^4 A^2}{(2\omega\mu''/\mu'^2)|\mu_e|^2 N^2 AC(1-D)/l}$$

$$\doteq \frac{10\omega^3 \mu\epsilon^2 Al}{C(1-D)\mu''}\mu'^2$$

$$\doteq \frac{10}{120\pi}\beta^3 V \left[\frac{\mu'^2}{\mu''(1-D)C}\right]$$

where V is the antenna volume and μ' and μ'' are the relative values of permeability. The last bracketed factor describes the effect of the ferrite loading on the efficiency. For a length-to-diameter ratio of 10, the demagnetization factor is about 0.02; therefore $(1-D)C \doteq 1$. The factor μ'^2/μ'' is the so-called μQ product characterizing the ferrite material. One of the better ferrite materials for the lower MHz frequency range has a μQ product of about 4×10^4. This is certainly a substantial factor by which to increase the efficiency.

However, in spite of this increase in the efficiency of the small loop, the efficiency is still almost negligible. That is, for the 1 in. × 10 in. ferrite loop postulated earlier, the factor $(10/120\pi)\beta^3 V$, at 1 MHz, is very small indeed:

$$\eta = \frac{10}{120\pi}\beta^3 V \frac{\mu'^2}{\mu''}$$

$$= \frac{20\pi^3}{12}\frac{(2.54)^3 \times 10^{-6}}{27 \times 10^6}(4 \times 10^4)$$

$$= 1.26 \times 10^{-6}$$

For 2 MHz, the value is eight times this; thus the value of efficiency assumed earlier is justified.

8.7 RECEIVER NOISE

Although it might be said that the subject of receiver noise is outside the scope of a book on antennas, a brief review of the main ideas will nevertheless be presented here; the purpose is to aid the antenna engineer in his decisions concerning how much sweat and money should be expended in improving his designs of receiving antenna systems.

The receiver itself is made up of various combinations of multiport electronic devices (such as amplifiers, detectors, and converters), each of which adds noise to the signal it processes. A common figure of merit for the noise performance of a device is known as the *noise figure F*. The noise figure is defined as the ratio of the signal-to-noise ratio at the input to the signal-to-noise ratio at the output, when the input is maintained at a reference temperature T_0:

$$F = \frac{(S/N)_{in}}{(S/N)_{out}}$$
$$= \frac{N_{out}}{GkT_0\,df}$$

where G is the gain of the device, k is Boltzmann's constant, df is the bandwidth, and T_0 is the reference temperature, set by Institute of Electrical and Electronic Engineers standards to 290°K. The noise figure is then the ratio of the noise output of a device to the noise output of a perfect (i.e., noiseless) device of the same type. Given the noise figure, determined experimentally, the noise output is easily determined; if the input-circuit temperature is T_0, the noise output is

$$N_{out} = FGkT_0\,df$$

from the definition. If the input circuit is at temperature T, the output noise is the sum of the noise that would be generated with input at temperature T_0 plus the "excess" noise:

$$N_{out} = FGkT_0\,df + Gk(T - T_0)\,df$$
$$= GkT_0\,df\left(F - 1 + \frac{T}{T_0}\right)$$

In systems work, frequent use is also made of the concept of *effective temperature T_e* of a network. The effective temperature is defined to be the input temperature that would account for the noise that is generated internally by the device. The internally generated noise is given by the previous equation in the event that the input-circuit temperature is 0°K. The effective temperature must be such that $GkT_e\,df$ gives the internally generated noise; that is, $GkT_e\,df = GkT_0\,df\,(F-1)$, or

$$T_e = T_0(F - 1)$$

In terms of the effective temperature of the network, the noise output is

$$N_{out} = Gk\,df\,(T + T_e)$$

which of course is the sum of the input noise and the noise generated by the network.

Having procedures for the determination of the noise added by a single device, the next item of interest is the noise output of two devices connected in cascade. Let the devices be characterized by gains G_1 and G_2 and effective temperatures $T_{e,1}$ and $T_{e,2}$. If the input to the first device is at temperature T, the noise output from the first device is

$$N_{\text{out},1} = G_1 k \, df \, (T + T_{e,1})$$

Thus the noise output of the second device is

$$N_{\text{out},2} = G_2 G_1 k \, df \, (T + T_{e,1}) + G_2 k T_{e,2} \, df$$

It is of interest to identify an effective temperature and a noise figure for the cascaded system. The effective temperature is to be such that the quantity $G_1 G_2 k T_e \, df$ gives the noise added (noise output with input temperature of 0°K), that is,

$$T_e = T_{e,1} + \frac{T_{e,2}}{G_1} \qquad \text{cascaded devices} \tag{17}$$

The noise figure is such that $F = N_{\text{out}}/G_1 G_2 k T_0 \, df$, where

$$N_{\text{out}} = N_{\text{out},1} + \text{noise added by device 2}$$
$$= F_1 G_1 k T_0 \, df \, G_2 + G_2 k T_0 \, df \, (F_2 - 1)$$

that is,

$$F = F_1 + \frac{F_2 - 1}{G_1} \qquad \text{cascaded devices} \tag{18}$$

Equations (17) and (18) both reveal a first principle in receiving-system design: Provide a low-noise device (preamplifier) with reasonable gain at the antenna terminals. If this is done, then the noise properties of subsequent devices in the system are rendered less critical. The quality of the transmission line leading to the main receiving electronics can be lower if it is preceded by a good preamplifier. For example, it was demonstrated earlier that the noise added by a matched transmission line at temperature T is $N_{\text{added}} = kT \, df \, (1 - G_L)$, where G_L is the gain of the line (it is less than one). The effective temperature of the line is then

$$T_{e,L} = \frac{T(1 - G_L)}{G_L}$$

The effective temperature of a system consisting of a preamplifier (characterized by G_p and $T_{e,p}$) and a transmission line is then

$$T_e = T_{e,p} + \frac{T(1 - G_L)}{G_p G_L}$$

If the output of this preamplifier and line system is put through another amplifier (characterized by G_a and $T_{e,a}$), the total effective temperature $T_{e,0}$ for the system is

$$T_{e,0} = T_{e,p} + \frac{T(1 - G_L)}{G_p G_L} + \frac{T_{e,a}}{G_p G_L} \tag{19}$$

316 ANTENNA ENGINEERING

This equation is helpful in tradeoff studies whose object is to determine the most economical way to minimize the noise added by the receiving system.

Example An antenna has an efficiency of 90 percent, a directive gain of 20 db, and an effective temperature $T_a = 200°K$. It is connected to a preamplifier with a gain of 100 and a noise figure of 6 db. The preamplifier is followed by 100 ft of transmission line with a loss of 3 db. The ambient temperature is 300°K. The transmission line leads to another amplifier which has a gain of 200 and a noise figure of 10 db. Find the signal-to-noise ratio at the output of this system, assuming an incident field strength of 20 μv/m, rms, at 2 GHz. The bandwidth is 10 MHz.

The signal power delivered by the antenna is

$$P_{\text{del}} = \frac{(20 \times 10^{-6})^2}{377} G_d \frac{\lambda^2}{4\pi} \eta = \frac{4 \times 10^{-12}}{47} \times 100 \times 0.15^2 \times 0.9$$

the gain of the system is $100 \times \frac{1}{2} \times 200 = 10^4$; therefore the signal level at the output is about 1.7×10^{-9} watts. The noise output of the antenna is

$$P_{\text{del},n} = k \cdot 200° \cdot 10^7 \cdot 0.9 + k \cdot 300° \cdot 10^7 \cdot 0.1$$
$$= 1.38 \times 10^{-16}(180° + 30°)$$

The noise at the output is

$$N_{\text{out}} = P_{\text{del},n} G_p G_L G_a + \text{noise added by the system}$$
$$= 1.38 \times 10^{-16}(210) \times 10^4 + 10^4 k T_{e,0} \, df$$

The effective noise temperature of the combination of preamplifier, transmission line, and post-amplifier is given by Eq. (19). The effective noise temperatures of the amplifiers are

$$T_{e,p} = 290(3.98 - 1) = 868°K$$
$$T_{e,a} = 290(10 - 1) = 290 \cdot 9$$

Thus

$$T_{e,0} = 868° + \frac{300°(\frac{1}{2})}{100(\frac{1}{2})} + \frac{290° \cdot 9}{100(\frac{1}{2})}$$
$$= 868° + 3° + 52.2° = 923°K.$$

The noise output from the system is then

$$N_{\text{out}} = 1.38 \times 10^{-16} \times 10^4(210° + 923°)$$
$$= 1.38 \times 10^{-12} \times 1.133 \times 10^3$$
$$= 1.56 \times 10^{-9} \text{ watts}$$

The signal-to-noise ratio in this case is then about unity. Note that in this case, most of the noise originates in the receiving system, in particular in the preamplifier. This fact could justify the expense of a larger antenna having a greater directive gain.

Exercise Consider a system identical to the one described in the preceding example, except without the preamplifier. Calculate the signal-to-noise at the output.

8.8 DIVERSITY RECEPTION

Radio signals that are propagated over the surface of the earth are altered to some extent by the time variations of the earth environment. At the very low frequencies, the signals are remarkably stable. On the other hand, the HF band, with propagation via the ionosphere, the signals exhibit wide variations in phase, amplitude, and polarization; these variations lead to signal fades and distortions at both slow and rapid rates. Even in the UHF band, the changing properties of the atmosphere often significantly influence the field at a given location. The adaptive systems, described in Chap. 3, hold promise for coping with some of the problems. Of course, the distortions and fading of the radio signals reduce the rates at which information can be transmitted and introduce errors.

To alleviate this problem, the idea of *diversity* systems was introduced early. The basic idea here is to duplicate some part of the signal or system in the hope that even if one part experiences a deep fade or distortion, the other parts may not. There are four common forms of diversity reception: (1) time diversity, (2) frequency diversity, (3) space diversity, and (4) polarization diversity. In a time-diversity system, the signal is sent through the channel more than once. In a frequency-diversity system, the same information is transmitted and received on two or more different frequencies. Although both of these systems are used, they have obvious disadvantages in that time diversity reduces the traffic capacity of the circuit, and frequency diversity increases the frequency band that is required to transmit a given amount of information. It will be our purpose here to discuss only space diversity and polarization diversity, since these are more directly related to antenna engineering.

Let us consider first the idea of space diversity. A signal fade at a given location frequently results from an interference between the radio waves reaching the location by different propagation paths. At the time of the fade, the electric path difference between paths from the transmitter to the receiver is in the vicinity of 180°. It follows that at another receiving point, at some distance from the first, since the path lengths are different, the electric path difference will be different from 180°. It might even be 360°; or, it may be uncorrelated since the actual paths to the two locations may be entirely different. From these considerations, it appears that if antenna systems are erected at different locations, with a fair separation between, then when one antenna system is in a fading field, one or more of the others may not be. Clearly, there is a minimum antenna separation for effective spatial diversity. The minimum separation is hard to estimate, but it has been found experimentally[1] that separations of 300 to 600 m with frequencies in the HF band lead to

[1] G. L. Grisdale, J. G. Morris, and D. S. Palmer, *Proc. IEE*, **104B**:39 (January, 1957).

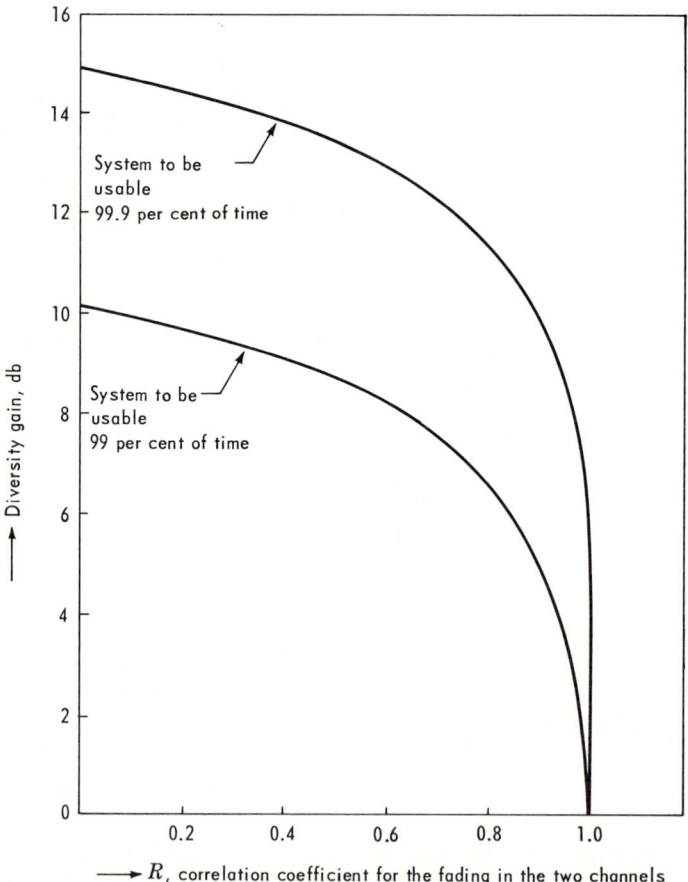

Fig. 8.6 Double-diversity gain as a function of the correlation between signals, assuming Rayleigh law fading.

effective diversity systems. In the UHF band, vertical separations of about 10 to 20 m are often used. In any case, the separation should be many wavelengths.

In the HF band, the very large separations required for effective space diversity are inconvenient. Moreover, because of the magnetoionic effects of the signals traveling via the time-varying ionosphere, the polarization is time-varying. This means that a single vertical or horizontal antenna will be improperly oriented at least part of the time. The obvious solution is to design an antenna system with two orthogonal states of polarization, such as horizontal and vertical, and to connect an output channel to each state. This arrangement is known as polarization diversity; it has the advantage that the two parts of the diversity system can have essentially the same location. The work of Grisdale et al., referenced above, has shown that in the HF band, polarization diversity can be as effective as space diversity.

Figure 8.6, taken from the referenced paper by Grisdale et al., is included to indicate the level of advantage to be gained through the employment of diversity systems. The plot shows "diversity gain" in decibels versus the correlation coefficient R between the channels in a diversity system with two antennas. The diversity gain measures the amount by which the power transmitted could be decreased while maintaining the same signal-to-noise ratio (comparing single-channel and the double-diversity system). The correlation coefficient is such that $R = 1$ corresponds to complete correlation, and $R = 0$ corresponds to zero correlation between the fading in the channels. The curves are plotted on the assumption of Rayleigh law fading.

The results differ according to the amount of "out" time that can be tolerated in the communication system. The two curves shown apply respectively to situations in which the channel can be unusable 1 percent of the time and 0.1 percent of the time. Note especially the shape of the curves; it appears that there will be a significant gain even for correlation coefficients as high as .8 but relatively little additional gain for correlation coefficients below about .5. Measurements indicate that in the HF band, polarization-diversity or space-diversity systems usually operate with correlation coefficients below .5.

It is also worth noting that in the HF band, a vertical and horizontal dipole system, when set up as a polarization-diversity system, also has what might be called a "path" diversity feature. That is, the patterns over ground of the horizontal-dipole system and of the vertical-dipole system are different, because of the difference in the imaging process over ground. Because of this, the horizontal and vertical dipoles are responsive to different vertical angles of arrival, another feature that varies with the state of the ionosphere.

9. ANTENNA MEASUREMENTS

A great deal of antenna engineering is subject to mechanical and environmental constraints which are very difficult to handle theoretically. As a result, much antenna development is accomplished experimentally. The antenna engineer must be well informed and skilled in the techniques of measurements, for in the course of his work he will frequently find it necessary to measure the radiation patterns and input impedances of various kinds of antenna structures. In addition, he will sometimes be called upon to measure near- and far-field amplitude and phase, polarization, gain, directivity, radiation efficiency, power-handling capacity, and antenna noise temperature. The following discussion of these measurements is largely adapted from the IEEE Standards Publication "IEEE Test Procedures for Antennas," as published in *IEEE Trans.*, **AP-13**:437 (May, 1965).

As is true for any product of engineering, an antenna design may be considered satisfactory only if the structure functions throughout a prescribed period of time in the environment for which it is designed. We cannot hope to consider all aspects of this problem here. Instead, we will list a few of the factors which must be considered, to spark the imagination of the reader. For example, antennas must withstand the mechanical loading imposed by high winds and ice. They must withstand the vibration and shock characteristic of their environments. They must survive the effects of all kinds of storms, and often must incorporate special features so that they will survive and pass to ground a direct lightning strike. Some antennas must be designed to survive an event as drastic as a nuclear blast; as an illustration, Fig. 9.1 shows such an antenna.

There are many special requirements: shipboard antennas must withstand water-wave impact and saltwater corrosion. Antennas on hyper-

Fig. 9.1 A blast-hardened monopole antenna. (*Collins Radio Company.*)

sonic vehicles must withstand very high temperatures and pressures. Antennas for space vehicles must withstand intense ionizing radiation, hard vacuum, and extreme temperatures.

Certain antenna applications require tests involving quite ordinary aspects of the physical environment. For example, antennas involving ferrite components are sometimes especially sensitive to changes in ambient temperature. Antennas in which intermodulation distortion must be minimized may have difficulty with nonlinear corrosion films. In precision tracking antennas, a pointing error can be caused by deflections resulting from nonuniform heating of the structure by the sun. Large movable antennas may develop errors because of unwanted deflections of all kinds.

In many cases the electrical environment of an antenna is transient and may cause problems of a more elusive nature. Moisture, when combined with impurities, can create troublesome conducting paths; water itself, because of its high dielectric constant, may change performance significantly. Antennas for missiles must contend with the ionized gases that exist in the exhaust or may be generated by the high vehicle velocity. Certain antennas may be susceptible to precipitation static, which is a series of noise pulses created when charged particles, such as raindrops, strike the antenna.

Frequently, an antenna must have a cover to protect parts of it from the physical environment. Needless to say, final tests for the antenna should be performed with this cover, often called the radome, in place.

Now let us consider some of the details of the testing procedures.

9.1 SCALE MODELS

Much preliminary antenna development is based on scale models. Scale models are also often used when the measurement of the original antenna in its original environment is impractical or inconvenient. Antennas for ships, aircraft, and space vehicles often fall into this category. Physically large antennas are also often modeled. The main motives for scale modeling are to obtain improved control over the conditions of measurement or to cut measurement costs.

Although the model is usually smaller than the original, a model may be either larger or smaller provided the following requirements for exact simulation are satisfied:

The linear dimensions of the model are $1/n$ times those of the full-scale antenna.

The operating frequency and conductivity of the materials used in the model are n times those of the full-scale antenna.

The complex electric permittivity and magnetic permeability of the materials used in the model are the same as in the full-scale antenna.

In the above, n is an arbitrary number which determines the scale of the model.

In a practical model, it is usually not feasible to satisfy exactly all of the above requirements. However, in antennas which are not highly resonant, the error will usually be small if good conductors are used to simulate good conductors, even though the conductivity is not scaled as it should be. Care should be taken to be aware of the frequency dependence of dielectric and magnetic materials so that the proper values are presented at the model frequency. As far as possible, the environment as well as the antenna itself should be modeled.

If the exact scaling procedure described above is followed, all fields are reproduced exactly in shape. Thus, the radiation pattern, power gain, directivity, radiation efficiency, input impedance, boresight error, and in general all properties dependent only on field ratios are preserved. If the exact scaling requirements are not followed, efficiencies will not be reproduced, but the remaining properties may be reproduced accurately enough for most purposes; especial care should be taken when the antenna has regions of extremely high current concentration or mismatch. Certain antenna characteristics, such as the power level for high-voltage breakdown and the noise temperature, do not scale because of the frequency-dependent nature of the mechanisms involved.

In the measurement of radiation patterns, it is likely that cables to the scale model will perturb the patterns. To overcome this, the scale model may transmit from a battery-operated transmitter, or alternatively, the scale-model antenna may contain a receiver from which the demodulated signal can be removed by means of high-resistance wire leads which minimize the disturbance of the radio-frequency field. The support for the model should receive particular attention if the pattern null structure is to be accurately determined.

9.2 MEASUREMENT OF RADIATION–FIELD AMPLITUDE

In general, the relative amplitude of the field as a function of angular bearing depends on the distance from the antenna. This is so because the relative phase and amplitude of the contribution to the field from different parts of the antenna depend upon their respective distances to the field point. However, in free space, if the distance to the observation point is great enough, the angular distribution of the field is essentially independent of distance. The minimum distance to this "distant-field region" is obviously not sharply defined. The usual criterion for the

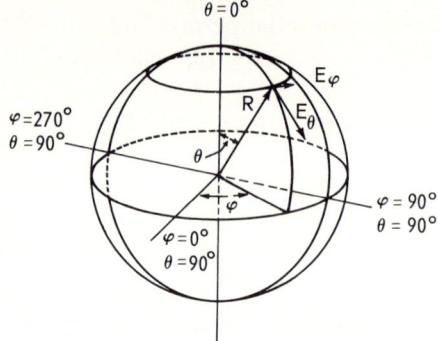

Fig. 9.2 System of spherical coordinates.

minimum distance between transmitter and "distant-field" receiver is

$$R = \frac{2D^2}{\lambda}$$

where D is the largest dimension of the radiating portion of the antenna. The criterion in general is set by specifying that the path difference from the center of the radiating aperture and from the edge of the aperture should be less than some small fraction of a wavelength. The criterion above makes the path difference less than $\lambda/16$. For accurate information concerning the depth of nulls or the relative size of minor lobes, the distance required may be several times larger than this.

Special care must be observed when the environment plays a significant part in the formation of the radiation pattern. For example, the "antenna" may have to be defined as the radiator plus all its supports and environment, such as an entire aircraft or other vehicle, or as the ground system and terrain in a ground-based installation.

A system of coordinates often employed for the two-dimensional radiation pattern of an antenna is shown in Fig. 9.2. In such a coordinate system, the radiation amplitude is determined on the surface of a large sphere whose center is located at the antenna. The coordinates are the usual angle coordinates in a conventional spherical system θ, φ. The antenna may be oriented at the origin in any convenient way, but of course the patterns should indicate the choice clearly. Then, the radiation patterns can be described as, for example, $E_\theta(\theta, \varphi_0)$ (which means the θ component of E as a function of θ for a fixed value of φ, φ_0) or $E_\theta(\theta_0, \varphi)$ or $E_\varphi(\theta, \varphi_0)$ or $E_\varphi(\theta_0, \varphi)$. Or, if the antenna is designed to radiate circular or some specified elliptic polarization, this state of polarization, call it $E_c(\theta, \varphi)$, may be measured and displayed in the form $E_c(\theta, \varphi_0)$ or $E_c(\theta_0, \varphi)$.

The measured field strength is usually recorded immediately on some type of coordinate paper. For simple antennas, amplitude information is fed to a pen-drive system, which positions a pen at a distance from an origin which depends on the field strength received. In one system, a turntable beneath the pen is synchronized in angle with a rotating system supporting the antenna under test. A sheet of polar-coordinate paper, attached to the turntable, is then marked by the moving pen as the antenna rotates. The result is a permanent record of the experiment. The polar plot is not satisfactory for highly directional antennas nor for detailed information on sharp lobes. For such antennas, a linear- or strip-chart recorder is commonly used so that the indication of angular bearing may be spread out to some convenient scale along a linear chart (for example, 1° may be spread out to, say, 1 in. along the linear chart, if desired). Also, commercial equipment is available which takes the received field-strength data and presents it in the form of a two-dimensional contour plot, with digital display of relative field strength, on a single sheet of paper.

In recording radiation patterns, the signal must inevitably pass through some type of detection and amplification system. For meaningful pattern information, the transfer properties of this detection-amplification system must be accurately known and controlled over the complete dynamic range of interest. Sometimes, the dynamic range of interest in radiation-pattern measurement is large, which means that the circuits must be carefully calibrated and compensated. In the simplest (and often the most accurate) type of measurement, a bolometer detector is used which is known to be accurately square-law in its response. This means that the output of the detector is proportional to E^2, or the power density in the radiation pattern. If this detector is followed by an amplifier which is accurately linear over the dynamic range of interest, the recorded pattern is a *power* pattern. A special circuit or amplifier with an accurately logarithmic response can be included, so that the power pattern is displayed directly with a decibel scale. For many applications, however, it is desirable to obtain a plot of relative field strength rather than relative power. To accomplish this, the square-law detector must be followed by a circuit whose output is accurately proportional to the square root of its input. This circuit can either precede or follow the linear amplifier.

9.2.1 ON-SITE MEASUREMENTS

The measurement of the complete two-dimensional radiation patterns of a full-scale antenna located on its ultimate site is a laborious and expensive task. Nevertheless, it is necessary to make such measurements when the antenna radiation is significantly affected by the site on which

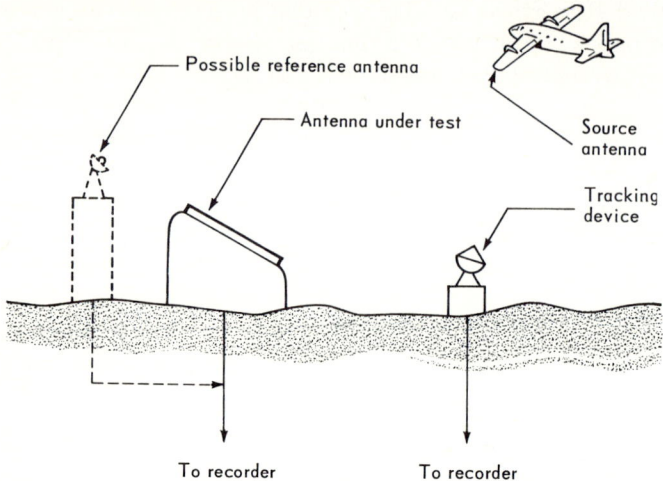

Fig. 9.3 System for on-site measurements of amplitude patterns.

it is located or when construction of the antenna is practical only at full scale on the site. Furthermore, on-site measurements of an antenna may be required as a conclusive demonstration that the delivered antenna is constructed properly, is correctly excited, and interacts with its environment in the predicted manner.

A wide variety of techniques have been employed for the on-site measurements of the amplitude patterns of an antenna. The procedures outlined in the following discussion are commonly used ones, and serve to illustrate the problems involved.

Figure 9.3 indicates the essential parts of the typical measurement system. A distant source is carried by a vehicle which is maneuvered through the space surrounding the antenna to provide a series of plane waves incident from all the directions of interest. The direction to the source is obtained from a tracking device, and this information serves to position the chart of a recording device. The response of the antenna, which is operated as a receiver, controls the position of the pen of the recorder. These data may then be processed to present the antenna patterns in the desired form.

To carry the source, airborne vehicles, such as conventional airplanes, helicopters, blimps, and free and captive balloons, have been employed on various occasions. The source should be in the far-field region of the antenna system being measured. When the far-field distance is greater than the maximum height obtainable with airborne vehicles, man-made earth-orbiting satellites have some usefulness. Even the sun and radio stars have been considered, since they provide a natural source. Of all these possible vehicles, light airplanes have proven most generally suitable.

If the attitude of the source antenna relative to the antenna under test changes, a change in the received signal is likely to occur. To minimize this, the source antenna should be oriented so that its peak is in the direction of the antenna being measured, and the useful portion of the source pattern should be as uniform as possible. In addition, the course flown by the aircraft should be chosen to minimize changes in altitude; a favorable course for this purpose lies along a circle centered on the antenna being measured and contained in a plane perpendicular to a vertical axis through the antenna. Since the radiation pattern of the source antenna is subject to modification by the aircraft on which it is carried, the design and placement of the source antenna must include the effects of this environment. This factor is greatly dependent on the relative sizes of the operating wavelength and the aircraft; the source antenna may be separated into two classes according to this distinction.

In the lower-frequency class (HF and VHF), the selection of a source antenna depends on polarization. For horizontal polarization, a light sleeve-dipole antenna trailed behind the aircraft has proven satisfactory. This type can be readily fabricated from a length of standard coaxial cable by removing the shielding braid for a distance of a quarter wavelength, and it can be supported in a horizontal position by a miniature parachute. For vertical polarization, useful results have been obtained with a cage-type monopole where the framework of the aircraft serves as the "ground." Since this antenna system is essentially an electrically short dipole, its radiation pattern has a null in the vertical direction; if measurements are made at appreciable elevation angles, the nonuniform nature of the source-antenna pattern should be recognized.

In the higher-frequency class (microwave region), wing-tip antennas have been successfully utilized. This placement minimizes excitation of the airplane structure and provides a source radiation pattern which is as close as possible to the pattern of the isolated source antenna.

As indicated in Fig. 9.3, a tracking device is required to establish the direction of the source. This process defines the direction of the aircraft with respect to the tracker; the direction of the aircraft relative to the antenna under test must then be determined by correcting the parallax error introduced by the known separation between the antenna under test and the tracker. In addition to source direction, it is often necessary for the tracking system to determine range to the source. This information may be needed in the parallax correction and, if the aircraft does not fly in a perfect circle about the antenna to be measured, may also be needed for an inverse-distance correction to the received signal level.

Two types of tracking instruments are in common use: optical trackers and radar trackers. An important distinction between the two is that the ordinary optical tracker furnishes only the direction of the source, but radar furnishes range as well. To determine range when an optical system is employed, aircraft altitude may be measured by an instrument in the aircraft; this information must then be transmitted to the test site so that range can be calculated.

For measuring the on-site amplitude patterns of an antenna, the system described in the preceding discussion may be adequate. However, there are occasions when this system is not sufficiently accurate, because the amplitude or polarization of the wave radiated by the source toward the antenna under test is too variable; this is particularly likely at microwave frequencies. In these cases, a reference antenna must be introduced in the system, as indicated in Fig. 9.3. This reference antenna should be placed close enough to the antenna under test so that the effect of variations in the strength of the wave from the source can be substantially eliminated by normalizing the signal received by the antenna under test with respect to the signal received by the reference antenna. In addition to the shape of the amplitude pattern, the reference antenna also may permit a measurement of power gain, as discussed in Sec. 9.4. Finally, the reference antenna can sometimes be designed to determine the polarization of a set of source antennas; this in turn may permit the polarization of the antenna under test to be determined. It should be recognized, however, that unless the reference-antenna system is carefully designed, it may introduce errors as great as those it is intended to cancel.

At microwave frequencies, it is desirable to make the response of the reference antenna substantially independent of the ground and surrounding structures. This dictates the use of a reference antenna with a narrow beam and low minor lobes and may also require treatment of the ground and other structures near the reference antenna with absorbing material or special fences to minimize reflections. The narrow beam in turn imposes a pointing requirement on the reference antenna which is usually satisfied by slaving the reference antenna to the tracking device. At frequencies well below the microwave range, the pattern of the reference antenna may be so broad that ground reflections must be accepted as part of the mechanism of pattern formation of the reference antenna, and the reference antenna may be practical only when it is fixed in position rather than pointed at the distance source. If the pattern of the reference-antenna system can be accurately determined, then it is still possible to employ its output as a means for correcting variations in the wave radiated by the source during measurement of the pattern of the antenna under test. However, at the lower frequencies such as HF, it is customary to use the reference antenna only during the measurement of power gain at one angle. Regardless of the frequency being employed, the reference-antenna system should be designed and located so that its effect on the patterns of the antenna under test is negligible.

The process of pattern measurement and recording may involve either a point-by-point or a continuous method; the latter is preferable. Commercially available continuously recording equipment can often be adapted to introduce automatically the various corrections which are needed in both the signal and angle inputs. The original data are recorded in a form dependent on the particular measurement geometry; when the aircraft flies in circles centered on the antenna under test, the patterns are plotted in the form of antenna response versus azimuth direction for a series of elevation directions. The final antenna patterns can be presented in different forms; the most comprehensive form is a contour plot.

9.2.2 OFF-SITE AND MODEL MEASUREMENTS

When the antenna pattern is essentially independent of the environment, a simple measurement technique can be used. This consists of operating the antenna under test as a receiving antenna, rotating it, and measuring the signal it receives from a suitably located fixed-source antenna. The method can be directly applied to most microwave antennas and has been used with success at any frequency which does not lead to impossible mechanical restrictions; it is also most suitable for scale-model measurements.

This type of measurement tends to be divided into two classes, one appropriate to antennas having electrically large apertures and the other appropriate to antennas having dimensions of the order of a wavelength or less. For those antennas with dimensions much less than a wavelength, the far-field region starts near $\lambda/2\pi$. This is such a short distance that the practical consideration of producing a uniform field over the region of rotary motion of the test antenna leads one well into the desired far-field region for the placement of the source antenna. Even for antennas having dimensions of the order of a wavelength, the far-field region is usually conveniently close to the test antenna. In addition to convenience, it is desirable to have a short distance between a small test antenna and the source antenna in order to minimize the spurious power that may enter the test antenna by wide-angle scattering from obstructions in the test range. Because antennas having dimensions of the order of a wavelength or less have wide radiation patterns, the rotating mechanism should provide 360° rotation of the test antenna and may have relatively low precision. An example of a test range for such antennas is indicated in Fig. 9.4.

For antennas having electrically large apertures, the $2D^2/\lambda$ criterion for the far-field region becomes most important. The distance between the source antenna and the antenna under test should equal or exceed

330 ANTENNA ENGINEERING

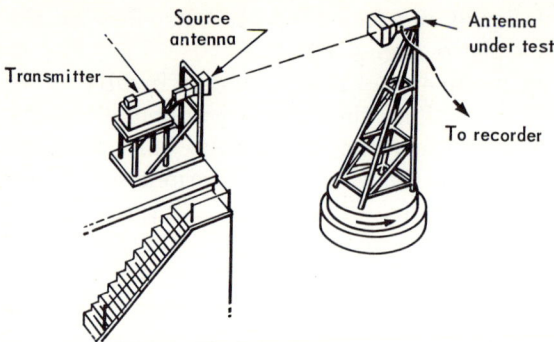

Fig. 9.4 Off-site pattern range for antennas having dimensions of a wavelength or less.

this value. If, as is usually the case, the maximum aperture dimension of the source antenna is equal to or less than that of the test antenna, then the $2D^2/\lambda$ criterion also fairly well ensures that the essential variation in amplitude of the incident field across the aperture of the test antenna is small enough to have a negligible effect. Thus, the measured pattern taken at a $2D^2/\lambda$ separation at least differs from the far-field pattern by amounts usually within the limits of instrumental error. If necessary, useful measurements can be made with the separation between source and receiving antennas as little as D^2/λ or even less. However, the disparity between the measured pattern and the far-field pattern becomes more marked as the separation decreases, and measurement results rapidly become less meaningful.

With an electrically large aperture, the radiation pattern is narrow and the antenna may be physically large and heavy; the rotating mechanism should therefore be a rugged platform of relatively high angular precision. Provision should also be made for precise alignment of the antenna in the coordinate orthogonal to the one being varied for the pattern measurement. An example of a test range for antennas with electrically large apertures is indicated in Fig. 9.5.

With either large or small antennas, measurements of patterns with low side lobes require a test range which is free of reflections whose amplitudes are greater than a small fraction of the side-lobe level to be measured. Techniques for minimizing undesired reflection effects include the following:

1   Mounting the source and test antenna on towers or adjacent hills or buildings. This arrangement reduces the ground reflections, since the radiation directed downward is in the side-lobe region of the test and

source antenna, and this radiation propagates along a greater path length between the two antennas than does the direct radiation.
2. Employing a source antenna with maximum permissible directivity, consistent with maintaining a sufficiently uniform illumination across the aperture of the receiving antenna.
3. Designing the source antenna so that the ground area causing reflections lies in the null region of the source. This technique is sometimes frequency-sensitive, however, and is often incompatible with 2.
4. Utilizing diffraction fences or absorbing baffles to either scatter or absorb the reflected radiation. A series of such fences may be required for precision measurements.

A less common alternative to minimizing ground reflections is to smooth the intervening ground between the source and test antenna to produce specular reflections. The source- and test-antenna heights are then adjusted in conjunction with the antenna spacing to yield an interference-pattern maximum in the vicinity of the test antenna. If the test antenna is small relative to the period of the interference pattern, it will intercept a quasi-uniform plane wave. No matter which of the above methods is employed for reducing the harmful effects of ground reflection, it should be considered standard procedure to probe the field across the aperture of the antenna under test with a small sampling antenna. By this means the uniformity of the field can be determined directly, and the expected accuracy of the antenna-pattern measurement can be estimated.

For very large antennas having aperture dimensions many times greater than the wavelength, the inner boundary of the far-field region

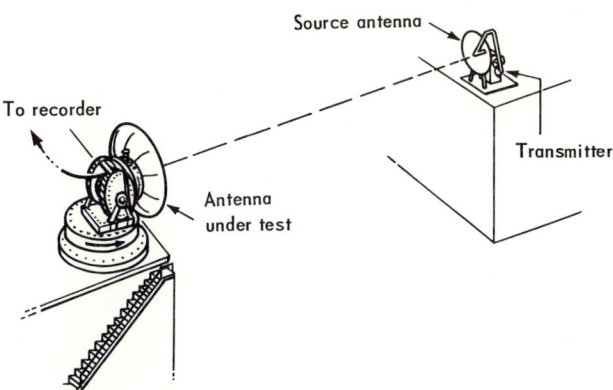

Fig. 9.5 Off-site pattern range for antennas having electrically large apertures.

may extend too far to allow a practical measurement to be made by the methods so far described. However, the far-field pattern, which exists at very large distances from large-aperture antennas, can sometimes be determined to a good approximation by refocusing the antenna to a focal region within the radiating near-field region, much closer to the antenna. As in analogous optical systems, focusing at a finite distance rather than at infinity requires the generation of a convergent spherical wavefront rather than a plane wave across the antenna aperture. With phased-array antennas, the phasing of the elements can in general be adjusted to produce faithfully the far-field pattern distribution in the near-field focal region. For the special case of arrays requiring in-phase excitation, it may sometimes be convenient to physically adjust the radiating elements on a concave spherical arc centered or focused at a usable distance. Even on a paraboloidal reflector, a moderate motion of the primary feed away from the focus along the reflector axis will produce a nearly perfect concave spherical wavefront on the aperture. To a good approximation, the far-field radiation-pattern characteristics (major lobe and near-in minor lobes) of a large aperture antenna as a function of angle remain constant as the antenna is refocused at a finite distance, provided that this distance is very large compared with the antenna aperture. In general, this equivalence does not apply to minor lobes far from the major lobe.

Microwave patterns of antennas which are not physically large can also be measured in an anechoic chamber or "microwave darkroom." Such chambers are lined with absorbing walls or baffles which greatly reduce the reflected radiation and may be designed so that reflected rays can travel from one antenna to the other only by multiple bounces. Anechoic chambers are generally valuable for design studies, but are usually not suitable for accurate measurements of low minor lobes. The frequency response of the absorbing material; reflection characteristics of walls, ceiling, and floor; and the influence on these properties of humidity, temperature, and traffic must be considered prior to use.

9.3 MEASUREMENT OF OTHER PATTERN CHARACTERISTICS

9.3.1 PHASE MEASUREMENT

The phase characteristic of an antenna radiation pattern often provides important information for antenna design and use. In the near-field region or at the aperture of an antenna, a knowledge of phase and amplitude as a function of location around the antenna can permit accurate prediction of the far-field patterns. The phase center of an antenna feed, which must be accurately located in focused antennas, may be determined by a measurement of the phase of the feed radiation pattern as a function of angle or displacement. Phase measurements may also be useful in determining distortions caused by radomes, lens stops, and the like.

The measurement of far-field phase is usually made at a constant distance from the antenna under test; it may be thought of as being measured over a spherical surface, where the antenna under test is located at the center of the sphere. By analogy with the more common amplitude radiation pattern, this form of measurement of phase information is referred to as the phase pattern of the antenna. The measurement of near-field phase requires more attention to the definition and control of the measurement parameters because of the very rapid and often unpredictable phase variations in this region. If the location of interest is known, such as a planar surface parallel to an antenna array, or a line parallel to a line source, the phase sampling should be done at that location. If constant-phase contours in a certain region are to be determined, several "slices" can be taken at convenient increments of distance; in the case of measurements on spherical surfaces, the location of the actual center of rotation of the patterns should be known with respect to the antenna itself. The contours of constant phase may be determined by interpolation of the measured values of phase. At times it may be possible to measure a constant-phase contour directly, by means of suitably designed apparatus.

All of these measurements are made in a similar fashion. Since phase is a relative quantity, a reference signal must be provided at all times for comparison. For measurements made at short distances, the antenna under test may be used as the transmitting antenna and a simple receiving antenna or probe used to sample the radiated field, as shown in Fig. 9.6a. The reference signal is coupled out from the transmitter line, and compared with the received signal in a suitable bridge circuit. For measurements at distances too long to permit direct connection of the reference and sampled signal, the arrangement of Fig. 9.6b may be used, wherein the signal from a distant source is received simultaneously by the antenna under test and by a fixed reference antenna; the antenna under test is rotatable in the usual manner for measuring radiation patterns. The distant source is also useful in the arrangement of Fig. 9.6c for the measurement of relative phase between two ports of a multiport antenna as a function of angle.

A number of techniques are described in the literature[1,2,3] for perform-

---

[1] H. Jasik, "Antenna Engineering Handbook," pp. 34-26—34-28, McGraw-Hill Book Company, New York, 1961.

[2] M. Wind, "Handbook of Electronic Measurements," vol. II, chap. 7, pp. 7-37 and 7-38, Polytechnic Institute of Brooklyn, N.Y., 1956.

[3] R. A. Sparks, "Microwave Phase Measurements," *Microwaves*, January, 1963, pp. 14-21.

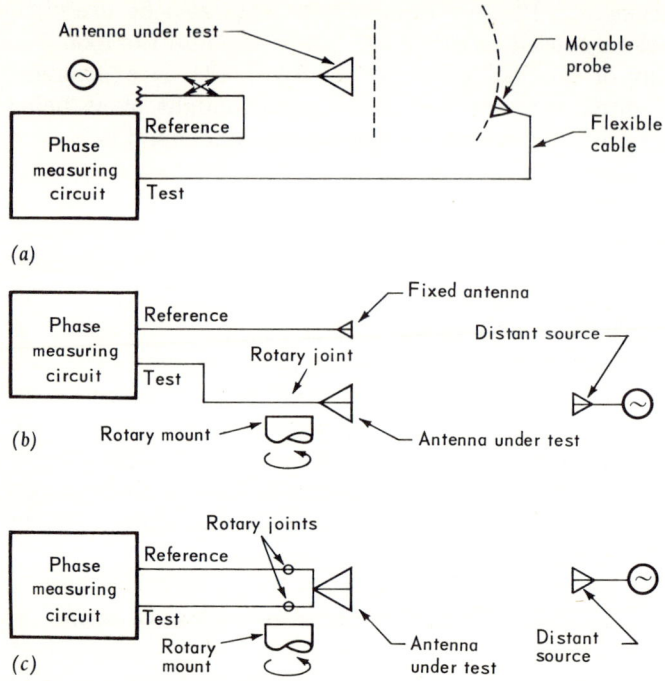

Fig. 9.6 Arrangements for measuring-phase patterns: (*a*) near-field, or short-distance, pattern phase; (*b*) far-field pattern phase; (*c*) far-field relative phase between two patterns.

ing the actual phase-comparison measurement; a typical method is described below. Among the necessary precautions in phase-pattern measurements are the preservation of constant-phase lengths in the reference and test-signal paths, in view of the presence of a movable joint or flexible cables in either or both lines, and the avoidance of distortion of the radiated field by reflections from the sampling probe, the reference antenna, or any support structures.

Figure 9.7 shows a very simple phase-measuring circuit which may be used in any of the arrangements described above. The hybrid junction is a microwave four-port component analogous to a bridge circuit. When equal signals are fed into conjugate arms of the hybrid, the signal emerging from each of the other ports is a function of the relative phase between the two inputs. By properly adjusting the phase of one of the input signals, a null can be produced at one of the output ports.

The test signal, which is assumed to be weaker than the reference signal, is brought through a low-loss calibrated phase shifter to one port of the hybrid junction. The reference signal passes through a variable attenuator having constant or calibrated phase shift and enters the conjugate port of the hybrid. (If the test signal is stronger than the reference signal, as is likely on the major lobe of a highly directive antenna, then either the variable attenuator or a fixed attenuator may be placed in the test-signal branch.) A detector and a termination are connected to the remaining ports of the junction. For accurate results, all reflections should be minimized. To make a measurement, the phase shifter and attenuator are varied until a null is observed at the detector; the phase-shifter setting, corrected for the attenuator phase if necessary, is noted. In this manner, a point-by-point phase pattern may be obtained as the antenna position is varied and the circuit rebalanced.

Automatic recording of antenna phase patterns may be accomplished in a number of ways by more elaborate phase-measuring circuits.[1] Since the test signal will usually vary widely in amplitude with respect to the reference signal, either an amplitude-insensitive method of phase sensing (such as a quadrature null detector) or amplification and limiting preceding the phase discriminator must be used. Circuits having a quadrature-signal null detector usually use it to servocontrol an adjustable phase shifter in the reference signal channel; the phase-shifter setting is the recorded indication of phase.[2] In such circuits, it is sometimes convenient to apply audio modulation to only one of the comparison channels before

[1] W. A. Cumming, Radiation Measurements at Radio Frequencies, a Survey of Current Techniques, *Proc. IRE*, **47**:705–735 (May, 1959).
[2] W. F. Gabriel, An Automatic Impedance Recorder for X-band, *Proc. IRE*, **42**:1410–1421 (September, 1954).

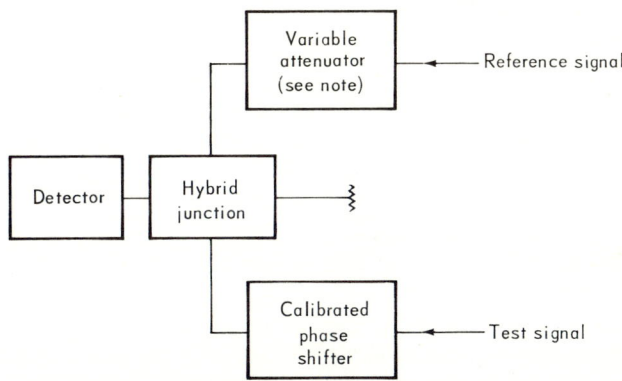

Note: Transmission phase of attenuator to be constant or calibrated

Fig. 9.7 Phase-measuring circuit.

combination in the null detector. The "homodyne" technique[1] generates suppressed-carrier sideband signals in one channel; the "subcarrier" technique[2] employs simple amplitude modulation.

9.3.2 POLARIZATION MEASUREMENTS

The radiation pattern of an antenna is not completely determined unless, in addition to amplitude and phase, polarization is also measured for all directions of interest. The nominal polarization of the antenna is expected to predominate near the peak of the major lobe, but accurate determination of the cross-polarization in this direction is often essential. In other directions, the polarization usually departs appreciably from the nominal value, and it is important in many cases to obtain comprehensive information about this polarization variation.

In any specified direction, the polarization of an antenna can be measured by sensing the relative amplitude and phase of two orthogonal components of its radiation field. This is usually done in the far-field region of the antenna, where the electric field vector has no significant radial component, so that it may be sampled two-dimensionally. Two orthogonally polarized but otherwise identical test antennas, such as crossed linear or opposite-handed circular, may be used, and their corresponding responses compared in amplitude and phase to determine the unknown polarization. If it is not feasible to make phase measurements, measurements of amplitude response with four test antennas having certain known polarizations can permit a complete determination[3] of polarization by a graphical computation.

An incomplete, but nevertheless informative, measurement of polarization is often made by observing the amplitude of response of a single rotating test dipole or other linearly polarized antenna. If the observed response is plotted as a function of the rotation angle Ψ, a dumbbell-shaped curve may be obtained, as indicated in Fig. 9.8. When sampling a linearly polarized field in this manner, the plot has two deep minima and follows a cosine variation in polar coordinates. When sampling a circularly polarized field, a circular plot is produced. For the general case of an elliptically polarized field, shown in Fig. 9.8, the axial ratio of the polarization ellipse is given by:

$$\text{Axial ratio} = \frac{|E_{\max}(\Psi)|}{|E_{\min}(\Psi)|}$$

---

[1] S. D. Robertson, A Method of Measuring Phase at Microwave Frequencies, *Bell System Tech. J.*, **28**:99–103 (January, 1949).

[2] G. E. Schafer, A Modulated Subcarrier Technique of Measuring Microwave Phase Shifts, *Trans. PGI*, September, 1960, pp. 217–219.

[3] G. A. Deschamps, Part II—Geometrical Representation of the Polarization of a Plane Electromagnetic Wave, *Proc. IRE*, **39**:540–544 (May, 1951).

ANTENNA MEASUREMENTS 337

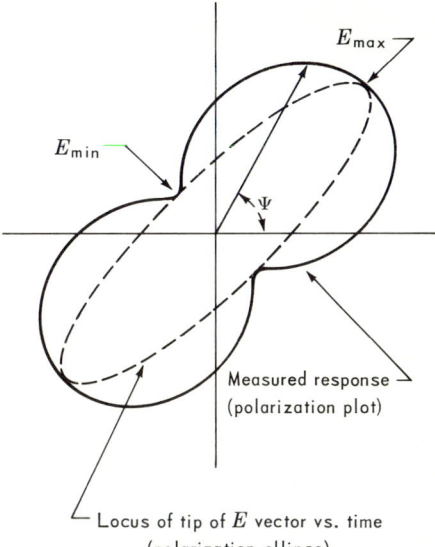

Fig. 9.8 Relative amplitude of elliptically polarized field as detected by a rotating linearly polarized antenna.

where $E_{\max}$ and $E_{\min}$ are the maximum and minimum electric fields sampled by the rotating dipole. The orientation angle of the major axis of the polarization ellipse is the angle of maximum response with the rotating dipole. Thus the two main parameters of the polarization ellipse are determined by this simple measurement. The handedness of the elliptical polarization, however, is not determined unless the phase of the signal from the rotating dipole is also measured. However, if the antenna under test is nominally circularly polarized with a known handedness, it is usually safe to assume that the measured elliptical polarization has the same handedness.

9.3.3 SPECIAL MEASUREMENTS FOR BORESIGHT AND ANGULAR SENSITIVITY

Directive antennas often require precise determination of the direction of the beam or the tracking axis, based on an electrical indication from the antenna system; such a direction is called the electrical boresight. This direction is determined with respect to some reference direction, called the reference boresight, which may be a specified stationary direction, or may be obtained from the antenna by a physical indication such as an optical or mechanical axis, or by a prior electrical indication. Measurements of boresight error are concerned with the angular deviation of the electrical boresight of an antenna from its reference boresight. Based on such measurement, it may be necessary to adjust the antenna system to minimize boresight error, or to place the electrical boresight in alignment or perpendicularity with mechanical axes or other physical reference.

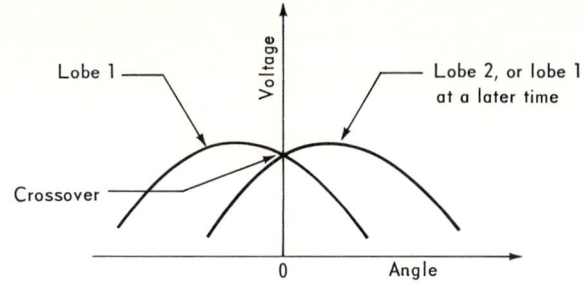

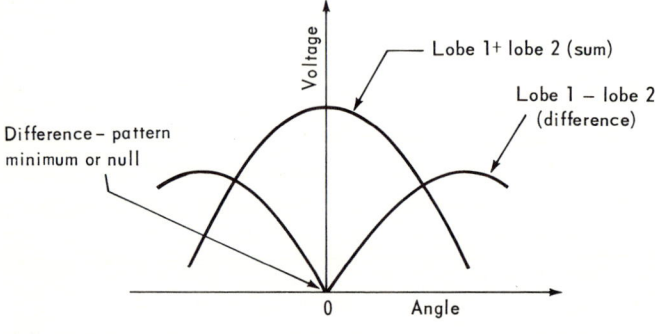

Fig. 9.9 Signals received by tracking antenna versus angle of target: (a) lobe patterns, monopulse or conical scanning; (b) monopulse sum and difference patterns.

The beam direction of an antenna with a single major lobe is usually determined by noting the direction of maximum response, or the direction halfway between equal responses either side of the peak. The precision of such determination is usually in the order of one-tenth of the half-power beamwidth.

Much greater precision of direction is demanded of tracking antennas, where two or more overlapping beams or lobes may be provided as indicated in Fig. 9.9. These lobes are compared to indicate a common direction by equality of amplitude or phase. When the comparison is done sequentially it is termed sequential lobing and typically compares amplitude; two examples of this class are conical scanning and lobe switching. When the comparison is done simultaneously it is called simultaneous lobing or monopulse; in this case either amplitude or phase may be compared.[1] The precision of determination of electrical bore-

[1] D. R. Rhodes, "Introduction to Monopulse," McGraw-Hill Book Company, New York, 1959.

sight with such antennas can be better than about one-hundredth of the beamwidth of an individual lobe.

Conical scanning antennas use a single beam which traverses a conical scan pattern which is usually circular; Fig. 9.9a represents a cross-section through the pattern in any plane. The response to an off-axis target varies with beam position, so that an audio error signal with a fundamental frequency equal to the scan rate is generated. The strength of this audio signal, as a percentage modulation of the rf carrier, is proportional to the target angular displacement from electrical boresight, and the phase of the audio signal is an indication of the direction of the angular displacement.

Monopulse tracking antennas, either amplitude comparison or phase comparison, commonly have separate ports for three responses: the sum, the azimuth difference, and the elevation difference. The sum (or range) pattern has a single major lobe in the boresight direction. The two orthogonal-difference patterns (elevation and azimuth) each have a minimum response in the boresight direction, and amplitude proportional to angle off axis over a small central range. Figure 9.9b shows the sum and a difference pattern in one plane; comparison with Fig. 9.9a indicates that they may be considered as the sum and difference of the overlapping lobes. In general, an off-axis target generates a difference signal in each difference channel; these signals, when referenced or normalized to the sum signal, provide error signals whose amplitudes are proportional to the target displacement from electrical boresight in the appropriate azimuth or elevation plane, and whose phases indicate the sense of displacement.

The test setup for determining the electrical boresight of a tracking antenna comprises the antenna under test, the proper associated circuits for performing the appropriate signal processing to obtain an error signal, a distant source, a precise means for continuously adjusting antenna direction (or source location), and a highly accurate optical or mechanical indication of antenna direction (or source location). The antenna (or source) is adjusted for the condition of minimum output observed at the appropriate error-signal port of the circuit. This antenna direction (or source location) is then noted as the electrical boresight under the particular condition of frequency, environment, or other variable under test; the direction may be compared with a reference direction to determine a boresight error. If the antenna system output cannot be noted continuously as a function of source angle, the minimum point may be interpolated. Since the absolute direction of the electrical boresight during test is often of less interest than its variation with system parameters, the chief requirement of the antenna direction-indicating mechanism may be high precision, over a very small angular range. A telescope rigidly fixed to the antenna mount may be used to sight on a calibrated optical target at the distant source, or a dial indicator on a long-radius arm may be used. A camera may be similarly employed to measure dynamic errors between electrical and mechanical axes when tracking a moving target.

Another measurement typical of tracking antennas is that of the appropriate output error signal (percent modulation, or normalized difference signal) as a function of angular displacement from electrical boresight. The test setup described above may be used, with particular attention to any circuit properties affecting the output level, such as automatic gain controls or limiters. The angular sensitivity of the antenna is measured as the rate of change of output error signal with angle in the vicinity of electrical boresight, and is related to the slopes of the patterns indicated in Fig. 9.9. For conical-scanning antennas, this property is sometimes termed "modulation slope"; for monopulse antennas, it may be called "error slope." For antennas in which the error-signal magnitude is used to determine angular location of an off-axis target, the linearity of error-signal output with angle is of interest, as well as the range over which the linearity holds. Another property of interest is the region of angular pull-in, within which the error-sensing circuits will generate an error voltage acting to reduce the angular displacement from electrical boresight.

The measurements described above for boresight and angular sensitivity involve the error signal, as derived from the complete system including the circuits associated with the antenna. While such measurements yield the true performance of the system, some significant contributions of the antenna to system performance may not be evident. For this reason, the signals obtained directly from the antenna are sometimes used to determine an electrical boresight and an angular sensitivity. In a conical-scan antenna, such an "antenna" boresight would be defined by equality of the lobe signals ("crossover" in Fig. 9.9a); in a monopulse antenna, the difference-signal minimum (see Fig. 9.9b) would define the "antenna" boresight. For this latter case, the additional boresight errors contributed by interaction between the antenna and the associated circuits can be determined from additional measurements and a simple graphical computation.[1] As regards angle sensitivity, the information usually desired is the slope of the lobe-signal patterns at crossover or the asymptotic slope of the difference-signal patterns. During measurement, these slopes should be referred to an absolute level of signal voltage, such as that which would be produced by an isotropic radiator or some other appropriate standard,[2] in order to

[1] H. W. Redlien, The Monopulse Difference Chart, *IEEE Intern. Conv. Record*, pt. I, Antennas and Propagation, pp. 129–131 (1963).
[2] R. R. Kinsey, Monopulse Difference Slope and Gain Standards, *IRE Trans. Antennas Propagation*, **AP-10:**343–344 (May, 1962).

permit a significant evaluation of the antenna design and the system signal-to-noise ratio.

9.4 MEASUREMENT OF POWER GAIN AND DIRECTIVITY

9.4.1 GENERAL

The power density in the distant field of an antenna depends upon angle and range. To remove the range variable from power considerations, it is convenient to consider the power in a unit solid angle in a given direction. The power in a unit solid angle may be found from the power density in a given direction and at a given range r by multiplying the power density by the square of the range r.

Thus, the *directive gain* of an antenna in a specified direction may be defined as 4π times the ratio of the power radiated per unit solid angle in that direction to the total power radiated by the antenna.

The *power gain* of an antenna in a specified direction is defined as 4π times the ratio of the power radiated per unit solid angle in that direction to the net power accepted by the antenna from its generator. This term differs from directive gain, because it includes antenna dissipation losses. The power gain is an inherent property of the antenna, and it does not include or involve system losses such as those arising from mismatches of impedance or polarization. Such system losses are accounted for separately. The ratio of the power gain to the directive gain in the same direction is the *radiation efficiency* of the antenna.

Directive gain and power gain are often stated relative to a hypothetical isotropic radiator which is lossless and has unity gain.

The gain of an antenna is usually measured in the direction in which it is a maximum; the gain in other directions may then be inferred from the radiation pattern. In the direction of the maximum value, the power gain is called the *maximum power gain*, and the directive gain is called the *directivity*. The maximum values are often referred to merely as "antenna gain," but this usage is deprecated because it is not definitive.

9.4.2 MEASUREMENT OF MAXIMUM POWER GAIN

In practice, the maximum power gain of an antenna is almost always determined from a measurement of relative gain involving a standard-gain antenna. The power received by the antenna under test is compared with the power received by the standard-gain antenna when both antennas are placed in a uniform plane-wave field and oriented for maximum received power. The maximum power gain of the antenna under test is determined from the known maximum power gain of the standard-gain antenna as follows:

$$G_{\max} = \frac{P_r \text{ (test antenna)}}{P_r \text{ (standard antenna)}} \times G_{\max} \text{ (standard antenna)}$$

where P_r is proportional to the received power in each case. It is common practice to express $G_{\max}$ in decibels.

Two precautions must be observed in the above measurement. First, account must be taken of the impedance mismatch which may exist between each antenna and its associated receiver. It is usually preferable to eliminate such mismatches by means of tuning devices which are adjusted to provide conjugate match in each case. Care should be taken to account for any dissipation present in the tuners. If the impedances cannot be matched, then the mismatches should be separately determined from impedance measurements of each antenna and its associated receiver. The losses caused by such mismatches may then be calculated and included in the computation of antenna gain.

The second precaution to be observed involves the polarizations of the antenna under test and the standard-gain antenna. It is generally desirable to make these polarizations identical in order to simplify the gain computation. Usually this situation is closely approximated when the antennas are nominally linearly polarized and are carefully aligned during the measurement. If the polarizations of the two antennas are not the same, as in the case of elliptically polarized antennas having different axial ratios, then it is necessary to account for the difference. In general, the polarization of both antennas should be measured separately and each compared with the measured polarization of the field produced by the source antenna. The losses caused by polarization mismatch may then be calculated and included in the computation of antenna gain.

Accurate knowledge of the maximum power gain of the standard-gain antenna is essential to the measurement just described. This gain can be calculated when the standard antenna has negligible dissipation and a simple configuration such as a half-wave dipole or a large horn. However, it is desirable to measure its gain by a fundamental method, and this can be done as follows when two identical antennas are available. One antenna is connected to a transmitter and the other to a receiver; both antennas are impedance-matched to their respective loads. The two antennas are placed a distance R apart (where R is beyond the inner boundary of the far-field region) and are aligned for maximum power transfer. The maximum power gain is then:

$$G_{\max} \text{ (standard antenna)} = \frac{4\pi R}{\lambda} \sqrt{\frac{P_r}{P_t}}$$

where P_r and P_t are the received and transmitted power, respectively.

The ratio P_r/P_t must be measured; this may be done by employing a calibrated receiver which is transferred from the receiving antenna to the transmitter, and rematched.

Various sources of error must be considered when attempting to achieve accuracy in the measurement of maximum power gain. For example, at microwave frequencies the antenna under test and the standard-gain antenna may react differently to irregularities which are often present in the field of even the most carefully designed test site. Several techniques are available to minimize such errors; for instance, if the standard antenna is small compared with the antenna under test, it should be moved across the aperture of the antenna under test and the average power recorded.

At much lower frequencies, where the ground must be regarded as an essential part of both the standard antenna and the antenna under test, the measurement technique depends on the polarization employed. For horizontal polarization, if the height above ground of the antenna under test is set for maximum power gain at a certain elevation angle, the reference antenna may be a half-wave dipole at the same height. In this case, the gain of the reference antenna in the direction of interest is taken as 8 db, which represents the approximate sum of the 2.15-db free-space gain of a lossless half-wave dipole and the approximate 6-db augmentation due to reflection from the assumed perfectly conducting earth. For an accurate measurement of gain of a vertically polarized low-frequency antenna, perfect reflection from the ground cannot always be relied on. If the ground were a perfect conductor, a reference antenna comprising a quarter-wave monopole would have a gain of about 5 db in the horizontal direction, representing the approximate sum of the 2.15-db free-space gain of a lossless half-wave dipole and the 3-db augmentation due to imaging of the monopole by the assumed perfectly conducting earth; for directions above horizontal, this gain would be reduced according to the standard E-plane pattern of a half-wave dipole. However, for an imperfectly conducting ground, the gain at low elevation angles may be seriously reduced. Even with the use of a large counterpoise, the pattern may deviate appreciably from the ideal one, because of diffraction and reflection from the edge of the counterpoise.

9.4.3 DETERMINATION OF DIRECTIVITY

When the complete radiation pattern of an antenna is available, it may be used to determine the directivity of the antenna. The particular quality of the pattern that is employed is the radiation intensity, which is the power radiated from the antenna per unit solid angle in a given direction. The directivity of an antenna is the maximum radiation intensity divided by the average radiation intensity. The latter quantity multiplied by 4π is the total power radiated. To compute the directivity the following relation may be employed:

$$G_{d,\max} = \frac{4\pi \times \text{maximum radiation intensity}}{\int_{\varphi=0}^{2\pi}\int_{\theta=0}^{\pi} \text{radiation intensity} \times \sin\theta\, d\theta\, d\varphi}$$

where θ and φ are the spherical coordinates of the radiation pattern, as indicated in Fig. 9.2.

Except for the simplest of antenna patterns, the integral must be evaluated by graphical numerical methods. There are two in general use: the "orange-slice" and the "conical-cut" methods. In the orange-slice method a set of patterns is obtained by measuring radiation intensity versus θ, for a number of discrete values of φ. Each pattern must be multiplied continuously by the $\sin\theta$ weighting factor and then integrated. The integrated values for the several patterns are then added according to the last equation above. In the conical-cut method a set of patterns is obtained by measuring radiation intensity versus φ for a number of discrete values of θ. Each pattern is integrated and the integrated values are each multiplied by the appropriate $\sin\theta$ weighting factor and added.

The number of patterns that should be measured increases as the pattern shape becomes less uniform. For pencil-beam antennas, it is desirable that the pole of the spherical coordinate system of measurements coincide with the beam so that this important part of the pattern is adequately covered; however, this is not always easy to accomplish. In addition, an antenna with a narrow major lobe is likely to have a large number of narrow minor lobes which must be included in the measurement and integration. Generally, it is practical to determine directivity accurately only on those antennas having radiation patterns which are not highly directive.

It is important that the radiation patterns which are measured for determining directivity include the effect of cross-polarization. Radiation intensity involves not only power radiated in the normal polarization but cross-polarized power as well; the latter component is often surprisingly large in directions away from the peak of the pattern. To determine radiation intensity, two sets of patterns may be measured using orthogonal polarizations, such as θ and φ linear, or right and left circular. Each pair of patterns, plotted in terms of power, is added together to yield the radiation intensity.

For estimation purposes, a rough value for directivity is sometimes determined from a measurement of only the major lobe, for only the normal polarization. When the major lobe is reasonably narrow, the following relation is employed:

$$G_{d,\max} = \frac{k}{\mathrm{BW}_1 \times \mathrm{BW}_2}$$

where BW_1 and BW_2 are the half-power beamwidths in degrees in the orthogonal principal planes through the major lobe, and k is a factor which may range from below 25,000 to about 40,000. The lower value of k is typical for microwave reflector-type antennas where spillover usually represents an appreciable fraction of power radiated into wide-angle minor lobes; the upper value can be approached in certain antennas having aperture excitations with appreciable amplitude tapering and no spillover. It is clear that this simplified method should be used only with great care because unexpected antenna defects are always likely to cause excessive minor lobes and cross-polarization, each of which may greatly reduce the directivity.

9.4.4 DETERMINATION OF RADIATION EFFICIENCY

The radiation efficiency of an antenna is the ratio of the total power radiated by the antenna to the net power accepted by the antenna at its terminals during the radiation process. The difference between these two powers is the power which is dissipated within the antenna. Radiation efficiency is an inherent property of an antenna and is not dependent on system factors such as impedance match or polarization match.

A fundamental method for determining radiation efficiency relies on the measurements described in Secs. 9.4.2 and 9.4.3. As noted in Sec. 9.4.1, radiation efficiency is equal to the ratio of power gain in any specified direction to the directive gain in that same direction. It is usually convenient to take the direction of maximum radiation for this determination of radiation efficiency; thus:

$$\text{Radiation efficiency} = \frac{\text{maximum power gain}}{\text{directivity}}$$

In measuring maximum power gain and directivity, all the precautions mentioned in Secs. 9.4.2 and 9.4.3 must be carefully observed. It is particularly important to include the power radiated in cross-polarization in the determination of directivity. Even when this is done, the results are not very accurate for low-loss highly directive antennas, because of the difficulty in calculating directivity from the measured patterns with sufficient accuracy.

Another method may be available when the antenna is electrically small and simple. In this case, an equivalent series circuit can frequently be found in which the real part of the input impedance, i.e., the antenna resistance, is equal to the sum of the radiation resistance and a loss resistance. The radiation resistance accounts for all radiated power, and the loss resistance accounts for all dissipation within the antenna. The antenna resistance is obtained from measurements of input impedance.

The radiation efficiency is then:

$$\text{Radiation efficiency} = \frac{\text{radiation resistance}}{\text{antenna resistance}}$$

This method is valid only if the antenna can be accurately represented as a series circuit. When the dissipation cannot be represented by a resistance in series with the radiation resistance, as in the case of an antenna coated with lossy dielectric or an antenna over a lossy ground, the method should not be used. Furthermore, the calculated radiation resistance and the measured antenna resistance must be referred to the same set of antenna terminals. It should also be noted that the input impedance of this type of antenna may present a large mismatch to the connecting transmission line, and a matching network having appreciable dissipation might be used. Such a loss is not usually included within the meaning of radiation efficiency, although it is clear that the loss would be important to the system as a whole.

9.4.5 GAIN OF ANTENNAS IN LOSSY ENVIRONMENTS

The measurements and application of the concepts of power gain, directivity, and radiation efficiency are particularly difficult when the antennas must operate in lossy environments. The definitions of these quantities were conceived initially to provide simplicity in free-space environments; they do not lend simplicity elsewhere. Honest and meaningful evaluations are also clouded by the divergent motives and objectives of the pure antenna engineer and the user. For, if system performance is degraded by an environment over which he has no control, the antenna engineer understandably would like to describe the gain and efficiency of his product as they would be in an ideal environment, since after all "his" part of the system is "working fine." On the other hand, the systems engineer and user are concerned with overall performance and, understandably, are not kindly disposed toward an antenna that would "work fine" in an ideal environment but fails to provide communication in the actual environment. So, to sell his product, the antenna engineer must evaluate his product as it would function in an actual environment.

The most common application in which these difficulties are manifest is that of antennas for operation on or close to the surface of the earth, at frequencies below, say, 30 MHz. Here, one of the most basic difficulties is that the definitions of power gain and directive gain have in them inherently the assumption that the fields vary as $1/r$ and that the power density varies as $1/r^2$. In the actual earth environment, this is not the case; in the directions near the ground, the field falls off faster than this, perhaps much faster. Thus with a direct application of the usual definitions, the gain depends on distance from the antenna. The radiation pattern may also depend on the distance from the antenna structure, even though the distance may be large compared with the free-space distant-field criteria. At large distances, even for vertical antennas, there is usually essentially a null at the horizon. This means that if the directivity is determined from the graphs of actual radiation-pattern measurements, as described in Sec. 9.4.3, the directivity over a lossy ground might be calculated to be greater than the directivity of the same antenna over a perfect ground. Moreover, the ratio of the directivity to the maximum power gain and the ratio of the "radiated" power to the input power are not necessarily the same; this means that the two definitions of radiation efficiency are not equivalent. Finally, the input impedance, and to a smaller extent, the patterns, both depend on the specific ground constants and the size and quality of the ground-screen system.

To determine the performance of an antenna near the earth, the engineer and user should agree as to whether the ground wave is to be considered as the important propagation mode or whether the radiation at angles above the horizon is the more significant (as in ionospheric circuits). Let us consider the latter case first. To obtain a meaningful gain measure in this case, the antenna should be set up in the actual environment. A set of patterns and field-strength measurements should be taken at a distance from the antenna so great that the field strength in the ground wave is negligible (or its contribution to the field subtracted away from the measured values). The average input power P_{in} should be carefully measured. Then, the power gain in a particular direction is

$$G(\theta,\varphi) = |E(\theta,\varphi)|^2 \frac{R^2}{30P_{\text{in}}}$$

where R is the range. With this type of measurement, the same antenna system will have different gains in different sites, depending on, among other things, earth conductivity at the site. The measurement evaluates the antenna for use in the particular site. In practice, a good measurement of E in an arbitrary location at an arbitrary frequency is hard to obtain. Thus it may be desirable to make a very careful determination of field strength at one location and thereafter make only careful relative measurements from the point of calibration.

To obtain a measurement of the performance of the antenna, rather than the antenna plus the propagation environment, comparative measurements may be performed. This may also be very useful from a systems viewpoint since the result is the comparison of the antenna in question with a standard antenna in a specified environment. To obtain such a comparison, a given power is fed into the antenna and an indication of field strength E_t at a given position is recorded. Next, the same power level is fed into a standard antenna, and using the same receiving and recording equipment, an indication of the field strength E_s is recorded. The ratio of the power gain of the antenna under test G_t to that of the standard G_s is then

$$\frac{G_t(\theta,\varphi)}{G_s(\theta,\varphi)} = \left| \frac{E_t(\theta,\varphi)}{E_s(\theta,\varphi)} \right|^2$$

Note that the relative gain depends only on the ratio of the field strengths; therefore an absolute measure of neither field strength nor power is required, only accurate relative values. Of course, if the gains of the test antenna and the standard are quite different, the dynamic range of the receiving equipment must be carefully verified or calibrated in advance. It is sometimes good practice to introduce precision attenuators, so that the receiving equipment operates with nearly the same input in both cases. If it is more convenient, the antennas under test may be employed as receivers and the relative power delivered to matched loads measured (due care being taken to account for any mismatches in the system) with a single transmitting antenna at the field point. The most common and convenient antennas at frequencies below, say, 100 MHz are quarter-wave vertical monopoles (or smaller) for vertical polarization, and half-wave dipoles (or smaller), mounted a half wavelength above the ground, for horizontal polarization.

When the important mode of propagation is or involves the ground wave, only a relative measure such as described in the last paragraph gives a unique measure of the gain of the antenna structure for the ground-wave field. The measure is then relative to the performance of, say, a quarter-wave vertical.

A fundamental problem which faces the antenna engineer and his customer, the systems engineer, is that of deciding in advance which of several possible antennas will perform most satisfactorily in a given function in a given earth environment. That is, in general it is not possible to install each of several possible antennas in each of many different locations in order to obtain, experimentally, a measure of the relative performance. To compare antennas in arbitrary earth environments, the following workable procedure is suggested: First, the input impedance of the antenna is carefully measured in an idealized environment. (It is then assumed that in the actual environment there is a sufficiently large ground screen so that the input resistance will be the same as in the idealized environment; if this is not the case, the input resistance in the actual environment must be calculated or measured.) Next, the current distribution, in both phase and amplitude relative to the input current, is carefully measured at all points on the antenna. Then, the input resistance of the standard antenna is measured or calculated. Finally, the electric field produced by a distant point source (plane wave or ground wave) with appropriate polarization is calculated in phase and amplitude for points in the region of the antenna structure. This calculation is done with actual earth parameters, with the proposed ground screen present but in the absence of the antenna structure. Except at very low frequencies and grazing incidence, the electric field will usually be simply the sum of an incident plane wave at the given angle plus a reflected wave calculated by means of the appropriate plane-wave reflection coefficient (see, for example, Chap. 2, p. 52) for the actual ground. (If the ground screen is very large, the diffraction field of the ground screen should also be included; or, alternatively, the contribution to the field strength by the ground-screen currents should be added in.) Let us call this field produced by the distant point source in the presence of the actual earth E_{i+r}. Then, the relative gain can be defined as the relative power delivered by each antenna to a matched load:

$$\frac{G_t(\theta,\varphi)}{G_s(\theta,\varphi)} = \frac{R_s}{R_t} \frac{|\iiint \mathbf{E}_{i+r} \cdot \mathbf{J}_t \, dv|^2}{|\iiint \mathbf{E}_{i+r} \cdot \mathbf{J}_s \, dv|^2}$$

In this formula, $\mathbf{J}_t$ is the current-density distribution on the test antenna, adjusted so that the input current is unity, and $\mathbf{J}_s$ is the current distribution on the standard antenna, adjusted so that the input current is unity. Of course for wire antennas, the volume integrals will reduce to integrals over the length of the wire (for example, $\int \mathbf{E}_{i+r} \cdot I(\xi) \, \mathbf{d}\xi$). This relative-gain measure may be defined for different, arbitrary modes of propagation. In the literature, this relative-gain measure, defined for a specific mode of propagation, has been called *Relative Communication Efficiency* (RCE).[1]

9.5 INPUT IMPEDANCE MEASUREMENTS

The input impedance of an antenna at the specified terminal pair (or port) affects the interaction between the antenna and its associated circuits. Antenna impedance can be an important factor in consideration of power transfer, noise, and stability of active circuit components. Frequently, it is the antenna impedance which limits the useful bandwidth of the antenna.

Measurement of input impedance is made at a single port of the antenna. For the usual problems and procedures related to this meas-

[1] W. L. Weeks and R. C. Fenwick, *IRE Intern. Conv. Record*, **I**:108 (March, 1962); also, *IEEE Trans.*, **AP-11**:296 (May, 1963).

urement, reference should be made to the appropriate Standard.[1] There is, however, a particular problem inherent in radiating structures, since the input impedance is modified by the environment of the antenna. For this reason, the antenna should be placed in a replica of its operating environment before the measurement is made. Usually, this requirement is easily approximated for narrow-beam antennas that can be pointed away from reflecting obstacles, but may be more difficult for omnidirectional-type antennas where much of the surrounding structure affects the input impedance.

At low frequencies, the two components of input impedance may be measured directly by a radio-frequency bridge. These instruments are commercially available up to about 500 MHz. The frequency range for direct measurement may be extended to about 3 GHz, by admittance meters or reflectometers which are arranged to measure the components of admittance or impedance.

At microwave frequencies, nominally above 1 GHz, the slotted section is the usual instrument for impedance measurement. With this instrument, the measurement of standing-wave ratio and position of the standing-wave minimum is equivalent to an impedance measurement. Often, this slotted-section data are plotted directly on a polar chart of complex reflection coefficient; such a chart employs the relation between reflection-coefficient magnitude and standing-wave ratio. If desired, these data can be transformed to input impedance by means of simple graphical methods involving impedance coordinates drawn on the reflection-coefficient chart. Two such coordinate systems are in common use: one involves the resistive and reactive components of impedance, while the other involves the magnitude and phase of impedance. The impedance so determined is in a form which is normalized to that of the slotted-section transmission line; this is usually the convenient form at microwave frequencies.

9.6 POWER-HANDLING MEASUREMENTS

Antennas are often tested to determine their ability to handle the power generated by their associated transmitters. These tests may be concerned primarily with limitations imposed by metallic or dielectric heating at high average power levels or with limitations imposed by the arcing, voltage breakdown, or corona discharge associated with high electric fields at high-peak power levels. In cases where the antenna is subject

---

[1] IRE Standards on Antennas and Waveguides: Waveguide and Waveguide Component Measurements, sec. 2.8—Measurement of Input Impedance, sec. 2.9—Measurement of Normalized Input Impedance; IEEE no. 148 (formerly 59 IRE 2 S1); 1959.

to low atmospheric pressures, it is necessary to simulate the high-altitude conditions by means of a bell jar or some other suitable chamber in order to achieve satisfactory peak power-handling measurements, since extrapolation from tests conducted at sea level is not reliable. At very low pressures, the breakdown power level decreases with increasing pressure, reaches a minimum at a pressure that depends on the wavelength (the glow-discharge region), and then increases with pressure at high pressures. The use of a radioactive material to create a continuous supply of free electrons in the atmosphere, in the near vicinity of the antenna under test, is a useful means of providing reliability and consistency in breakdown tests, and does not lower the absolute breakdown threshold.

Because of the great variety of antenna configurations and environmental conditions, no attempt will be made to provide a specific procedure for conducting power-handling measurements.[1] For accurate results, it is important to ensure that the source of power for the measurements has the same characteristics as the actual transmitter with which the antenna is to be used; that is, the modulation, pulse shape, pulse width, pulse rate, etc., should be the same and should not change with the applied power level during the test. Temperature rise may be measured by means of thermocouples or temperature-sensitive paints applied to critical surfaces. Care should be taken that these added components are not themselves heated by the radio-frequency power. Depending on the type, breakdown may be detected by visual observation, audible indication, change in signal picked up by a transmission-monitor antenna, or change in a reflected-wave monitor signal as the applied power level is increased. It is sometimes important, as in the case of glow discharge, to determine the power level at which breakdown ceases once it has been initiated. Sharp corners, dirt, metal particles, and corrosion are often major offenders in lowering the breakdown power level.

It is sometimes desired to maintain transmission from an antenna in the presence of breakdown, such as when high-altitude glow discharge occurs with a missile antenna. In those cases, it is usually desirable to measure not only the breakdown power level but also the changes in input impedance, radiation pattern, power gain, and distortion of the transmitted signal for power levels above the breakdown level.

Whenever measurements are performed in which high-average power levels exist, it is essential that proper safety devices and procedures be employed for protection of the personnel in the vicinity.

[1] IRE Standards on Antennas and Waveguides: Waveguide and Waveguide Component Measurements, sec. 2.10—Measurement of Dielectric Voltage Breakdown, sec. 3.7—Measurement of Power-handling Capacity; 1959.

9.7 NOISE TEMPERATURE MEASUREMENTS

Measurements of antenna noise temperature (T_{ant}) are usually made by comparing the antenna noise output with that from a standard source. For greatest accuracy, it is desirable that the noise temperature of the standard (T_{std}) be close to T_{ant}. Values of T_{ant} in practice may range from 3 to 3000°K and further. Several standard sources covering this range are available. One type comprises a matched resistive termination and a means for controlling its physical temperature, which is also its equivalent noise temperature. Cooling agents such as liquid nitrogen (77.4°K) or liquid helium (4.2°K) may be used; the termination may be air cooled and have a thermometer for measuring ambient temperature; or artificial heating, for example, by means of boiling water, may be employed. The low- and high-temperature standards must be such that the impedance match is preserved at their operating temperatures. Another type of standard source involves a gas tube[1] such as an argon noise lamp having an equivalent noise temperature (about 10,000°K) much higher than ambient temperature. In addition to its use in measurements of high values of T_{ant}, a gas tube is applicable in low-temperature measurements by connecting it to the system through a directional coupler of known coupling loss.[2] Its effective temperature can also be reduced by flashing the tube on a low-duty cycle.

The circuit arrangements and procedures for measuring antenna noise temperature vary considerably, depending on the type of information required. In some cases, a simple substitution test is made in which two readings of power are sufficient. On the other hand, a complex radiometer system[3] in which many readings are averaged to achieve a greater precision is often employed in radio astronomy. Sometimes the circuit includes more than one type of standard source, and in many of the more accurate measurements a calibrated precision attenuator is adjusted to maintain constant output noise power in the two test conditions. In the latter case, the noise introduced by the attenuator, as well as its attenuation, must be included in the calculation. A simple example of this is shown in Fig. 9.10, in which it is assumed that the attenuator is a resistive (not reactive) type and that all components are matched to their transmission lines, which are lossless. With the switch in position 1 and the precision attenuator set to some convenient value of loss (ratio of power input to output) L_1, the noise temperature of the

---

[1] W. W. Mumford, A Broadband Microwave Noise Source, *Bell System Tech. J.*, **28**:608–618 (October, 1949).

[2] R. W. DeGrasse et al., Ultra-low-noise Measurements Using a Horn Reflector and a Traveling-wave Maser, *J. Appl. Phys.*, **30**(12):2013 (December, 1959).

[3] R. H. Dicke, Measurements of Thermal Radiation at Microwave Frequencies, *Rev. Sci. Instr.*, **17**:268–275 (July, 1946).

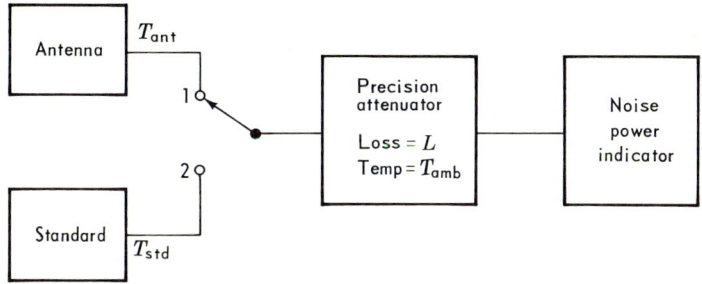

Fig. 9.10 Circuit for measuring noise temperature of an antenna.

noise power output to the indicator is:

$$T_{out} = \frac{T_{ant}}{L_1} + \left(1 - \frac{1}{L_1}\right) T_{amb}$$

where $T_{ambient}$ is the ambient temperature of the attenuator. With the switch in position 2, the attenuator is now adjusted to a new value L_2 such that the noise power to the indicator remains the same; the equivalent output temperature is then:

$$T_{out} = \frac{T_{std}}{L_2} + \left(1 - \frac{1}{L_2}\right) T_{amb}$$

Equating these two relations permits calculation of the antenna noise temperature:

$$T_{ant} = T_{amb} + \frac{L_1}{L_2} (T_{std} - T_{amb})$$

It should be noted that with the circuit of Fig. 9.10, an antenna temperature below ambient temperature requires a standard with a temperature also below ambient temperature, and vice versa. No matter what circuit arrangement is used to measure antenna noise temperature, reliable results are obtained only when care is taken to avoid the pitfalls which are present in any noise measurement of any component.[1]

The measurement of antenna noise temperature is complicated by the dependence of the noise on several variables. Two such variables are the direction of antenna pointing and the antenna location; these determine the particular external sources of noise which may couple to the antenna. The antenna noise temperature is also likely to vary with time, because of changes in the external noise sources.

---

[1] IRE Standards on Methods of Measuring Noise in Linear Twoports, 59 IRE 2S1, *Proc. IRE*, **48**:60–68 (January, 1960).

BIBLIOGRAPHY

Aharoni, J.: "Antennas," Clarendon Press, Oxford, 1946.

Fry, D. W., and F. K. Goward: "Aerials for Centimetre Wavelengths," Cambridge University Press, London, 1950.

Jahnke, E., and Fritz Emde: "Tables of Functions," 4th edition, Dover Publications, New York, 1945.

Jasik, Henry: "Antenna Engineering Handbook," McGraw-Hill Book Company, New York, 1961.

Jordan, E. C.: "Electromagnetic Waves and Radiating Systems," Prentice-Hall, Inc., Englewood Cliffs, N.J., 1950.

——— (Ed.): "Electromagnetic Theory and Antennas," Pergamon Press, New York, 1962.

King, R. W. P.: "Theory of Linear Antennas," Harvard University Press, Cambridge, Mass., 1956.

Kraus, J. D.: "Antennas," McGraw-Hill Book Company, New York, 1950.

Laport, E. A.: "Radio Antenna Engineering," McGraw-Hill Book Company, New York, 1952.

Moullin, E. B.: "Radio Aerials," Clarendon Press, Oxford, 1949.

Papas, C. H.: "Theory of Electromagnetic Wave Propagation," McGraw-Hill Book Company, New York, 1965.

Schelkunoff, S. A.: "Electromagnetic Waves," D. Van Nostrand Company, Inc., Princeton, N.J., 1943.

———: "Advanced Antenna Theory," John Wiley & Sons, Inc., New York, 1952.

——— and H. T. Friis: "Antennas: Theory and Practice," John Wiley & Sons, Inc., New York, 1952.

Silver, S.: "Microwave Antenna Theory and Design," McGraw-Hill Book Company, New York, 1949.

Smith, R. A.: "Aerials for Metre and Decimetre Wavelengths," Cambridge University Press, London, 1949.

Wait, J. R.: "Electromagnetic Radiation from Cylindrical Structures," Pergamon Press, New York, 1959.

Walter, C. H.: "Traveling Wave Antennas," McGraw-Hill Book Company, New York, 1965.

Watson, W. H.: "The Physical Principles of Waveguide Transmission and Antenna Systems," Clarendon Press, Oxford, 1947.

Weeks, W. L.: "Electromagnetic Theory for Engineering Applications," John Wiley & Sons, Inc., New York, 1964.

Williams, H. P.: "Antenna Theory and Design," Sir Isaac Pitman & Sons, Ltd., London, 1950.

GLOSSARY OF SYMBOLS

The number after the symbol refers to the chapter in which the symbol first has the stated meaning. If there is no number, the symbol has the same meaning throughout.

$\doteq$		Approximately equal to
A		Magnetic vector potential
A	5	Potential function for spherical coordinates $A = r \sin \theta \, H_\varphi$
A	6	Width of horn aperture
A_e	8	Effective area of receiving antennas
A_n	3	Complex number representing relative amplitude and phase of the nth element in a linear array
A_x	2	x component of magnetic vector potential A
A_y	2	y component of magnetic vector potential A
A_z	2	z component of magnetic vector potential A
AF	3	Array factor
B	6	Height of horn aperture
B	7	Required bandwidth in a log-periodic dipole array
B_s	7	Design bandwidth in a log-periodic dipole array
B_{ar}	7	Bandwidth of the active region in a log-periodic dipole array
B^t	1	Time-dependent magnetic flux density
C	6	One of the Fresnel integrals (p. 246) (the one involving the cosine function)
D^t	1	Time-dependent electric flux density
$D(\psi)$	3	Difference between the array factor of an array whose elements have small departures from equal

356   ANTENNA ENGINEERING

		spacing and that of the corresponding array with equal spacing
E	1	Complex vector representing electric field strength at circular frequency ω; it is a function of space but not of time
$\mathbf{E}^t$	1	Electric field strength, a function of space and time, volts/m
$\mathbf{E}_{inc}$	8	Electric field at the position of the receiving antenna with the receiving antenna structure removed
$\mathbf{E}_{ap}$	5	Applied electric field
$\mathbf{E}_{ind}$	5	Induced electric field
E_θ	1	θ component of electric field strength **E**
F	1	Electric vector potential
F	8	Noise figure
G_d	2	Directive gain
G_{di}	2	Directivity (maximum value of directive gain)
G_{dE}	6	Directivity of a sectoral horn flared in the E plane only
G_{dH}	6	Directivity of sectoral horn flared in H plane only
H	1	Complex vector representing magnetic field strength, at circular frequency ω; it is a function of space but not of time
$\mathbf{H}^t$	1	Magnetic field strength, a function of space and time, amp/m
H_φ	1	φ component of magnetic field strength **H**
H*	1	Complex conjugate of **H**
H	4	Half-length of a wire dipole antenna, or height of a monopole antenna
H_p	5	Hankel function or Bessel function of third kind of order p
$\hat{H}_n$	5	Spherical Hankel function or spherical Bessel function of third kind of order n
I	1	Electric current, amp
I_{sc}	8	Short-circuit current at the terminals of a receiving antenna
J_p	2, 5	Bessel function of first kind of order p
$\hat{J}_n$	5	Spherical Bessel function of first kind of order n
$\mathcal{J}$	1	Complex number representing the density of electric current sources at circular frequency ω; a function of space but not of time
$\mathcal{J}^t$	1	Density of electric current sources; a function of both space and time, amp/m²

GLOSSARY OF SYMBOLS

Symbol	Chapter	Description
$\mathcal{J}_t$	8	Current density distribution on the structure comprising the receiving antenna when that structure is energized as a transmitting antenna from the same pair of terminals
$\mathcal{J}_z$	1	z component of electric current source density
K	1	Complex vector representing single-frequency electric surface current density, amp/m
L	2, 4	Length of transmission-line section
L	3	Length of a continuous array
L	4	Length of a linear antenna
L	5	Radius of biconical antenna
$\mathfrak{M}$	1	Complex vector representing the density of magnetic current sources, volts/m^2
N_p	5	Bessel function of second kind of order p
$\hat{N}_n$	6	Spherical Bessel function of second kind of order n
$\mathfrak{N}$	1	Complex vector representing magnetic surface current source density, volts/m
$P_{\text{rad,av}}$	2	Average radiated power
P_n	5	Legendre polynomial of order n
P_n^1	5	Associated Legendre polynomial of order n and degree 1
R	1	Distance between two points in space
R	3	Ratio of the strength of the main beam to the side-lobe level in a Chebyshev array
R_1	6	Slant length of a sectoral horn
R_a	8	Input resistance of an antenna
R_r	2	Radiation resistance
R_{12}	4	Mutual resistance between two antennas
$R(r)$	5	Separable part of $A(r,\theta)$, depending only on r
Re	1	Real part of
S	1	Area
S	6	One of the Fresnel integrals (p. 246) (the one involving the sine function)
T	8	Absolute temperature, °K
T_a	8	Receiving antenna temperature (equivalent)
T_b	8	Brightness temperature
T_e	8	Effective temperature of a network or system
T_M	3	Chebyshev polynomial of mth order
$U(z - z')$	3	Unit step function starting at $z = z'$, so that $U(z + z_2) - U(z - z_1)$ is a unit pulse function of width $z_1 + z_2$
V	1	Electric potential function (scalar)
V_{ab}	1	Voltage between points a and b

V_g	4	Strength of a voltage generator
V_{oc}	8	Open-circuit voltage at receiving antenna
X_{12}	4	Mutual reactance between two antennas
Y_{in}	5	Input admittance
Y_0	5	Characteristic admittance
Y_t	4	Terminating impedance for biconical antenna theory
Z_a	4	Inverse radiation impedance
Z_{in}		Input impedance
Z_0		Characteristic impedance of transmission line
Z_{12}	4	Mutual impedance between two antennas
$\hat{Z}_n$	5	Any of the three kinds of spherical Bessel functions
a	3	Constant to be determined in Chebyshev array theory
a	5	Radius of cylindrical post or antenna
a	6	Radius of cylinder; also, radius of circular aperture
a_m	5	Coefficients in the power series expansion of the function Θ
b	3	Constant to be determined in Chebyshev array theory
d	3	Spacing between elements in a linear array
d_y	3	Spacing of elements along y axis in a two-dimensional array
d_z	3	Spacing of elements along the z axis in a two-dimensional array
dA		Differential area
df	8	Incremental bandwidth of receiving system
dl		Differential length
dv		Differential volume or elemental volume
f		Frequency of oscillation
g	6	A potential function such that $\mathbf{E} = \nabla \times g\hat{\mathbf{z}}$
h	8	Planck's constant
$i(z)$	3	Current distribution in a linear antenna normalized so that the current at the input is unity
j		imaginary unit $(-1)^{1/2}$
k	3	Phase shift per unit length of the sources in a continuous linear array
k	5	Separation constant or eigenvalue for the differential equations in spherical coordinates
k	8	Boltzmann's constant
$\hat{\mathbf{n}}$	1	Unit vector normal to a surface
p	1	Current moment or dipole moment

GLOSSARY OF SYMBOLS 359

r	1	Distance from origin in spherical coordinates
$\hat{r}$	1	Unit vector in r direction
t	1	Time
v	5	Alternative coordinate variable in the spherical coordinate system, $v = \cos\theta$
x		One of the variables in a rectangular coordinate system
$\hat{x}$		Unit vector in the x direction
y		Rectangular coordinate variable
$\hat{y}$		Unit vector in the y direction
z	1	Rectangular coordinate variable
$\hat{z}$		Unit vector in z direction
z	3	A variable in the polynomial representation of linear arrays, $z = e^{j\psi}$
α	3	Phase shift angle from one element to the next in a linear array
α	7	Half-angle of apex angle in a log-periodic dipole array
β		Propagation (phase) constant, $\beta = \omega\sqrt{\mu\epsilon} = 2\pi/\lambda$
γ	1	Complex propagation constant ($= j\omega\sqrt{\mu\epsilon}$)
γ_i	5	Complex propagation constant for the region inside the conducting material
γ_z	5	Propagation constant along z direction in cylindrical system; separation constant for z-dependent part of the differential equations
Δ_n	3	Error in positioning of nth element in an unequally spaced array, measured from its position in an equally spaced array
$\delta(x)$		Impulse function, $\int_{-\infty}^{\infty} \delta(x)\,dx = 1$
ϵ		Permittivity of a medium, usually a complex number
$\Theta(\theta)$	5	Separable part of $A(r,\theta)$ depending only on polar angle variable in spherical coordinate system
θ		Polar angle variable in spherical coordinate system
$\hat{\theta}$		Unit vector in the θ direction
η		Radiation efficiency
λ		Wavelength
μ		Permeability of a medium, often a complex number
ξ	3	A variable in the Fourier transform theory of linear arrays, $\xi = \beta\cos\phi + k$
ρ	1	Electric charge density

ρ	2	Reflection coefficient
ρ	5	Coordinate variable in cylindrical coordinate system, (ρ,φ,z)
σ	1	Electrical conductivity, mhos/m
σ	7	In a log-periodic dipole array, the ratio of the element spacing to twice the length of the next larger element
τ	7	Ratio of successive distances between the apex and the elements in a log-periodic antenna, $\tau = R_{n+1}/R_n$
φ		Azimuthal angle variable in spherical coordinates and in cylindrical coordinates
$\hat{\varphi}$		Unit vector in the φ direction
ϕ ·		Polar or conical angle with respect to the line of the array
ϕ_m		Value of ϕ in which the radiation pattern is maximum (i.e., the direction of the main beam)
ψ	3	A variable used in discrete array theory, $\psi = \beta d \cos \phi + \alpha$
ψ_1	3	A variable used in the theory of continuous arrays, $\psi_1 = \tfrac{1}{2}(\beta \cos \phi + k)L$
Ω	3	Polar angle measured from the y axis (Fig. 3.2a)
ω	1	Circular frequency, $2\pi f$
∇		Vector operator; $\hat{x}(\partial/\partial x) + \hat{y}(\partial/\partial y) + \hat{z}(\partial/\partial z)$ in rectangular coordinates

INDEX

INDEX

Abraham, M., 3
Absorption in water vapor and oxygen, 306
Active region, 279
Adams, R. T., 123
Adaptive arrays, 177
Admittance, biconical antenna, 212
Aharoni, J., 354
Amplitude taper, 256, 345
Andre, S. N., 128
Anechoic chamber, 332
Antenna, applications, 4
  feeders, 179
  history, 1
  small, 27
  temperature, 303, 308–309, 352
  voltage, 44
Aperture, effective, 297
Aperture antennas, 228
Aperture blocking, 260
Aperture distributions, 256
Aperture theory, 228
Array factor, 62, 64–65
Arrays, 62, 64–65
Artificial dielectrics, 262
Ashmead, A. S., 118
Asymmetric feeding, 161
Atmospheric irregularities, 123
Atmospheric noise, 304
  measurements, 306–307
  grades, 307

Attenuation, coaxial cable, 5
  waveguide, 5
Axial mode helix, 144
Axial ratio, 336

Babinet's principle, 232, 273
Backfire radiation, 279
Baillin, L. V., 118
Balance problems, 167
Balanced to unbalanced converters (*see* Baluns)
Baluns, 167–180, 270
  broadband, 275
  frequency range, 172
  transformers, 176
Bandwidth of small antennas, 52
Bazooka balun, 180
Beam direction, measurement, 338
Beamwdith, Chebyshev array, 108
Bell, R. L., 291
Berry, D. G., 275
Bessel functions, 60, 236
  spherical, 204–206
Bibliography, 354
Bickmore, R. W., 122
Biconical antennas, 38–39, 268
  impedance calculation, 209
  input impedance, 150–151
  small, approximate reactance, 149
  terminal admittance, 212

Biconical antennas, theory, 199
Biconical transmission lines, 145
Binomial array, 85
  factors, 85
Bolljahn, J. T., 265
Boltzman's constant, 301
Booker, H. G., 232, 268
Boresight error, 337
  measurement, 337–340
Boundary conditions, 77
Bow-tie antenna, 268, 270
Bracewell, R. N., 118
Brewster angle, 262
Bridge circuit, 334
Brightness temperature, 303–305
Broadband antennas, 265
Broadcast antennas, 86
  towers, 164
Broadcast applications, 62
Brown, G. H., 47
Brown, J. L., 105
Bueller, L. L., 118

Capacitor plate antenna, 28
Cardioid pattern, 193
Carrel, R. L., 277, 283
Carson, J. R., 4
Carter, P. S., 4
Cavity backing for slot antenna, 232
Characteristic impedance, average, of
    antenna elements, 286
  biconical transmission line, 148
  parallel rod line, 287
Chebyshev arrays, 101
Chebyshev polynomials, 102–103
Cheng, D. K., 116
Christiansen, P. L., 116
Chu, L. J., 4
Circular aperture antennas, 254
  excitation, 258
Circular array, 221
Circular polarization, 145, 269, 300, 337
Coaxial cable losses, 5
Collin, R. E., 93, 116
Collinear array, 64

Communication efficiency, 349
Complex permittivity, 6
Conductivity, earth, 49
Conical horns, 254
Conical-scanning antennas, 339
Continuity of charge, 9
Continuous linear arrays, 66, 78
Cornu spiral, 246
Correlation array, 133
Corrosion, 320–321
Cosine integral function, 160
Cross polarization, 345
Cumming, W. H., 335
Current density, electric, 6
Current distribution, integral equation for, 224–225
  wire antennas, 137
Current element, 16
Current moment, 16
Cutler, C. C., 128
Cutler, Maine, VLF installation, 52
Cylindrical antenna theory, 222
Cylindrical monopole antenna, input
    impedance, 152–153
Cylindrical post, impedance, 217–221
Cylindrical structures, theory of, 214

Deep space communication, 123
Deflections, mechanical, 321
DeGrasse, R. W., 352
Deschamps, G. A., 267, 336
Dicke, R. H., 352
Directive gain, 30, 341
  capacitor plate antenna, 30
  limitations, 208–209
Directivity, 30, 292, 341
  circular aperture, 256
  common antennas, table, 298
  end-fire array, 81
  horns, 244–248
  measurement, 343–344
  two-element array, 191–192
Director, 265
  wire, size, and spacing, 196
Distant field, 17, 323–324
Distortion, intermodulation, 321

Diversity, frequency, 317
  gain, 319
  polarization, 317
  reception, 317
  space, 317
  system, 317
  time, 317
Dolph, C. L., 102–103
Dolph-Chebyshev arrays, 102
Duals, 22–23, 232
DuHamel, R. H., 90, 105, 115, 270, 275
Duncan, J. W., 179
Duncan, R. H., 222
Dunn, G. R., 44
Dwight, H. B., 102
Dyson, J. D., 267, 268, 270

Earth, effect on radiation patterns, 52
Earth conductivity, 49
Earth environment, 346–349
Effective aperture, 297
Effective area, 297, 300
Effective permeability, 311
Effective temperature of a network, 314
Efficiency, communication, relative, 349
  receiving antenna, 308
  (See also Radiation efficiency)
Ehrlich, M. J., 118, 251
Electric vector potential, 236
Electrical storms, 164
Electromagnetic modeling, 289
Electromagnetic scaling, 289
Elfving, C. T., 291
Elliott, R. S., 118
Elliptically polarized antennas, 299
Emde, F., 354
End-fire array, Chebyshev, 113
  directivity, 81
  Schelkunoff's, 89
Epstein, J., 47
Equation of continuity, 9
Equiangular spiral, 268

Equivalence theorem, 228
Errors, effect on pattern, 118
Excitation errors, 118
External impedance, 217

Fading, Rayleigh, 319
Fast waves, 81
F/D ratio, 260
Feeders, 179
Fenwick, R. C., 349
Ferrite loop antennas, 309
  efficiency, 310–311
Field, collinear array, 69
  current element, 16
  linear array, 68
  point current, 16
Field strength measurement, 325
Flush-mounted antenna, 233
FM antenna, 291
Folded dipole, 180, 265
Folded monopole, 184
Fong, T. S., 126
Foster, D., 142
Fourier representations, 92
Fourier series, 94
Fourier transform, 92
Franks, R., 291
Frequency-independent antennas, 263, 267
Fresnel integrals, 246
Fry, D. W., 354

Gabriel, W. F., 335
Gain, directive, 30
  diversity, 319
  effect of errors on, 118
  in lossy environments, 346
  limitations, 208–209
  system, 297
Galactic noise, 306
Gangi, A. F., 44
Gilbert, E. N., 118
Goldstone, O. L., 253
Goward, F. K., 354
Grating lobes, 76, 81, 118

Green's function, 15
Grisdale, G. L., 318
Ground, 41
Ground stake resistance, 47
Ground systems, radial wire, 47
Gruenberg, E. L., 128

Hallen, E., 4, 222
Hansen, R. C., 122
Hansen, W. W., 81
Hansen and Woodyard condition, 81
Hardened antennas, 320
Harper, A. F., 142
Harrington, R. F., 23, 118, 121, 253
Harrison, C. W., 34, 184
Helical wire antennas, 143–145
Hertz, H., 1
Hertz potentials, 9
High-gain arrays, 123
Hinchey, F., 222
History, antenna, 1
Hogg, D. C., 305
Holey waveguide, 253–254
Honey, R. C., 253
Horn antennas, 240
  directivity, 244–248
Huygens' principle, 26, 228, 240
Huygens' sources, 229
Huygens' surface, 229–230
Hybrid junction, 334
Hyperboloidal lens, 261

Illumination, E-plane and H-plane, 260
Images, 39
Impedance, of antennas over ground, 194
  calculation, wire antennas, 158
  cylindrical post, 217–221
  mismatch, 342
  transformers, 173
Impulse function, 13
Induced emf method, 153
Induction field, 17
Input admittance, shunt-fed antenna, 165–167

Input impedance, 28, 29
  biconical antennas, thin, 150–151
  cylindrical-monopole antennas, 152–153
  measurement, 349–350
  rhombic antennas, 142
Integral equations, 224–225
Internal impedance, cylindrical post, mast, or wire, 218–219
Intrinsic impedance, 20
Inverse radiation impedance, 149–150
Ionosphere, 319
Ionospheric circuits, 55
Isbell, D. E., 273, 275, 277
Isotropic radiator, 30

Jacomimi, O. J., 305
Jahnke, E., 354
Jasik, H., 52, 105, 118, 139, 333, 354
Johnson, C. M., 128
Jordan, E. C., 54, 92, 267, 354

King, D. D., 118
King, L. V., 4
King, R. B., 47
King, R. W. P., 34, 162, 164, 184, 222, 227, 354
Kinsey, R. R., 340
Kompfner, R., 128
Kraus, J. D., 145, 354
Ksienski, A., 133
Kummer, W. H., 126

Laplace transform, 116
Laplacian operator, 13
Laport, E. A., 354
Large arrays, 118
Leaky-wave antennas, 253
Legendre polynomials, 210
  associated, 203
Lenses, 254, 260–262
  metallic, 262
  reflections, 262
  waveguide, 262

Leonard, D. J., 128
Lewis, R. F., 47
Lightning, 164
Lightning strike, 320
Linear antenna, 39
 radiation patterns, 137–140
 theory, 222
Linear-array theory, 62
 polynomial representations, 84
 uniform, 74
Litz wire, 44
Lo, Y. T., 118
Log-periodic antennas, 271–272
Log-periodic cells, 289
Log-periodic concept, 289
Log-periodic dipole array, 275–288
 voltages and currents, 278, 280
Log-periodic dipole design, 283–288
Logical switching, 131
Loop antennas, 56
 circular, 59
 efficiency, 59
 impedance, 58
 radiation resistance, 58
 rectangular, 56
 power radiated, 58
Losses, coaxial cable, 5
Love, A. E. H., 228
Low side lobes, 260

Ma, M. T., 116
MacDonald, H. M., 228
Magnetic current equivalent, 231
Magnetic current source, 6, 154
Major lobes, 75
Maley, S. W., 47
Man-made noise, 304
Marconi, G., 2
Marker beacon, 128
Matt, S., 305
Mattingly, R. L., 132
Maxwell, J. C., 1
Maxwell's equations, 5, 6, 8
 in spherical coordinates, 146
Mayes, P. E., 267

Measurements, 320
 of dynamic field quantities, 7
Mechanical loading, 320
Mills, B. Y., 118
Minerva, V. P., 179
Model measurements, 329
Modeling, 322
Monopulse tracking antennas, 339
Morgan, S. P., 118
Morris, J. C., 318
Moullin, E. B., 354
Multiple tuning, 37
Mumford, W. W., 305, 352
Mushiake, Y., 184
Mutual effects, 184
Mutual impedance, calculations, 184
 dipoles with sinusoidal current,
   189–193
 formula, 187
 short antennas, 188

Near field, 17
 of sinusoidal current, 156
Noise, added by lossy transmission
   line, 307
 added by receiving antenna, 308
 atmospheric, 304
 external, 303–304
 extraterrestrial, 305
 galactic, 306
 internal, 307
 man-made, 304
 precipitation, 304
 receiver, 313–314
 sky, 306
Noise considerations, 301
Noise delivered by receiving antenna,
   301–302
Noise figure, 314
Noise output devices in cascade, 314–
   315
Noise power, radiated by a black
   body, 301
Noise temperature, measurement,
   352–353
Normal mode helix, 143

Norton, K. A., 55
Norton equivalent for receiving antenna, 273
Nuclear blast, 320
Nulls in radiation pattern, 320

Oliner, A. A., 249, 253
O'Neill, H. F., 118
Ore, F. R., 275

Palmer, D. S., 318
Papas, C. H., 354
Paraboloidal reflectors, 254
　feeding, 258–260
　gain, 258
Parasitic elements, 184, 196
Partial sleeve monopole, 266
Pattern multiplication, 132
Pattern range, test, 330
Pattern synthesis, 92
Pencil beam antennas, 344
Permeability, 6
　effective, 311
Permittivity, 6
Phase center, 332
Phase disturbances, 123
Phase errors, 117
Phase measurement, 332–333
　circuit, 334–335
Phase shift, element-to-element, 77
Phased arrays, 71–72, 76–77
Phillip, E. M. R., 128
Planck distribution law, 301
Poisson's equation, 11
Polarization, ellipse, 336
　measurement, 336–337
　orthogonal, 299
Polynomial representation, linear-array theory, 84
Pon, C. Y., 128
Power delivered by receiving antenna, 292
Power gain, 341
　measurement, 341–343

Power handling, measurement, 350–351
Power pattern, 18, 325
Power rating, 44
Power transfer between elliptically polarized antennas, 299
Poynting vector, 26, 29
Preamplifiers, 315–316
Preferred coverage, 62
Primary feed, 259
Primary source, 21
Principal planes, 18
Pyramidal horns, 240

Quarter-wave monopole, 309
Quasi-static field, 17

Radar antenna, 239
Radiation efficiency, 31, 341
　loop, 59
　measurement, 345–346
Radiation field, 17
Radiation pattern, 17
　center-driven dipoles, 139
　circular aperture, 255–257
　effect of the earth, 52–53
　horn antennas, 240–246
　measurement, 323–324
　recording, 325
　straightwire antennas, 137–140
　zeros, 84
Radiation resistance, 29
　(*See also* Impedance; Input impedance)
Ramo, S., 60
Random errors, 117
Range reflections, 330
Rayleigh fading, 319
Rayleigh-Jeans distribution law, 301
Receiver noise, 313–314
Receiving antennas, 292
　efficiency, 308
　open-circuit voltage, 293–296
　short-circuit current, 293–296
Receiving pattern, 25, 292

Reciprocity theorem, 23
Redlien, H. W., 340
Reference antenna, 328
Reflection coefficient, complex, 350
Reflections, range, 330
Reflector, 265
  wire, size, and spacing, 196
  (*See also* Paraboloidal reflectors)
Relative communication efficiency, 249
Resistance, ground stake, 47
Resonance absorption, water vapor
    and oxygen, 306
Retrodirective arrays, 128
Rhodes, D. R., 243, 338
Rhombic antennas, 142–143
Riblet, H. J., 103
Robertsone, S. D., 336
Rondinelli, L. A., 118
Rumsey, V. H., 253, 267
Ruze, J., 118

Sampling theorem, 96
Satellite communications, 123
Saturn SA-202 radar antenna, 239
Scale models, 322
Scaling, 267, 322–323
Scanning, phase shift, 77
Schafer, G. E., 336
Schelkunoff, S. A., 4, 84, 150, 213,
    229, 354
Schelkunoff's end-fire array, 89
Sectoral horns, 240
  E-plane, 247
Self-complementary structures, 268
Self-phased arrays, 124
Sensiper, S., 44
Series slots, 252
Shaped beam, 260
Shock, 320
Shunt feeding, 164
Shunt slots, 249
Side lobes, 76
  levels, 118
  reduction, 121
Signal-processing antennas, 122
Signal-to-noise ratio, 292, 306, 310

Silver, S., 254, 354
Simmons, A., 252
Simulation, 322
Sine integral function, 160
Sine-tapered amplitude distribution,
    81–82
Sinusoidal current, near field of, 156
Skolnik, M. I., 128
Sky noise, 306
Sleeve antenna, 163
  dipole, 327
  monopole, 163–164
Slot antennas, annular or circular, 233
  in cylinder, 235
  impedance, 233
  rectangular, 230
  waveguide, 249–253
Slotted-cylinder antenna, 235
  radiation patterns, 237–238
Slotted-line section, 350
Slow waves, 81, 253
Small antennas, bandwidth, 52
Smith, R. A., 354
Sommerfeld, A., 3, 55
Space factor, 62
Spacing of the zeros, 89
Sparks, R. A., 333
Spherical antenna, 207
Spillover, 260, 345
Spiral antenna, feeding, 270
Split-coax balun, 179
Spot-band antennas, 263
Standard-gain antennas, 342–343
Standing-wave ratio, 350
Stratton, J., 4
Sunde, E. D., 47
Supergain array, 92
Superposition integrals, 19
Synthesis, antenna pattern, 92
  continuous array, 96
  by sampling theorem, 99
System gain, 297

Tchebysheff arrays (*see* Chebyshev
    arrays)
TE, TEM, and TM fields, 147

Telemetry antenna, 36
Television antenna, 265, 291
Temperature, antenna, 303
    brightness, 303–305
    effective, of a network, 314
Terminal admittance, biconical
    antenna, 212
Terrio, F. G., 126
Thevenin equivalent for receiving
    antenna, 293, 295–296
Thinned arrays, 118
Thompson, J. J., 3
Tillotson, L., 128
Time-modulated arrays, 126
Tolerance problems, 117
Top hat loaded dipole, 28
Tracking antennas, 338
Tracking device, 327
Trailing-wire antenna, 163
Transmission-line antennas, 34
Transmission-line loading, 31, 37
Transponder, 128
Trapezoidal toothed structure, 273
Traveling-wave array, 141
Traveling-wave slot arrays, 252–253
Tuning coil, 44
Turner, E., 268
Twin-lead transmission line, 184
Two-dimensional array, 69
Two-way response, 133

Umbrella-loaded monopole, 44
Unbalanced systems, 167
Unequally-spaced arrays, 117
Uniform-linear array, 74
University of Illinois, 265

V antennas, 141
Van Atta, L. C., 128
Van Duzer, J., 60
Vector potential, electric, 236
    magnetic, 9, 12, 19
Vibration, 320
Villeneuve, A. T., 126
Visible range, 73, 81, 88, 107
VLF installation, Cutler, Maine, 52
Voltage, antenna, 44
Voltage generator, field theory equivalent of, 154
VSWR, small horns, 244

Wait, J. R., 47, 55, 354
Walter, C. H., 253, 354
Walters, L. C., 55
Watson, W. H., 354
Waveguide, attenuation, 5
    radiators, 239
    slots in, 249–253
Weeks, W. L., 23, 54, 101, 232, 349, 354
Whinnery, J. R., 60
Williams, H. P., 354
Wind, M., 333
Woodward, P. M., 98
Woodyard, J. R., 81
Wu, T. T., 162

Yagi antenna, dual-channel, 264
    two-element, 264
Yagi array, 196
Yaru, N., 92

Z transforms, 115–116